ERGONOMIC TRENDS FROM THE EAST

PROCEEDINGS OF ERGONOMIC TRENDS FROM THE EAST, JAPAN, 12–14
NOVEMBER 2008

Ergonomic Trends from the East

Edited by:

M. Kumashiro
University of Occupational and Environmental Health, Kitakyushu, Japan

 CRC Press
Taylor & Francis Group
Boca Raton   London   New York   Leiden

CRC Press is an imprint of the
Taylor & Francis Group, an **informa** business

A BALKEMA BOOK

CRC Press/Balkema is an imprint of the Taylor & Francis Group, an informa business

© 2010 Taylor & Francis Group, London, UK

Typeset by MPS Ltd. (A Macmillan Company), Chennai, India

Printed and bound in Great Britain by Antony Rowe (A CPI Group Company), Chippenham, Wiltshire

Published by: CRC Press/Balkema
                P.O. Box 447, 2300 AK Leiden, The Netherlands
                e-mail: Pub.NL@taylorandfrancis.com
                www.crcpress.com – www.taylorandfrancis.co.uk – www.balkema.nl

ISBN: 978-0-415-88178-4 (Hbk)
ISBN: 978-0-203-83334-6 (eBook)

Ergonomic Trends from the East – Kumashiro (ed)
© 2010 Taylor & Francis Group, London, ISBN 978-0-415-88178-4

Table of Contents

Chapter 16　*Ergonomics in manufacturing*

Preface

The First East Asian Ergonomics Federation Symposium was held from November 12–14, 2008, at Ramazzini Hall on the campus of the University of Occupational and Environmental Health, Japan. Five ergonomics organizations from Japan, South Korea, Taiwan, Hong Kong, and China (The Society for Occupational Safety, Health and Ergonomics, Japan; the Chinese Ergonomics Society, China; the Hong Kong Ergonomics Society, Hong Kong; the Ergonomics Society of Korea, Korea; and the Ergonomics Society of Taiwan, Taiwan) teamed up with the objective of improving ergonomics research in East Asia and to further strengthen their ties. Organizations from a total of 13 countries participated, including nine countries outside East Asia. The theme of the symposium is "Ergonomic Trends from the East."

Japan was the first to create such an organization when it formed the Japan Ergonomics Society in 1964. It was followed in order by South Korea, Taiwan, China, and Hong Kong. In addition, the Society for Occupational Safety, Health and Ergonomics, which focuses on occupational safety and health, was established in Japan in 1996, and the ergonomics research competition between these two ergonomics societies is expanding the breadth of ergonomics research. In this way, ergonomics research in East Asia is reaching maturity. Further, the purpose of this forum is to increase the level of ergonomics research in East Asia.

Finally, East Asian ergonomists created a venue in 2008 where they all could share information.

There is a limit to what can be accomplished in one country for ergonomics research, the industrial applications of ergonomics, and the utilization of ergonomics in daily life. Today, in the early 21st century, it is difficult to produce good research in the closed environment of one country. This is the age of communication using the Internet, and videoconferencing is also possible. We can further ergonomic research using this IT environment. A lack of face-to-face communication, however, can lead to a serious situation with adverse effects.

That's why this symposium is greatly needed. When the regions are connected by a relatively close distance, all of the aforementioned communication channels can be easily established. Now, both industrial and policy circles are dealing with subjects from the point of view of a bloc in East Asia. Why ergonomists did not seek to achieve regional solidarity was an important question for me.

I began developing a great interest in this issue in 1988. After I went to Beijing to attend an international conference on occupational safety and health at China's invitation, the Chinese consulted me about creating an organization and research methods to start their own ergonomics studies. This was a rather large burden for me to bear alone, so I sought the help of neighboring countries. That was the start of the Pan-Pacific Council on Ergonomics (PPCOE). It first met here in Ramazzini Hall in 1990.

I followed that up by working hard to create ties between the Japan Ergonomics Society (JES) and the Ergonomics Society of Korea (ESK). The Department of Ergonomics, University of Occupational and Environmental Health (UOEH), Japan was the site of several meetings convened to bring about those ties. My third project involved efforts to create ties between the Society of Occupational Safety, Health and Ergonomics (SOSHE), and the Ergonomics Society of Taiwan (EST).

Today, the ties between SOSHE and EST are strong. And now we have achieved this symposium that is an East Asian conglomeration transcending bilateral ties. The means I used to convene this symposium were to invite the ergonomists of the East Asia region individually. I asked the ergonomics society of each country to participate. I established 11 areas for presenting research results in light of the current state of ergonomics research in Asia. These are cognitive ergonomics, ergonomics in manufacturing, ergonomics modeling and usability evaluation, ergonomics in occupational health, physical ergonomics, safety ergonomics, social and organizational ergonomics, aging and work, gender and work, universal usability design, and kaizen. As a result, the symposium has been favored with the active participation of many countries despite the global economic downturn.

Discussions were held regarding a road map outlining how ergonomics activities in East Asia should progress, as well as the ideal form of a practical standard for qualifications in the East Asian conglomeration. Indeed, a truly appropriate volume of research information was provided for the launch of this symposium that aims for the unity of ergonomics research in East Asia. This book consists of 56 papers carefully selected from among those presented at the symposium. I hope it will be useful for the research of East Asian ergonomists in the future.

In conclusion, I would like to express my thanks to President Osami Hagiwara and Akiko Inage of Alphacom for their great assistance in publishing this book.

Masaharu Kumashiro,
Professor of University, Occupational and Environmental Health, Japan
Chairman of the 1st East Asian Ergonomics Federation Symposium

Special lectures

Ergonomic Trends from the East – Kumashiro (ed)
© 2010 Taylor & Francis Group, London, ISBN 978-0-415-88178-4

Will visual discomfort influence on muscle load and muscle pain for Visual Display Unit (VDU) workers?

A. Aarås, G. Horgen, M. Helland & T.M. Kvikstad
Department of Optometry and Visual Science, Buskerud University College, Kongsberg, Norway

ABSTRACT: In order to study more in detail the correlation between visual stress and muscle load, two laboratory studies were carried out. Further, to study the correlation between visual discomfort and muscle pain four prospective epidemiological studies were performed.

In the laboratory studies, visual stress was induced by the size of characters on the screen (8 points and 12 points Times New Roman) and the luminance levels in the surroundings of the screen (between 1500 and 2300 cd/m^2) versus (between 70 and 100 cd/m^2). The results showed that the smallest characters 8 points and the highest luminance levels had no significant influence on the muscle load in neck and shoulder regions. However, the productivity was significant lower when using 8 points characters compared with 12 points. There was also a tendency to an increase in the number of errors made.

In three different prospective epidemiological studies, correlations between visual discomfort and average pain intensity in the neck and shoulder, were $0.30 < r < 0.72$ for VDU workers. In the first study, correlation between visual discomfort and pain in the neck and shoulder was $0.30 < r < 0.40$. In the second study, visual discomfort correlated to neck pain ($r = 0.40$, $p = 0.003$). In the third study, correlation were for neck pain ($r = 0.69$, $p = 0.000$) and shoulder pain ($r = 0.72$, $p = 0.000$). In the fourth study, the correlation for neck pain was ($r = 0.44$, $p = 0.038$).

1 LABORATORY STUDIES

Two papers of these studies are published by Horgen *et al.* (2005 and 2007). The aims of the studies were to evaluate how the luminance levels of the surroundings of VDU and the size of the characters on the screen effect the muscle load.

1.1 *The design and method of the study*

The design and the methods of the study are described by Horgen *et al.* (2005). The "glare" luminaries gave a luminance between 1500–2300 cd/m^2 (measured across the screen). Two "glare" luminaries were mounted vertically on the right side of the VDU, at approximately 45° horizontal angle from the sightline to the centre of the screen, simulating windows as they very often appear in a normal work station set up. The work task was interactive work on a 15 inch LCD screen. The test set up is shown in Fig. 1. The lowest luminance level of the surroundings of the screen, (between 70 and 100 cd/m^2), and the normal size of the characters on the screen, (12 points New Roman), were defined as baseline. This baseline was recorded for each participant at the start and the end of the trial. The mean of these two measurements was used as a baseline in the statistical analysis (Horgen and Aarås 2003). The smallest text size was 8 points Times New Roman. The combination of high luminance/normal character size, high luminance/small character size and low luminance/small character size was tested according to the orthogonal Latin square design. The postural load on the neck and shoulder muscles was quantified by electromyography (EMG) using the Physiometer. Surface electrodes were used (Aarås *et al.* 1996). The load in m. trapezius (descending part) and m. Infraspinatus was used as indicators of load on the neck and shoulder areas. To perform continuous measurement of postural angels, three dual axis inclinometers were used. Angles were measured relative to the vertical by these inclinometers attached to the upper

Figure 1.   The Workplace, with glare source.

arm, head and back. The angle measurements were mainly used to control the work posture during the VDU work. The EMG and the postural angle methods are described and the methodological limitations discussed (Aarås *et al.* 1996; Hagen *et al.* 1994). There were five test sessions. Each session lasted 10 minutes of active recording, with a period of rest in between. The rest period was about 5 minutes. 16 subjects were needed in order to detect a difference in muscle load between 0.5 to 1% Maximum Voluntary Contraction (MVC) at a power level of 80%. (Horgen *et al.* 2005).

1.2   *Results*

The size of the characters and the glare condition had small influence on the muscle load.

M. trapezius activities did not show significant differences when comparing the mean of the two baseline measurements with muscle activities when working with small characters and glare. This was true both for static ($p = 0.21$) and median values ($p = 0.07$) (Horgen *et al.* 2005). This was opposite what to be expected. For the median muscle load, there was significant higher activity at baseline than when working with small characters with glare ($p = 0.008$) and small characters without glare ($p = 0.015$). M. infraspinatus was in most cases relatively heavy loaded. There were no significant differences when comparing the static value of the baseline measurement, working with small characters with glare ($p = 0.11$) and small characters without glare ($p = 0.14$). However, when similar comparison for median muscle load was done, there were significant higher activity at baseline then when working with small characters with glare ($p = 0.008$) and small characters without glare ($p = 0.015$). Erector spina lumbar part, at L3 level did not show significant differences between the baseline and the three test situations. This was true for both static and median values ($0.13 < p < 0.96$). Productivity, in terms of amount of text processed was significantly reduced when working with 8 points characters. In addition there was a tendency to increased number of errors when working with glare.

1.3   *Conclusions and recommendations*

Working with small characters and glare did not increased static muscle load for Trapezius, Infraspinatus and Erector spina. M. infraspinatus was relatively heavy loaded during this type of computer work due to high precision-dependence during tracking work. For presbyopic VDU workers, the character size should be more than 8 points letters.

2   LONGITUDINAL PROSPECTIVE EPIDEMIOLOGICAL STUDIES

The aims of these studies were to investigate the correlation between visual discomfort and pain in the upper part of the body. Longitudinal epidemiological studies were performed to evaluate the aims (Aarås *et al.* 1998; Aarås *et al.* 2005, Helland *et al.* 2008 and 2008).

Table 1. Correlation (r) between Visual discomfort and body pain.

First study (N = 150)		Second study (N = 90)		Third study (N = 34)		Fourth study (N = 32)	
Body area	r-value	Body area	r-value	Body area	r-value	Body area	r-value
Neck	0.30 < p < 0.40	Neck	r = 0.40, p = 0.003	Neck	r = 0.64, p = 0.000	Neck	r = 0.44, p = 0.038
Shoulder	0.30 < p < 0.40	Headache	r = 0.34, p = 0.01	Shoulder	r = 0.56, p = 0.001		
				Forearm	r = 0.35, p = 0.04		

2.1 The first study

This was a prospective epidemiological study where VDU workers were followed for a period of six years. Visual discomfort showed a relationship with pain intensity in the neck and shoulder $(0.30 < r < 0.40)$ (Aarås *et al.* 2001). The level of discomfort/pain was assessed on a Visual Analogue Scale (VAS). Visual discomfort was 29.9 (21.7–38.09) and shoulder pain 23 (15.3–30.7) as group mean with 95% Confidence Interval (CI). Zero was no pain 100 indicated extreme or unbearable pain. However, such studies have a lot of confounding factors such as organizational and psychosocial factors. For the psychosocial factors, there was no statistical intervention effect or time effect and no interactions between time and intervention were found.

2.2 The second prospective study

This study was a multidisciplinary multinational ergonomic study MEPS (musculoskeletal-eyestrain – psychosocial – stress). The objective of the study was to examine the effects of various kinds of ergonomic interventions including corrective lenses on a combination of musculoskeletal, postural, and psychosocial outcomes among VDU workers. In this study, visual discomfort was related to neck pain, $r = 0.40$, $p = 0.003$; regression coefficient 0.37 with CI of 0.18–0.57. Neck pain was also related to burning and itching of the eye $(p = 0.004)$. Headache was related to visual discomfort, $(r = 0.34, p = 0.01)$ (Aarås *et al.* 2005).

2.3 The third epidemiological study

This is a follow up period which covers from 6 to 13 years of the same study as described in 2.1, The results showed a significant correlation between visual discomfort and neck pain $(r = 0.64, p = 0.000)$ as well as shoulder pain $(r = 0.056, p = 0.001)$. For the forearm this correlation was weaker, but still significant $(r = 0.35, p = 0.04)$. In a multivariable regression model when lighting and glare were excluded, visual discomfort explained 53% of the average of the neck and shoulder pain (Helland *et al.* 2008).

2.4 The fourth epidemiological study

The design of this study was the same as the study described in 2.1. The follow up period was only one year. The results showed a significant correlation between visual discomfort and neck pain $(r = 0.44, p = 0.038)$ In a multivariable regression model containing headache, subjective tenseness and dry mouth as predictor for neck and shoulder pain, these independent variables explained 76% of the variance of neck and shoulder pain. By considering each of the three variables separately in a simple regression model the explanation percentages were subjective tenseness 57%, headache 44% and dry mouth 28% (Helland *et al.* 2008).

2.5 Conclusion

Three different prospective epidemiological studies have shown that there is a clear indication of a relationship between visual discomfort and pain in the neck and shoulder. In a laboratory study visual stress had small influence on the muscle load. A reasonable explanation of the differences in the results between the epidemiological and the laboratory studies may be that in the laboratory study, the visual stress in terms of small character and glare reduced the productivity. Reduced productivity may reduce the static muscle load and pain. According to a study by Helland *et al.* (2006) glare had a significant correlation to visual discomfort, $r_s = 0.35$, $p = 0.040$. They showed also that visual discomfort explained 53% of the variance of the neck and shoulder pain in VDU workers.

REFERENCES

Aarås, A., Veierød, M. B., Ørtengren, R. and Ro, O. 1996. Reproducibility and stability of normalized EMG measurements on musculus trapezius. *Ergonomics* 39 (2), 171–185.

Aarås, A., Horgen, G., Bjørset, H-H., and Ro, O. 1998, Musculoskeletal, visual and psychosocial stress before and after multidisciplinary ergonomic interventions. *Applied Ergonomics* Vol. 29, No 5, 335–354. Elsevier Science Ltd, Great Britain.

Aarås, A., Horgen, G., Bjørset, H-H., Ro, O. and Walsøe, H. 2001. Musculoskeletal, visual and psychosocial stress in VDU operators before and after multidisciplinary ergonomic interventions. A 6 years prospective study – Part II. *Applied Ergonomics* 32, 559–571.

Aarås, A., Horgen, G., Ro, O., Mathiasen, G., Bjørset, H-H., Larsen, S. and Thorsen, M. 2005. The effect of an Ergonomic Intervention on Musculoskeletal, Psychosocial and Visual Strain of VDT Data Entry Work: The Norwegian Part of the International Study. *International Journal of Occupational Safety and Ergonomics (JOSE)* Vol. 11, No 1, 25–47.

Hagen, K. B., Sørhagen, O. and Harms-Ringdahl, K. (1994). Influence of weight and frequency on thigh and lower-trunk motion during repetetive lifting employing stoop and squat techniques. In K. B. Hagen (Ed.), Physical work load and percieved exertion during forest work and experimental repetetive lifting – *thesis.Stokholm: Karolinska Institutet – Stokholm.*

Helland, M., Horgen, G., Kvikstad, T. M., Garthus, T., Bruenech, J. R. and Aarås, A. 2008. Musculoskeletal, visual and psychosocial stress in VDU operators after moving to an ergonomically designed office landscape. *Applied Ergonomics* 39, 284–295

Helland, M., Horgen, G., Kvikstad, T. M., Garthus, T. and Aarås, A. 2008. Will Musculoskeletal, Visual and Psychosocial Stress Change for Visual Display Unit (VDU) in VDU Operators When Moving from a Single-Occupancy Office to an Office Landscape? *International Journal of Occupational Safety and Ergonomics (JOSE)* Vol. 14, No.3, 259274

Horgen, G. and Aarås, A. (2003). Visual discomfort among vdu-users wearing single vision lenses compared to vdu-progressive lenses. *Paper presented at the Human Computer International 2003, Crete.*

Horgen, G., Helland, M., Kvikstad, T. M., Aarås, A. and Bruenech, J. R. 2005. Do luminance Levels of the Surroundings of Visual Display Units (VDU) and the Size of the Characters on the Screen effect the Accommodation, the Fixation Pattern and the Muscle Load during VDU Work. *HCI International Conference July 22–27, 2005 Las Vegas, Nevada USA.*

Horgen, G., Helland M., Kvikstad, T. M. and Aarås, A. 2007. Do the Luminance Levels of the Surroundings of Visual Display Units (VDU) and the Size of the Characters on the Screen Effect the Accommodation, the Muscle Load and productivity during VDU Work? *HCI International Conference July 22–27, Beijing China SpringerCD-Rom ISBN 978-3-540-73738-4.*

Differences in thinking and reasoning in between East Asian and Western approaches and their impact on organizational issues related to the field of ergonomics

K.-D. Fröhner & K. Iwata-Fröhner
System- und Orgplanungen, Bad Oldesloe, Germany

1 METHODOLOGICAL REMARKS

A topic with a wide scope asks for laying open definitions, procedures and ideas being applied especially when different scientific approaches and methodologies are involved. These prerequisites for scrutiny are important steps to an issue venturing generalization.

When dealing with organizational issues with a wide scope the term of macroergonomics is often used, considering the way organizations are designed and managed. Macroergonomics views the organization as a sociotechnical system and incorporates concepts and procedures of sociotechnical systems theory in the field of ergonomics. It enables to approach interdisciplinary environments (Brown, 1990).

The organizations, being dealt with and conclusions are being derived from, are in a first phase of empirics small and medium sized companies of metal processing industry. In the second and third phase companies of different background are included.

Starting point to examine production of metal processing industry in the end of the 80-s, was the common interest of German and Japanese Technical Universities and their cooperating industrial partners in the field of reducing set-up times of machines. Two symposia were held in Japan and in Germany presenting the theoretical ideas pursued in the field followed by company visits of the industrial partners. Accompanying confrontational comparisons of the involved companies could be realized.

In a second phase of field research in the 90-s longitudinal in deep studies were carried out. By this intensified field research it became evident that culture plays an important role when coming up with explanations not only how but also why things were/are done in a particular way.

In a third phase of studies in the first decade of the second millennium the findings were evaluated and enhanced under economic conditions that had changed considerably. New combinations of organizational techniques based on Japanese culture could be found, exemplarily pointing out the complex approach in one company, showing the ability to cope with changing economic conditions.

Finally, it will be explained why the depicted cultural approaches are unique on the one hand side but offer high potentials for generalization towards Western and East Asian approaches on the other hand side.

2 RESULTS OF FIELD RESEARCH IN WEST-GERMANY AND JAPAN IN THE LATE -80S

In metal processing industry the added value is realized by organization, personnel and technology. No production technology differing from country to country could be found in the field studies. Therefore, no emphasis was put on technology in the survey covering the shop floor levels of ten Japanese and West-German companies. A half-structured questionnaire was used (Fröhner, 1990).

When looking closer at shop floor levels for parts production it became obvious that the coordination of production was accomplished in completely different ways in both countries. West-German companies relied on standardized software for production planning and control. There were many computer terminals on shop floor level to coordinate personnel, machine and material. Mainly the blue-collar workers had to inform the computer system when they started or finished a job. Some systems also offered support to the workers to realize a more sophisticated time schedule due to their needs and that of production. Mostly direct production, especially parts production, was the object of an accurate planning. The targets of production were very much related with the individual work place. A high standard of information of not only when but also how the work had to be realized was accomplished. Very accurate and precise information for operations planning concerning operation descriptions and standard times were performed.

When coming up to shop floor levels for parts production in Japan, it was obvious that there were no terminals for production planning and control and the workers got only little information concerning operation descriptions and the time planned for specific operations. But there was less coordinative necessity, because the workers had sound knowledge of many work places and received information concerning the strategy and the goals of the company.

The elements for coordination 'personnel' and 'organization' were marked in different ways in both countries (figure 1). Looking at West-German companies, there was frequent use of standard programs for production planning and control. Because of the function oriented layout the packages could be used in many companies, containing demanding models being adaptable to various organizational settings. This layout corresponds with the principles of education in West-Germany stressing the individual ability in special fields. Workers and employees mostly stay in the same or an adjacent field even when they change the company. This is stabilized by nationwide professional organizations, which support to create professional ethics in special fields. The comprehensive information concerning orders were/are laying ground for a middle-management orientated control strategy.

In Japan professional organizations were not strongly developed. Special professional ethics were and still are not marked in detail. Workers and employees have a high general qualification and very often move from field to field in one company. The coordination is arranged by groups also because it was nearly impossible to give voluminous written information to the shop floor level until the beginning of the 80-s. The complicated Kanji, Hiragana, and Katakana writing with round about 2000 Kanji (pictographic signs) and 100 syllable signs could practically not be written with typewriters. Until then, work orders, routing plans and bills of material e.g. had to be written by hand.

The barriers in the Kanji, Hiragana, and Katakana writing had to be overcome by the layout of production and product, which could be managed with little written information as KANBAN, TQC and KAIZEN, grounded on elements of Japanese culture. So, Japanese companies transformed a cultural barrier into an accelerator for the efficiency of production systems as the temporarily world-wide leading nation in production. This meant that Japan was able to modernize after World War II and to come up as a leading industrial nation incorporating Non-Western approaches.

WEST-GERMANY		JAPAN	
Organization	Personnel	Organization	Personnel
– many standard programs – many software adaptations – function-oriented specialization – planning of many functions – middle managemento control strategy – very specific written orders	– specialists – individual motivation – extern oriented labour market – extern oriented ethics – extern oriented trade unions	– very few standard programs – accurate planning of machine tool areas – focussed on the product – only little written information – integration by general qualification – coordination by groups	– general qualification – no individual motivation – very equal comp. efficiency wages – internal labour market – no specific ethic – intern oriented trade unions

Figure 1. Marking of organization and personnel in West-German and Japanese small and medium sized metal processing industry in the late 1980-s.

3 IN DEEP LONGITUDINAL STUDIES IN THE EARLY 90-S AND THE IMPORTANT ROLE OF CULTURE CONSTITUTING THE WAY OF THINKING, REASONING AND REFERRING

The investigations in the second phase of surveys covered the whole scope of production starting with design and ending up with assembly (figure 2).

Two companies more closely examined were a Japanese machine tool company and a German manufacturer of electromechanical products. Both companies had round about two hundred employees. Their products consisted of up to several thousand parts or components.

The German company realized out of a newly established CAD-design a very extensive production planning and control in combination with a data capturing system on shop floor level. Material control and capacity planning were realized by means of demanding standard programs with precise cost-oriented targets. In account of a long tradition and since there did exist no problems with writing, there is a high functional specialisation in this field in Germany. One of the main topics in this field is economic lotsizing, with the consequence that parts of different custom orders form different lot sizes, joined together in production planning and control for parts production. After parts production, they are being allocated to the custom orders. The result is a kind of piece work wages in parts production. Reasoning and inferring on shop floor could only be realized on a very mechanistic level serving the functions. The linking of the highly effective and specialized functions is looked for to be realized by software.

The Japanese company focussed on the design department concerning information technology since many years. To support new ideas was in the centre for the use of new information technology. A department for operations planning did not exist and therefore no routing plans and standard times were given. The operations planning had to be generated on the basis of the drawings from the design department by the groups on shop floor level. The deadlines for producing the modules of the customers' products were fixed by hand on the basis of an outline plane of the customer order. The parts needed for the modules of the custom order were kept together physically and produced together. No economic lotsizing was realized. The connection of the custom order was visible from the beginning to the end of production. The extent of usage of information technology concerning the technology in parts production was the same in the Japanese as in the German company. Both companies did not use special software for coordination in the assembly departments.

For a long time European managers identified well functioning industrial relations and the flexibility in working hours as the main causes of Japanese effectiveness and efficiency to explain the rapid growth of Japanese production abilities. But as depicted so far, it is obvious that on all levels of production cultural influences have a high impact on how production is generally laid out. So it is interesting to look at not only how but also why things are done in different ways. For scrutiny special cultural elements that are seen to have a high impact on the way things are done are being selected. As culture and the natural environment are connected through intellect, mind and brain it is shown that this leads to a set of elements determining thinking, reasoning and inferring. The cultural elements being selected are the Hiragana, Katakana, and Kanji writing, the tradition of thought, and familism with groupism (figure 3).

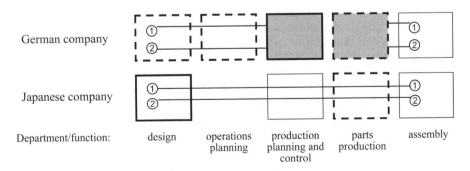

Figure 2. Use of information technology (extensive, partial, slight) and coherence of customer order (two orders) information in selected departments of a German and a Japanese company.

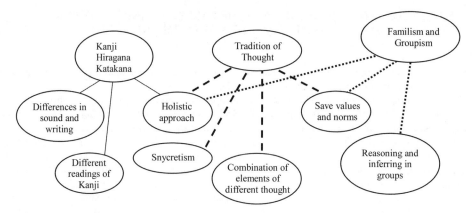

Figure 3. Elements of Japanese culture and their relations to some elements of thinking, reasoning and inferring.

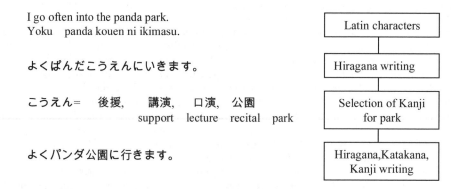

Figure 4. Writing a sentence on a computer with a Japanese word processor.

3.1 *Hiragana, Katakana, and Kanji writing*

Japanese writing is a pictographic writing (Kanji) combined with two syllables (Hiragana and for foreign words Katakana). It consists of approximately 2000 Kanji and 100 syllable signs. Therefore, typewriters did not really exist in Japan. Written information had to be written by hand and hence was very often reused.

Today, word processors, which are installed on computers, support the Hiragana, Katakana, and Kanji writing and laser printers are able to produce very difficult pictographic and syllable signs. The starting point of writing with a computer is normally to start with Latin characters and then to transform the writing into Hiragana syllables (figure 4), the phonetic transcription of Japanese. The next step is to transform it with the word processor into Hiragana, Katakana, and Kanji. But the phonetic writing can be depicted by different Kanji, called homonyms, as shown in figure 4 for the sound kouen. The meanings of the different Kanji can be a generic-subordinate, analogical, and opposite terms or a limited thought or area. Furthermore, in Japanese the number of homonyms does not decrease, it increases (Suzuki, 1990, pp. 83). So the proper Kanji have to be selected to depict the correct Hiragana, Katakana, and Kanji writing, as shown for the sound kouen (figure 4).

This leads to the conclusion that the specific meaning of a sentence is only revealed in the context of the full text. The hearer has possibly to bear in mind the spoken text so that he, by inferences, may be also of several sounds, can conclude the meaning. That's why Japanese people are often not sure whether to have understood a spoken text or not. Furthermore, different people may come to different conclusions with regard to the content.

Therefore, ambiguity, vagueness and contextual reasoning are basic elements of the Japanese language, also because Kanji can be read in Japanese or Sino-Japanese, and there can be several ways to read a Kanji.

The Japanese borrowed the Kanji from the Chinese language, made up only out of Kanji, also being ambiguous because of different Kanji for one sound. The problem of writing did/does exist also for the Chinese language. In Korea however, for the most part Korean is written in Hangeul, an alphabet having been developed in the 15th century consisting of 24 characters. The usage of Chinese characters has decreased in Korea, diminishing writing problems.

Looking at languages with Latin characters from the point of sound, writing and meaning, one can state a close connection in between sound, writing and meaning, except for few homonyms (in English e.g. night and knight). Problems producing written information did not exist since long because of typewriters.

3.2 *Familism and groupism*

There are two organizations of Japan's premodern social tradition that are seen to have exerted on modern Japanese behaviour (Hall, 1985). One is the remnant of the old Japanese family system (the ie), the hierarchical samurai-style corporate family incorporating new members by adaption and enrolment into the household. This meant that membership could be either by kinship or by contract. The ie was abolished with the Japanese constitution in 1947. The second organization is the consensus-type community exemplified in the rural village (the mura). The mura is seen as a type of organization based on consensual patterns of decision making. These two organizations have been part of the Japanese society since hundreds of years having subordinated individual interests to those of the state or the group for fulfilling common functions (Hall, 1985). Out of this tradition, today organizations still care for the development of the manpower and guarantee employment for their members setting great store by equal wage structures seen by German standard. Therefore, Japanese employees easily identify with the company and perform jobs and tasks they had not been educated for giving little space for professional ethics being influenced by external organizations. They often thrive for becoming an accepted member of the company and thereby their requirements, wishes, and talents being respected.

The consensus type of community in the rural village is seen as a necessity of rice cultivation, ubiquitous in many areas of East Asia, because of common needs for irrigation and the maintenance of irrigation systems. Both types of organization are not known in Western countries, at least not in modern times.

3.3 *Tradition of thought*

In contrast to Europe, there have been only few elementary changes in the orientation of thinking in the last centuries in Japan (Maruyama, 1988). For example, the archaic Japanese religion of Shinto is still dominant in society although it was influenced in the past centuries by several strong religions (Buddhism, Konfuzianism and Christianity). There is no example of an archaic religion having such influence on an industrialized country (Immoos, 1990). Moreover, religions that came to Japan in the last centuries were finally accepted in the society and even enriched the Shinto dogmatically (Maruyama, 1988). Religions are really coexisting in Japan and today people do not have ethical problems to adopt different religions for different reasons: Shinto to ask for successful business, Buddhism for the funeral ceremony and Christian church for wedding, called syncretism. The religions adapted to the sociocultural conditions and even have no teaching supervision of priests (Miyasaka, 1994). Pragmatism is exhibited by the world religions in Japan. Going along with that, combining elements of completely different thought is normal in Japan.

There had been more elementary changes in the orientation of thinking in the last centuries in China and Korea, also because of Japanese occupations. But still, archaic religions are present, Taoism and Shamanism, seen as part of the culture or a philosophy, giving way to syncretism.

Religions having originated and carrying out worshipping in the West cannot be used according to different situations of life. The creed of one religion restricts all possibilities of worshipping to this religion. Not only because of this, is tradition of thinking completely different esp. in Europe and the West, strictly sticking to the exemplified issues.

11

Do the ways of thinking going along with the writing system etc. have a deeper influence on the principles that govern the thought and the behaviour? Kumon does believe it and gives the following explanations in respect to Westerners and to Japanists, the latter ones being people applying contextual ways of thinking (Kumon, 1982).

By his interpretation recognition processes start and proceed in differentiated ways when mastering a complex whole. "In view of the fact that recognition processes of the Japanese begins with a vaguely grasped whole, we say that they are holists. At the same time it is obviously that they also can, at least tentatively, think dualisticly or pluralisticly by applying their power of analytic reason. They are aware of the possible conflict between the 'whole' and the 'parts'. . . .. However, Japanese tend to see the dualistic components being complementary, not conflicting". The personality of contextualists is seen to adapt to surrounding and to personal contacts. An essential element in the contextualistic order is the idea that in the world with its complexities it is impossible to name the single contributions of a context's member. He sees the recognition processes of Westerners when mastering a complex whole differently. Westerners are seen as tending to divide the whole into its parts and thus go on with procedures from the elements to the larger whole. Westerners are seen as individualists who come near to Weber's ideal type of an ascetic puritan acting as an independent individual out of an inner responsibility with regard to higher norms (Kumon, 1982).

4   ORGANIZATIONAL APPROACHES IN THE BEGINNING OF THE NEW MILLENIUM

The advantages of the Japanese approaches seemed to fade in the second part of the 90-s at the same time as these approaches and techniques were applied world-wide. With a tumbling banking sector and companies of the automobile sector, which could partly survive only by the help of their American or European competitors, the focus of international discussion on Japanese techniques disappeared. In the beginning of the new millennium Japanese companies came back with improved productivity showing that they are able to cope with changing economic conditions. The studies belonging to this phase already started in the second part of the 90-s involving four companies, exhibiting here a complex approach, closely linked with contextual thinking and reasoning in one company (Fröhner and Iwata-Fröhner, 2003).

Adjusting the strategic orientation of companies to the situation of the national economy is closely linked with the contextual thinking and reasoning of Japanese. Within the vague approach of adaptation it had to be considered that Japanese citizen have a high annual saving rate of about 8% of their annual income because expenditures for education are high and retirement income by the social system is low. Furthermore, the citizens were tangled by increasing unemployment rates. Because of holistic orientation the companies accepted the changed economic situation and as a first step developed products with low costs and high usability.

Japanese automobile industry had a good basis for the anticipated customer needs because of long experience with production of subcompact cars having special tax exemptions in Japan. Enhanced usability of these cars could be reached by placing the users as if sitting upright on a chair on the chassis of a subcompact. Thereby, the first microvans with a rectangular form offering seats for four persons and a small trunk were developed. Adapting the cars to customer usage, video sequences within field studies were realized. One outcome was e.g., to place the fuel tank under the front seat instead of the rear, realizing a quite voluminous trunk space. By an ergonomic point of view, the dimensions and the layout of microvans make it easy to get in and out and to place a baby on the back seat addressing the elderly and young families. Today Japanese car companies produce the third generation of microvans and additionally create editions of different subcompact cars for special niches. This can be seen as consistent consumer oriented product development.

But regaining high competitiveness cannot be explained by only one field of the holistic approach (figure 5). One more contextual element was the focussing of the company's culture on special belongings. Therefore, the national influence lasting for many years through ministries on employment strategies was reduced. But it became also obvious that contextual attitudes can lead to self-destruction, when the targets and the strategy of the management do not take into account higher norms and standards. But if the strategy and the targets are open and integrate higher norms, Japanese employees contribute their best and deepen the community.

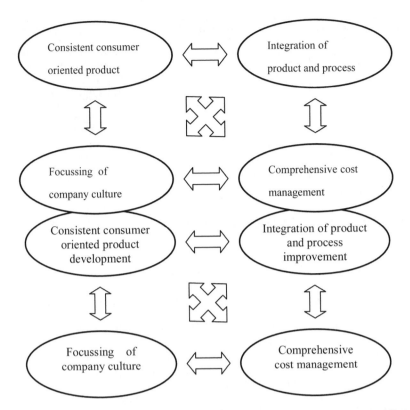

Figure 5.　Main elements of shaping and forming improvement processes in Japanese automobile industry.

The integration of product and process improvement was/is driven by the steady development of new variants, integrating gradual systematic improvement, not realizing the great design as realized by Audi A2 or Mercedes A-class. Consequently, the assembling lines are relatively unchanged for many years. The improvement processes aim at realizing the economy of scale and scope in one assembling line.

Improvement processes are driven by cost management incorporating all departments and all employees. It consists out of target costing and kaizen costing (Monden and Hamada, 2000). Target costing is an approach also realized in industrial sectors of Western industries because of the increasing change of products. Kaizen costing can be explained as a year by year target cost reduction for each cost element. The cost reduction targets are set and applied on a monthly level (Monden and Lee, 2000). So Kaizen was changed from a volatile to a target driven approach not taking into account the stress and strain induced in the employees. Establishing target and kaizen costing is facilitated by the fact, that the Shinto asks for yearly endeavours in life and supports joint working efforts. The first Japanese constitution by Prince Shotoku already excluded lonely decisions in article 17.

Figure 6 shows the evident results of the target and kaizen cost management in a car company concerning design with development and production. In this case the models Mira (in Europe Cuore), the microvan Move and the sports car Copen are assembled in one line at Daihatsu's central plant. It becomes evident that the resources in the car company are limited when taking the aesthetical building standards into account. But these standards emphasize the sparing use of resources to the customers. On the other hand there is a high customer commitment, when customers visit the company.

Therefore, the enhanced productivity in Japanese automobile industry is based on complex elements of the Japanese culture. By this approach copying is not as easy as with the organizational techniques depicted before.

13

Corporate Function	Ways of Acting	Intended Effects
Design and Development	Integrative development of sedans and sport cars: joint assembly	Considering many conditions: → many identical parts → high capacity load → flexibility of variant
Production	Extensive product mix in final assembly: small scale kanban	Considering many conditions: → many kanban → flexibility of scale

Figure 6. Evident results caused by the way of acting on behalf of target and kaizen cost management in automobile industry.

On the other hand side an evolutionary perspective may be supported as asked for by Hipsher *et al.*

5 DEVELOPMENT OF WESTERN AND EASTERN APPROACHES

It can be stated for both approaches that there had been and still are two consistent processes of development.

The German approach being depicted in the empirics seems to be on a first glance unique, but is being grounded on a basic element of Western culture. This element is the ability to produce written text with typewriters and thereby to centralize the knowledge of a company concerning production processes. The written information could easily be reproduced since the 20-s of the last century. With the rise of computers the access towards software was feasible for those using typewriters with Latin characters, because of the layout of the keyboards. So, Westerners could write software and install databases, the latter because of tens of years of experience collecting and formalizing written information for production. With the increase of information and knowledge specialization increased also. So, a consistent development in standardized software can be seen by at first focussing on direct production as parts production and then moving to indirect processes ending up with software integrating production, sales and economics, called ERP = enterprise resources planning. This software is offered by German and American companies (e.g. SAP and Oracle) setting standards in information processing going along with high monthly fees, fostering principle orientation and function oriented specialization.

The Japanese approach can be seen as an access towards the problem of coordination by building up mental models for the employees of what could be done and how it should be realized. This is accomplished to a high extent by talking and consensual decision making. The mental ability of the employees is not only being stimulated and organized, it is also used to install esp. cross-functional improvements using and enhancing implicit knowledge. So the strategy is to look for new options from a holistic point of view and not to look for full written information. This approach could also be seen as an East Asian approach, because East Asians are open to holistic approaches as depicted in chapter 3, and can easily install this kind of approach for organizational issues. To build up the holistic approach to full extent two prerequisites have to be achieved in the future:

To build up information systems with limited information supporting and not displacing information and

To analyze the effect of stress and strain of the complex approaches induced in the employees.

To follow up with the successful implementation of techniques seeing the employees in the centre of production great challenges have to be mastered, to support a long lasting evolutionary perspective.

REFERENCES

Brown, O. Jr., 1990, Macroergonomics: a review. In Human Factors in Organizational Design and Management – III, edited by Noro, K. and Brown, O. Jr. (Amsterdam: North-Holland), pp. 15–20.

Fröhner, K.-D., 1990, Cultural barriers are accelerators for the efficiency of production systems. In Human Factors in Organizational Design and Management – III, edited by Noro,K. and Brown, O. Jr. (Amsterdam: North-Holland), pp. 423–426.

Fröhner, K.-D. and Iwata-Fröhner, K., 2003, Kulturelle Elemente als Verzögerer und Beschleuniger für die Effektivität und Effizienz von Produktionssystemen. Japanische Unternehmen erfolgreich im turbulenten Umfeld. In REFA-Nachrichten, Vol. 56, No. 4 (Darmstadt: Refa), pp. 4–10.

Hall, J.W., 1985, Reflections on Murakami Yasusuke's Ie Society as a Pattern of Civilisation. In Journal of Japanese Studies, Vol. 11, No.1 (Seattle: University of Washington), pp. 47–55.

Hipsher, S.A., Hansanti, S. and Pomsuwan, S., 2007, The Nature of Asian Firms. An evolutionary perspective, (Oxford: Chandos Publishing). Immoos, Th., 1990, Japan. Archaische Moderne, (München: Peter Kindt Verlag).

Kumon, S., 1982, Some principles governing the thought and behaviour of Japanists (contextualists). In Journal of Japanese Studies, Vol. 8, No.1 (Seattle: University of Washington), pp. 5–28.

Maruyama, M., 1988, Denken in Japan, (Nihon no shiso, De aru koto to suru koto, Nihon no chishikijin), (Frankfurt: Edition suhrkamp).

Miyasaka, M., 1994, Shinto und Christentum. Wirtschaftsethik als Quelle der Industriestaatlichkeit. Thesis Universität Würzburg, (Paderborn: Bonifatius).

Monden, Y. and Hamada, K., 2000, Target Costing and Kaizen Costing in Japanese Automobile Companies. In Japanese Cost Management, edited by Monden, Y. (London: Imperial College Press), pp. 97–122.

Monden, Y. and Lee, J., 2000, Kaizen Costing: Its Function and Structure compared to Standard Costing. In Japanese Cost Management, edited by Monden, Y. (London: Imperial College Press), pp. 229–242.

Suzuki, T., 1990, Eine verschlossene Sprache. Die Welt des Japanischen (Tozasareta gengo-nihongo no sekai), (München: iudicium Verlag).

Ergonomic Trends from the East – Kumashiro (ed)
© *2010 Taylor & Francis Group, London, ISBN 978-0-415-88178-4*

Do modern office workers need more stress at work?

Leon Straker
School of Physiotherapy, Curtin University of Technology, Perth, Australia

1 INTRODUCTION

Physical stress is a necessity for human health. Low physical stress, as in space travel, bed rest or joint immobilisation, results in short term harm. Long term inactivity is associated with a range of musculoskeletal, cardiovascular, metabolic, mental and other disorders. Societal trends are creating a larger group of office workers whose tasks require little activity. This is combining with trends for less activity in leisure, domestic duties and travel. Office workers therefore appear to be at risk due to insufficient physical stress at work (Straker and Mathiassen 2009). This paper outlines the trends for increased sedentary work and examines the opportunities and challenges for ergonomics to design office work which can have a positive effect on health, productivity and satisfaction.

1.1 *Changing employment and activity patterns*

Affluent countries, such as Japan, have seen a change in workforce employment patterns over the last few decades – to more sedentary employment. This has been associated with a reduction in people working in traditionally activity industries and a change in the physical demands in occupations within industries.

Figure 1 shows data from the Statistics Bureau of Japan (www.stat.go.jp/english) showing the number of workers in selected industries. The decline in the number of workers in traditionally physically demanding jobs in the agriculture, forestry and construction sectors is clear. This is contrasted with the rise in the number of workers in service industries. Whilst service industries include some physically demanding occupations, they also include office based occupations such as financial services. Office based occupations are less active, with Tudor-Locke and Bassett

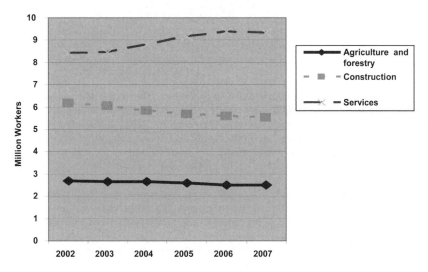

Figure 1. Japanese workers in agriculture, forestry, construction and services industry sectors (data from www.stat.go.jp/english).

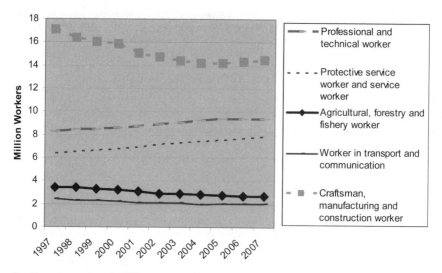

Figure 2. Japanese workers in different occupations (data from www.stat.go.jp/english).

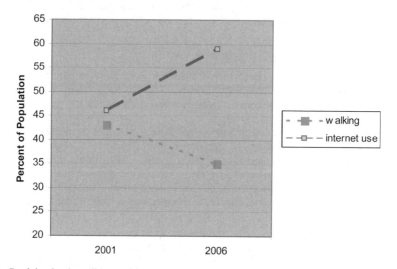

Figure 3. Participation in walking and internet use by Japanese (data from www.stat.go.jp/english).

(2004) reporting office workers accumulated only around 5,000 steps per day, around the half the commonly recommended 10,000 steps per day (Exercise and Physical Activity Reference for Health Promotion 2006).

Similarly, Figure 2 shows the decline in the number of workers employed in craftsman, manu-facturing and construction, agricultural, forestry and fishery occupations. Again this is contrasted with the growing number of workers within the professional, technical and service (largely office based) occupations.

Even within traditionally demanding occupations such as mining and forestry, engineering and ergonomics innovations have increased the mechanisation of physically demanding tasks. For many workers in these industries, most occupational tasks can now be performed sitting and operating a machine.

The implication of these employment trends is that many people are not getting the physical activity they used to from their occupation.

This comes at a time of decreasing physical activity during leisure time by Japanese. Results of the 2006 Survey on Time Use and Leisure Activities (www.stat.go.jp/english) from 80,000 household members showed a 7% reduction in participation in sport activities from 2001 to 2006. This included an 8% reduction in the most common activity, 'walking or light physical exercise'. Again this contrasts with a 13% increase in internet use of the same period, as shown in Figure 3.

2 POTENTIAL HEALTH EFFECTS OF INSUFFICIENT PHYSICAL STRESS

Insufficient physical activity is commonly linked with a range of cardiovascular, metabolic, mental and musculoskeletal disorders (Surgeon General 1996). Short term inactivity and insufficient physical stress, such as during space travel or bed rest, results in reduced bone density, muscle strength and endurance, coordination, cardiovascular fitness and a deterioration of mood (Greenleaf et al. 1989, Fitts et al. 2001). During joint immobilisation (such as after a fracture) muscle strength is lost as a result on local muscle and central nervous system changes (Pathare et al. 2005). The effects of insufficient physical activity/stress are likely to compound the effects of an aging workforce, resulting in even greater burdens on health systems.

Enhancing physical activity is therefore a major public health priority for many countries.

3 OCCUPATION AS AN OPPORTUNITY FOR PHYSICAL STRESS

Since it inception, ergonomics in the physical domain has had the approach of aiming to enhance health by reducing excessive physical demands. However, as demonstrated above, the risk for many workers is now insufficient physical demand.

Most public health initiatives to increase population physical activity have targeting discretionary activities such as leisure time sport participation. Not only does this rely on individual motivation, it misses the substantial time opportunity afforded at work. Most working adults will spend around 40 hours a week at work. On week days this is close to half their waking hours.

Ergonomics has tried to design work to enhance health. Whilst this was previously focussed on reducing physical stresses, in the psychosocial domain ergonomics has always aimed at not only removing negative stresses, but also providing positive psychosocial stresses. Physical ergonomics could adopt the same paradigm – and design work to have possible health effects rather than just minimising negative health effects.

In a physical context, for individuals at risk of insufficient physical stress (such as many office workers), this will require work to be designed to increase physical stress. Several examples of such initiatives have recently been reported.

Gilson et al. (2008) reported a pilot study of 58 female university workers allocated to control, 'walking routes' and 'walking in tasks' conditions. The 'walking routes' intervention was like the common public health approach and encouraged workers to walk particular paths. The 'walking in tasks' intervention encouraged workers to walk to perform tasks. Both walking groups showed an increase in step counts after 10 weeks, indicating office work can be modified to enhance physical activity.

Using a more radical approach, Levine and Miller (2007) tested 15 obese office workers while they worked at a computer workstation which enabled them to slowly walk on a treadmill whilst operating the computer. They reported mean (sd) energy expenditures of 191 (29) kcal/hour while using the 'walkstation' compared with just 72 (10) kcal/hour when sitting on a normal chair and using a computer. It was estimated that if these individuals replaced chair workstation work with walkstation work for 2–3 hours a day, they would loose 20–30 kg/year, if all other energy income and expenditure were held constant. In a follow-up study, Thompson et al. (2008) provided 3 walkstations to 8 office workers and measured their work steps for two weeks prior to provision of the new walkstations, over two weeks of acclimatisation and for a further 2 weeks. They found work step count increased from 2,200/day to 4,200/day with provision of the walkstation. No reminders were given to workers to walk, and yet all subjects walked at least an additional 30 minutes/day at work when the walkstation was available. However, the impact of these walkstations on productivity is unclear.

In fact the walkstation idea is not new and was promoted as early as 1989 (Edelson and Danoff 1989). Ideas for cycling workstations are also now appearing on the internet.

Taking a different approach, Meyer et al. (2006) evaluated use of a more active input device than the traditional keyboard and mouse – a dance mat. They developed software to work with mail and photographs based on foot movements and showed these could receive acceptance by users and increased their level of exertion. Others have evaluated more active seating designs which encourages more trunk activity (O'Sullivan *et al.* 2006).

Technological developments have reduced the physical activity and physical stress of many modern workers. However the examples above show that appropriately designed technology may also provide some solutions to the problem.

It may also be possible to design office work to provide higher physical stresses appropriate to maintain muscle strength, cardiovascular fitness and bone mineral density.

4 CONCLUSION

Ergonomics aims to design the world humans interact with to enhance their health, satisfaction and productivity. Physical ergonomics has previously focussed on enhancing health through the reduction of high physical stresses likely to cause musculoskeletal injury through an incident or cumulative trauma. However a new risk is now emerging for many workers – lack of sufficient physical stresses. Ergonomics therefore needs to adapt its paradigm and look for ways to design work to provide appropriate physical stress.

REFERENCES

Edelson, N. and Danoff, J., 1989, Walking on an electric treadmill while performing VDT office work, *SIGCHI Bulletin*, 21(1), 72–77.
Exercise and Physical Activity Reference for Health Promotion, 2006, Office for Lifestyle-related Diseases Control, Ministry of Health, Labour and Welfare, Japan.
Fitts, R., Riley, D., and Widrick, J., 2000, Physiology of a microgravity environment. Invited review: Microgravity and skeletal muscle, *Journal of Applied Physiology*, 89, 823–839.
Gilson, N., McKenna, J., Cooke, C. and Brown, W., 2007, Walking towards health in a university community: a feasibility study, *Preventative Medicine*, 44, 167–169.
Greenleaf, J., Bernauer, E., Ertle, A., Trowbridge, T. and Wade, C., 1989, Work capacity during 30-days of bed rest with isotonic and isokinetic exercise training, *Journal of Applied Physiology*, 67, 1820–1826.
Hagstromer, M., Oja, P. and Sjostrom, M., 2007, Physical activity and inactivity in an adult population assessed by accelerometry, *Medicine and Science in Sports and Exercise*, 39(9), 1502–1508.
Healy, G., Dunstan, D., Salmon, J., Cerin, E., Shaw, J., Zimmet, P. and Own, N., 2007, Objectively measured light-intensity physical activity is independently associated with 2-h plasma glucose, *Diabetes Care*, 30(6), 1384–1389.
Levine, J., and Miller, J., 2007, The energy expenditure of using a "walk-and-work" desk for office workers with obesity, *British Journal of Sports Medicine*, 41, 558–561.
Matthews, C., Chen, K., Freedson, P., Buchowski, M., Beech, B., Pate, R. and Troiano, R., 2008, Amount of time spent in sedentary behaviours in the United States 2003–2004, *American Journal of Epidemiology*, 167(7), 875–881.
Meyer, B., Bernheim Brush, A., Drucker, S., Smith, M. and Czerwinski, M., 2006, Dance your work away: exploring step user interfaces. In *Proceedings of CHI '06 Conference on Human Factors in Computing Systems*, Montreal, (New York:ACM), pp. 387–392.
O'Sullivan, P., Dankaerts, W., Burnett, A., Straker, L., Bargon, G., Moloney, N., Perry, M. and Tsang, S., 2006, Lumbopelvic kinematics and trunk muscle activity during sitting on stable and unstable surfaces. *Journal of Orthopaedic and Sports Physical Therapy*, 36:19–25.
Pathare, N., G. A. Walter, J. E. Stevens, Z. Yang, E. Okerke, J. D. Gibbs, J. L. Esterhai, M. T. Scarborough, C. P. Gibbs, H. L. Sweeney and K. Vandenborne. 2005, Changes in inorganic phosphate and force production in human skeletal muscle after cast immobilization, *Journal of Applied Physiology*, 98, 307–314.
Straker, L. and Mathiassen, S. 2009, Increased physical workloads in modern work – a necessity for better health and performance? *Ergonomics*, 52(10), 1215–1225.
Surgeon General, 1996, *Physical activity and health: A report of the Surgeon General.* (US Department of Health and Human Services, Atlanta, GA).
Thompson, W., Foster, F., Eide, D. and Levine, J. 2008, Feasibility of a walking workstation to increase daily walking, *British Journal of Sports Medicine*, 42, 225–228.
Troiano, R., Berrigan, D., Dodd, K., Masse, L., Tilert, T. and McDowell, M., 2007, Physical activity in the United States measured by accelerometer, *Medicine and Science in Sports and Exercise*, 40(1), 181–188.
Tudor-Locke, C. and Bassett, D., 2004, How many steps/day are enough? Preliminary pedometer indices for public health. *Sports Medicine*, 34(1), 1–8.

Chapter 1 The East Asian ergonomics roadmap

Ergonomic Trends from the East – Kumashiro (ed)
© 2010 Taylor & Francis Group, London, ISBN 978-0-415-88178-4

A roadmap of ergonomics in Korea

Kwan S. Lee
Department of Information and Industrial Engineering, Hongik University, Seoul, Korea

1 INTRODUCTION

In improving and sustaining advancement of quality of life of Korean people, it is vital to use ergonomics to respond to common ergonomics-related needs in daily life including working life. Although ergonomics has been introduced to Korean society for almost forty years, this has not been regarded as one of main scientific methods which need to be promoted in Korea. Thus, most of its effort has been more of academic or industrial level effort. One of reasons for this problem is that it has not been accepted as a society-wide tool which needs to be used in all aspects of human life in Korea. But, rapid and vast change of Korean society presents many social and industrial problems which need to be resolved using ergonomics. It is necessary for ergonomists to establish a clear strategy to meet needs of Korean society and industry to resolve problems and to improve quality of life in Korea in the future. Recently roadmaps have been used in establishing strategies for developing technologies. Thus it is good to try to develop ergonomics roadmaps in short and long-term perspectives to set up a strategy for the future. Horino et al. (2007) stated in their ergonomic road map study in Japan that "as we need to focus on meeting priority needs, it is important to identify practicable goals in main business management areas from people-centered points of view." Therefore, this paper intends to identify current and anticipated social and industrial problems in Korea and needs and to set priority in resolving problems and meeting needs from Korean society and industry.

2 ANTICIPATED AND CURRENT SOCIAL AND INDUSTRIAL CHALLENGES AND PROBLEMS

There are many current social and industrial challenges and problems which need to be resolved and/or overcome. Horino et al. (2007) suggested that ergonomics goals need to be achieved within the period by around 2020 and 2030 in the following areas; (1) production and systems, (2) support for community life, (3) products design and (4) transportation, (5) safety and health management, (6) coexistence with environment, and (7) communication networks. These areas should be the same in Korea although problems and challenges can be different. Thus, we will discuss these areas.

2.1 Developing people-centered production and work systems

Production and systems have been issues for traditional ergonomists for a long time starting from Taylor and Gilbredth. With the emergence of automation and robots, workers' role in production and systems seemed to be diminished in the earlier stages. But it was found that human ability appeared more often as a limiting factor for efficiency in production and systems rather than automated machines. Further, we expect that in the near future more elderly and women will be required to get jobs. They will encounter tasks of increasing productivity and efficiency in operating systems. However, Korean work environments are not very user friendly yet. Thus Korean ergonomists need to work to make work environment more user friendly. There has been saying for decades that the labor productivity of Korean workers is much lower than those of American and Japanese workers. But we found that most of work places, tools and facilities in Korea have not been ergonomically designed yet although a few companies have tried to apply ergonomics just to reduce incident rate of musculoskeletal disorders recently.

2.2   Support for community life and products design including transportation

It is impossible to think about meeting varying needs of people for the good community life without thinking about ergonomics. It is also important to promote people-friendly design and use of products. However, ergonomists have not been fully utilized opportunities to teach people ergonomics which should be a fundamental knowledge in designing things and systems not only in workplaces but also in our daily lives (Sanders and McCormick, 1993). Although safety and health professionals have worked to improve health and safety of workers at workplaces, safety and health are still big issues in most Korean workplaces. These will remain as big issues in the coming years since Korean people would concern more about their health and safety as their basic needs such as food, clothes and home would be fulfilled to a certain extent. Thus ergonomists need to work to enhance safety and health management for all people.

Korea became aged societies since the population of the elderly surpassed 7% of the whole population. It is expected that the population of the elderly will be more than 10% of the whole population very soon since the number of aged population increases drastically. Thus it will be very common and inevitable for the elderly to work until very old age. Thus it is important to check the workability at workplaces and to improve working conditions so that the elderly should not have much problem in working. Therefore, Korean ergonomists needs to find ways to improve workability of workers and improve workplaces to make working condition better. Further, there are still many problems exists for disabled persons as well as elderly in accessing and using transportation since ergonomics have not been widely used.

In industry, providing products which customers want is very important. Ergonomics has been used in designing workplaces for half a century, but since 1980, it has been widely used in designing products for consumers. Ergonomically designed products can improve community life as well as working life. One of examples of ergonomic application is the universal design for disables and elderly as well as minority and women. The capability of ergonomics to identify consumers' desire also blossomed during the last decades in 20th century under the name of sensibility engineering or Kansei engineering. Ergonomics has been promoting compatibility between human and systems. According to Sanders and McCormick (1993), ergonomics 'focuses on humans and their interaction with products, equipments, facilities, procedures, and environments used in work and everyday living'. Many researchers have shown that ergonomic design can improve quality.

2.3   Workplace design for workers' satisfaction, safety and health

Ergonomics has been pursued first to enhance effectiveness and efficiency with which work and other activities are carried out and second to enhance certain desirable human values. These require increasing convenience of use, reduction of errors, and increasing productivity. Further for workers, improving safety, reducing fatigue and stress, increasing comfort of working, and job satisfaction, and finally, but not the least important, improving quality of life are required. There exist many musculoskeletal stress disorders in workplaces. These cause high medical costs, employee complaints, and low productivity. Although ergonomists have tried to reduce work loads of workers by using guidelines and assessment tools such as NIOSH guidelines, ILO guidelines, RULA, OCRA, OWAS, etc., businesses still need to resolve these problems in the near future since workers want to work without the pain and fear of getting musculoskeletal diseases.

Modern management disciplines teach the importance of employees' satisfaction. Comfort at work, safety, and health which are within the domains of ergonomics are included in the criteria for employees' satisfaction. One of aims of ergonomics is to increase efficiency by improving compatibility between human and systems. Efficiency in this case includes productivity, ease of working, and comfort. These also imply quality of life which is quality of working life for workers These can protect workers' musculoskeletal system from fatigue, overexertion, and injuries such as carpal tunnel syndrome and lower back injuries, etc. Ergonomics has been regarded as a method which can provide employees' satisfaction and reduces medical cost and compensation cost.

2.4   Productivity improvement and working schedule design

Costs for production and costs to raise quality are regarded as inevitable costs but managers know that they should reduce as much as they can. TQM and production innovation are areas which can use ergonomics. However, production innovation is again a localized effort while TQM is, as the first initial indicates, dealing with almost all the departments in a company. If ergonomists can show that ergonomics can achieve significant gains in productivity and quality while providing safety and quality of work life, ergonomics can be adopted as an important tool in TQM. A case study by Helander and Burri (1995) can be used as an example in showing cost effectiveness and quality improvements of ergonomics.

It was known that works designed without considering ergonomics have usually lowered productivity (Sanders and McCormick, 1993). Therefore by providing working environment which requires work load within a worker's capability, the productivity can be improved. During my TQM consulting and field visits as a national quality award examiners, it was often observed in the improvement proposals from quality circles (QC) that they were using ergonomic approaches and Nagamachi (1991) also stated that ergonomics was used in Q.C. circles. Axelsson and Eklund (2002) presented in their report of researches that "participatory and holistic approach makes it possible to integrate knowledge and disciplines of ergonomics and TQM in order to create more efficient systems for business improvements. Its campaign to promote ergonomics is still in an early stage and but it is expected to be used more frequently in the future to determine working schedule since physical improvement alone could not fulfill workers' desire to work comfortably.

2.5 *Providing communication networks for achieving decent social life*

The internet and related technologies and application as well as communication technologies such as mobile phones and GPS (Global Positioning System) have changes the way people communicate each other and how information systems support business processes, decision making, and competitive advantage (O'Brien and Marakas, 2007). However, there are still many usability problems exists. Further, some jobs in information technology such as data entry are still quite repetitive and routine. And there is a frequent criticism of information system of their negative effect on the individuality of people. Computer-based systems are criticized as impersonal systems that dehumanize and depersonalize activities that have been computerized, since they eliminate the human relationships present in non-computer systems. Ergonomists should resolve these problems by either providing substitute mechanism or other means for personal interaction. We also need ergonomics to make information systems efficient and effective. One of roles of ergonomics in these hardware and software is called as human-computer interaction (HCI). We would also foresee a big role of ergonomics in ubiquitous environments. As modern technologies including computer technologies evolve fast, more role of ergonomics are expected in the future.

2.6 *Strengthening sustainable coexistence with global environment*

Recycling is almost a mandatory in Korea, but there are still many problems in practicing the application of ergonomics in resource-saving life styles and production. The procedures and containers or equipments for storing and processing waste food need attention by ergonomists since people would not easily behave in recycling if it is dirty or difficult.

3 HOW SHOULD ERGONOMICS BE APPLIED?

Although Kogi (2000) reported that ergonomics would be very effective in industrially developing countries, ergonomics has not been promoted well because of notion that it would cost a lot (Lee, 2005). To facilitate use of ergonomics, ergonomists need to show that it would return bigger benefit than cost. Ergonomics is not a tool which needs to be applied only once or a few times. Nagamachi (1991, 1998) used the idea of quality circle as a part of his participatory ergonomics approach and has reported that it was very effective. However, his quality circles were more of a project basis rather than regular everyday activities in companies. To be effective, it is important to keep ergonomics as a tool which should be used every day at workplaces and design departments.

The concept of quality has kept changing as time changes. Before 1980, providing products which fit to the specification and/or regulations was a concept of quality. In the 1990's, the concept of quality became that of providing products that satisfy customers' needs (ISO, 1987). In 2000, Kano, who is renowned as a quality expert claimed that this concept was old one since most products have already satisfied customers' explicit requirements. The new quality concept is to find latent or subconscious desires and realize these desires as features in products (Kano, 2000). Many Korean companies have successfully adopted ergonomics once but failed to keep it effective because there was not any scheme for incorporating ergonomics as continuing normal activities. However, competitive advantage gained by using ergonomics doesn't last long and is generally not sustainable over the long term unless ergonomics is used all times. Korean ergonomists need to promote ergonomics as the main techniques to be used in designing products and facilities for daily life as well as work places, ergonomics could be well used in Korea and result in improving quality of life for Koreans.

4 CONCLUSIONS

Ergonomics should be used more than ever in daily life and in industry in the forth coming several decades since people become more affluent and thus seeking for quality of life and effective and efficient business methods. Although it has been known for workplace improvement by considering the workers' characteristics, hardware and software requirements for the systems and the working methods together with ergonomic principles, there are many challenges which ergonomics should and would resolve. They ranged from the people-friendly design and use of products to enhancing competitiveness of business. They should make better working environment for all workers including elderly and women. This improvement can bring safety of workers and efficiency of work. It should also help quality of life by providing effective and efficient communication networks, increasing productivity and reducing costs.

REFERENCES

Axelsson, J. R. D. and Eklund J., 2002, Macro Ergonomic Management – The Integration and Applications of Participatory Ergonomics in Strategic Quality Management.

Helander, M.G. and Burri, G., 1995, Cost Effectiveness of Ergonomics and Quality Improvements in the Electronics Industry. *International Journal of Industrial Ergonomics*, Vol. 15, pp. 137–151.

Horino, S., Kogi, K., Sakai, K., Kishida, K, Mizuno, K., Ebara, T., HongSon, S. and Ohashi, O., 2007, *Practical Steps for Developing Practical Steps for Developing Roadmaps of Ergonomics Application in Major Technical Areas*, Proceedings of 13th Pan-pacific conference. Bangkok.

Kano, N., 2000, Historical Change of Quality as Competitive Edge, *Proceedings of Annual Meeting of Japanese Quality Control Society*, Tokyo, Japan.

Kogi, K., 2000, Low-Cost Ways of Reducing Musculoskeletal Risks at Work. Institute for Science of Labor, Kawasaki, Japan, pp. 216–8501.

Lee, K. S., 2005, Ergonomics in total quality management: How can we sell ergonomics to management? *Ergonomics*, Vol. 48, No. 5, pp. 547–558.

Nagamachi, M., 1991, Application of Participatory Ergonomics through Quality Circle Activities. In K. Noro and A.S. Imada (Eds.), *Participatory Ergonomics*, London: Taylor & Francis, pp. 139–164.

O'Brien, J.A. and Marakas G.M., 2007, *Enterprise Information System 13th* Ed. McGRAW-HILL, NJ, pp. 12.

Sanders, M.S. and McCormick, E.J., 1993, *Human Factors in Engineering and Design 7th* Ed. McGRAW-HILL, NJ, pp. 5.

Ergonomic Trends from the East – Kumashiro (ed)
© 2010 Taylor & Francis Group, London, ISBN 978-0-415-88178-4

Future directions in universal design of everyday products

Akira Okada
Graduate School of Human Life Science, Osaka City University, Sugimoto, Sumiyoshi-ku, Osaka, Japan

1 INTRODUCTION

The importance of universal design, including human-centered design for the elderly, the disabled, and others with special needs, is well known, but the focus on design has been misdirected for several reasons: (1) the appearance and mechanical function of a product receives higher priority than the usability of the product, (2) extra cost and time are required to produce such a design, (3) many users adapt to a product even if its usability is poor owing to the human's inherent adaptability, and (4) there is a remarkable lack of relevant data on human characteristics and human-centered design.

Knowledge of human characteristics alone is not sufficient to realize universal design. Methods to translate data on human characteristics into information valuable for design are also needed. Much of the basic data is available on the physiological and mental functions of users with normal health. However, the direct application of this information to the design of products is limited because it has been obtained under experimental conditions without consideration of a specific purpose. Moreover, those who participate in these studies are not always representative of the same populations that the designed product is targeted at, so a limited amount of the data is incorporated directly into the product design.

This paper shows the importance of the universal design philosophy for good designs, cites some examples of physiological characteristics, and presents the steps and problems involved in applying human characteristics to design in now and future from an ergonomist's perspective.

2 IMPORTANT IDEAS IN UNIVERSAL DESIGN

A universal design includes designs for the aged, the disabled, and any other special needs users. Therefore, a universal design is not a design just for a particular group, such as elderly or disabled people. It is impossible to design one product that meets the needs of all people; however, it is possible to design a series of products in one system that suits all people (e.g., size classes, adaptors).

A product or a system of products is universally designed only if all users are equally satisfied. A universal design does not function properly without considering the use of the product; therefore, to create a universally designed product, it is necessary to know how to create it as well as how to most effectively use it.

3 AGE-RELATED CHANGES AND DIFFERENCES IN PHYSIOLOGICAL FUNCTIONS

Universal design considers the physiological attributes and age-related changes of all users, so that they can most effectively use the product within their capabilities. It is important to incorporate specific human characteristics to design products for the aged. In this paper, I have followed several examples that include data on anthropometry, muscle strength, and sensory-cognitive characteristics.

Table 1.   Heights of Japanese youths and elderly (HQL, 2008).

	20–29 yrs.		70–79 yrs.	
	Male	Female	Male	Female (mm)
5th %ile	1625	1505	1532	1411
Mean	1719	1592	1625	1496
95th %ile	1812	1686	1725	1584

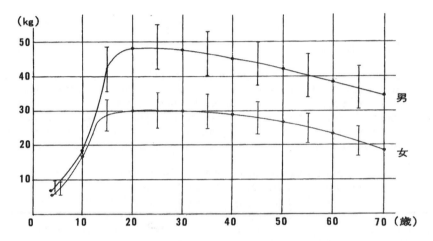

Figure 1.   Changes in maximum grip strength according to age (Himaru et al., 1991).

Consider the range in height of Japanese youth and elderly (Table 1). The difference in height can be explained by the fact that height decreases with age and the Japanese physique has grown rapidly in the past half-century. Suppose we intend to design a product based on the height of its users. If this product is intended for use by both older females with a small build (e.g., the 5th percentile among females in their seventies) and by large-build young males (the 95th percentile among males in their twenties), we must take into account a difference of 40 cm to satisfy both these user categories.

As an example of manually handled products, consider the changes in maximum grip strength that occurs with age in Japanese people (Figure 1). There is generally a large gender difference in muscle strength, including maximum grip strength. After adolescence, the mean maximum grip strength of women is approximately 50–70% that of men. Moreover, grip strength generally decreases with age. These two factors result in a great difference between the maximum muscle strength of the young male and that of the aged female. The operational force or the weight of a product designed by a young male designer based on his functional ability often leads to the inability of older females to use the product.

Changes in tactile senses also exert a large influence on motor performance. Age-related changes in tactile senses are not noticeable in daily life, but can be measured experimentally. In collaboration with a manufacturing company, we conducted an experiment in which participants were asked to recognize various types and sizes of convex symbols on an electric appliance button (Figure 2). The rate at which the participants failed to correctly recognize a symbol (error rate) was used as the index of tactile sensitivity. The error rate of the elderly group (60–70-years-old) was significantly higher than that of the youth group (20–30-years-old) and the middle-aged group (40–50-years-old) (Figure 3). This indicates a large decrease in tactile sensitivity with age, which may relate to the decrease in precise motor performance.

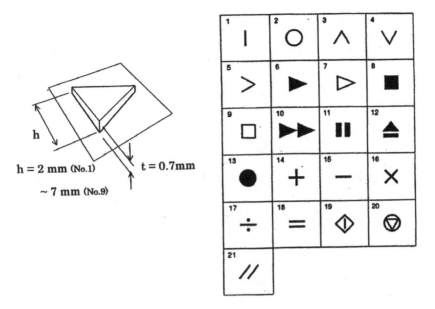

Figure 2.   Embossed symbols.

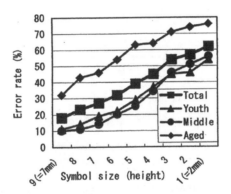

Figure 3.   Error rate in the recognition of embossed symbols (Marumoto & Okada, 1997).

4   DATA INTERPRETATION AND TRANSLATION

This type of physiological data can be cited from various reports; however, there is a big problem with its application to human-centered design.

The data is raw and cannot be adopted into a design without considering other factors. For example, it would not be appropriate to directly apply the maximum grip-strength data as the upper limit of approved operational force for a fire extinguisher because the grip shape of the experimental apparatus differs from the grip shape of a fire extinguisher. Furthermore, the user must adopt a different posture when using the measurement device than when using the fire extinguisher. Another example is that the height of a computer display is usually based on the eye level of its operators, but pupil height while sitting (pupil height, sitting) is more appropriate information to use to properly design a computer display. However, this height is measured while the subject is sitting erect with thighs fully supported, even though few people sit this way while using a computer (Figure 4).

Thus, to overcome such problems, raw data should be converted into information that can be used to properly design a universally usable product (Figure 5). This can be done by following three steps, data interpretation, translation and trade-off. To interpret the raw data we must consider the

Figure 4.   Experimental conditions are contrary to the conditions of use.

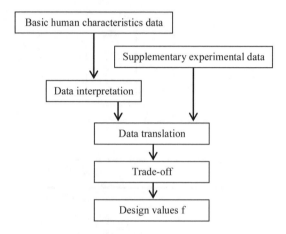

Figure 5.   Steps to interpret and translate data for application to the design.

4W1H (who, what, when, where, and how) of the measurement conditions and the 4W1H of the actual conditions of product use. The application of data will result in failure if it is not interpreted correctly.

How close the data translation is to the actual conditions of use contributes significantly to a good design. However, data translation is very difficult, and the method has not been sufficiently developed. In some cases, it is necessary to collect supplementary experimental data from conditions close to actual-use situations.

Even if the translation is successful, the characteristics may not be initially adopted as design characteristics because modifications may be necessary before production (trade-off). For instance, the recommended size of a mobile cellular phone may be so small that it conflicts with the usability of the buttons and display of the phone. These examples emphasize the need to effectively translate the experimental data and consider all of the factors related to the end use of a product for application to the design.

5   KEY POINTS OF UNIVERSAL DESIGN IN FUTURE

The application of human characteristics data to product design will increase along with progress in designing for all. The most fundamental steps for these applications include the acquisition of the appropriate data on human characteristics for the design, and the effective translation of these data into the product design. Because translation is generally difficult, the method must be systematized (Yamazaki, et al., 2006). In addition to experimental data, it is important to accumulate information from good and bad designs as well as to form an effective translation method.

It is the "real" universal design that all users are equally satisfied with, and that all users do not regard it as a universal design. It will be difficult to reach the level of such a design, but we can approach this goal without limit.

REFERENCES

Himaru et al. 1991: "Kenko tairyoku hyoka/kijunchi jiten" (in Japanese), Gyosei.
Marumoto, K., Okada, A., 1997: A study on cognitive characteristics in tactile display (in Japanese), Proceedings of the 38th Annual meeting of Japan Ergonomic Society, pp. 384–385.
Research Institute of Human Engineering for Quality Life (HQL), 2008: Japanese body size data book 2004–2006.
Yamazaki, K., Yamaoka, T., Okada, A., Saitoh, S., Nomura, M., Yanagida, K., Horino, S., 2006.: Practical universal design guidelines: A new proposal, Karwowski, W. (Edt.), Handbook of standards and guidelines in ergonomics and human factors., pp. 365–379, Lawrence Erlbaum Associates, Publishers.

Chapter 2   Information sharing about the factors
that cause trouble and human error

Ergonomic Trends from the East – Kumashiro (ed)

Prospects of expanding a user-involvement accident/incident sharing system to public facilities

Miwa Nakanishi
Graduate School of Engineering, Chiba University, Chiba, Japan

1 INTRODUCTION

In recent years, strategies for reducing accidents have shifted from "prevention of recurrence" to "prevention of occurrence." The concept of "prevention of occurrence" can be traced back to Heinrich's Law (1931) (Figure 1): "for one fatal or serious accident, there are 29 accidents and 300 potential incidents with a high possibility of injuries." Hence, it is important to take measures against potential incidents to prevent serious accidents.

Some industries have attempted to gather and analyze incident data using paper-based reports. However, few have been able to make effective use of the data. One of the reasons for this is that workers are not willing to report small unsafe incidents because they feel it is not worth the trouble, or because they are afraid of being rebuked. Another reason is that managers are unable to analyze the data properly or provide effective and swift feedback to the workers.

In response to this situation, we have developed a network-based accident/incident reporting system that enables workers to report accidents and incidents from computer terminals, and this gives swift feedback on their potential seriousness and priority of the measures that need to be taken. A field trial evaluation of the system at actual workplaces has been carried out. We presently aim to expand the system to public facilities such as stations and parks. A system that allows users to easily and quickly report accidents or incidents is expected not only to bring safety to such facilities but also enhance public awareness of safety.

In this paper, the key idea behind the system developed in a previous study by the present authors (Nakanishi and Okada, 2005) is explained in the next section. In section 1.3, the prospects and challenges of expanding the system to public facilities are discussed.

2 NETWORK-BASED ACCIDENT/INCIDENT SHARING SYSTEM

Figure 2 shows an outline of the system that we developed in a previous study (Nakanishi and Okada, 2005). When a worker experiences an accident or incident, he/she reports it from a computer

Figure 1. Heinrich's Law (1931): "for one fatal or serious accident, there are 29 accidents and 300 potential incidents with a high possibility of injuries."

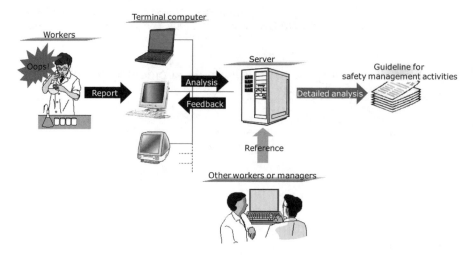

Figure 2.   Outline of the network-based accident/incident sharing system.

Table 1.   Eight significant factors that potentially cause fatal or serious accidents.

Category	Potential Factor
Software	Manual
	Label
Hardware	Display
	Controller
Environment	Space
	Indoor condition
Task	Difficulty
	Similarity

terminal. The reported data is sent to a central server and analyzed using the previously reported data. The result is immediately returned to the user. The database can also be used for further safety management activities.

Details of the reporting and feedback processes are provided below.

2.1   *Process of report*

A worker accesses the system and fills out some forms. In the first form, shown in Figure 3, he/she enters the date, place, affiliation, and position. He/she also provides a simple title and summary of the reporting data. The worker then chooses one of the provided options concerning the damage that has been caused by the accident or incident. In the second form, shown in Figures 4 and 5, he/she completes the checklist to determine the context of the accident/incident. The checklist includes questions to determine the conditions of eight significant factors (Table 1) that potentially cause fatal or serious accidents, factors which were selected according to our preliminary investigation. The worker evaluates the conditions of the factors from both subjective and objective viewpoints.

2.2   *Feedback process*

After the reported accident or incident is statistically analyzed on the central server, the following two results are returned to the worker's computer terminal. One result concerns the potential

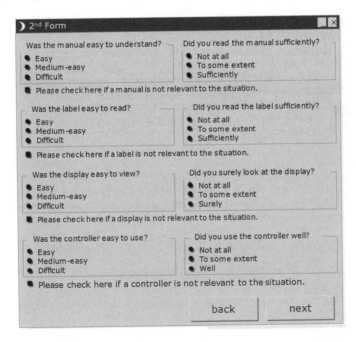

Figure 3.　First form, which is used to enter information on an accident/incident.

1st Form ✕

Date year `---- ▼` month `-- ▼` day `-- ▼` Place `------ ▼`

Affiliation `---- ▼` Position `---- ▼`

Title _____

Descriptive Question

What are the details of the incident?

_____

Why is it unsafe?

_____

How should it be prevented?

_____

Alternative Question

○ Someone got hurt.
○ An explosion or blaze occurred.
○ Something burned.
○ None of those happened.

back　　next

2nd Form ✕

Was the manual easy to understand?　　Did you read the manual sufficiently?

● Easy　　　　　　　　　　　　　　　● Not at all
● Medium-easy　　　　　　　　　　　● To some extent
● Difficult　　　　　　　　　　　　　● Sufficiently

■ Please check here if a manual is not relevant to the situation.

Was the label easy to read?　　　　Did you read the label sufficiently?

● Easy　　　　　　　　　　　　　　　● Not at all
● Medium-easy　　　　　　　　　　　● To some extent
● Difficult　　　　　　　　　　　　　● Sufficiently

■ Please check here if a label is not relevant to the situation.

Was the display easy to view?　　　Did you surely look at the display?

● Easy　　　　　　　　　　　　　　　● Not at all
● Medium-easy　　　　　　　　　　　● To some extent
● Difficult　　　　　　　　　　　　　● Surely

■ Please check here if a display is not relevant to the situation.

Was the controller easy to use?　　Did you use the controller well?

● Easy　　　　　　　　　　　　　　　● Not at all
● Medium-easy　　　　　　　　　　　● To some extent
● Difficult　　　　　　　　　　　　　● Well

■ Please check here if a controller is not relevant to the situation.

back　　next

Figure 4.　First part of the 2nd form, which is used to complete a checklist to determine the context of the accident/incident.

seriousness of the incident/accident, which defines the potential seriousness of the reported incident, even if it did not, by chance, become a major problem. The other result regards the priority of measures to be taken to effectively prevent large problems arising. The process to obtain these results is as follows.

37

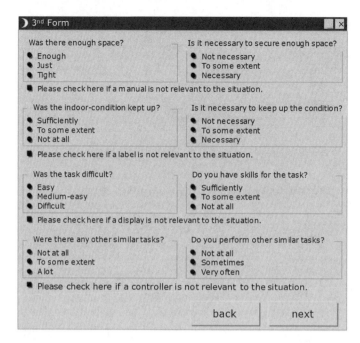

Figure 5.   Second part of the 2nd form.

Table 2.   Boundary judgments concerning the potential seriousness of an incident.

	y	JUDGMENT
	$y \geq 2$	Highly possible that a serious accident occurs.
	$1 < y < 2$	Possible that a serious accident occurs.
	$y \leq 1$	Little possibility that a serious accident occurs.

We carried out a pilot study to investigate the relationship between the eight potential factors of Table 1 and the level of damage caused by the reported accident/incident, and found that it could be expressed by the following multiple regression model.

$$y = 0.177x_1 + 0.18x_2 + 0.273x_3 + 0.239x_4 + 0.106x_5 + 0.102x_6 + 0.123x_7 + 0.184x_8 - 2.364 \tag{1}$$

$$(R^2 = 0.61),$$

where
   y: Level of damage
   x_1 to x_8: Scores of eight potential factors

Based on this, we developed algorithms to calculate the potential seriousness of incidents and the priority of required measures using the multiple regression model, which was built using the database of the previously reported data. Specifically, the potential seriousness is predicted by substituting conditions of the eight potential factors into the variables of x_1 to x_8 of the model. Table 2 shows the boundary judgments. Priority of the required measures is calculated using

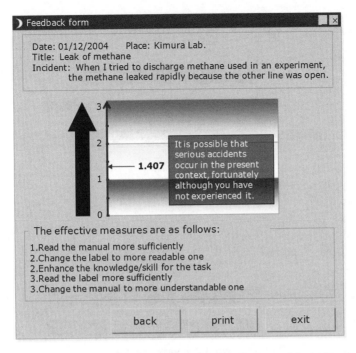

Figure 6. Resultant feedback dialog box displayed to the worker who reported the incident/accident.

partial correlation coefficients of the model. The partial correlation coefficients express the degree of correlation between the objective variable and each explanatory variable. If a condition of a potential factor is improved, reduction of the potential seriousness is determined by the following equation.

$$R_i = p_i(X_i - x_i) \quad i = 1 \text{ to } 8 \tag{2}$$

where
    R_i: Reduction of the potential seriousness
    P_i: The i-th Partial correlation coefficient
    x_i: Score of the ith potential factor
    X_i: Desirable score of the ith potential factor
    As R_i is larger, the priority of improving the potential factor P_i is expected to be higher.
    Through the above analysis, a feedback is displayed on the worker's computer terminal (Figure 6).

2.3 Evaluation of the system

We asked the actual workers, 33 workers of a chemical laboratory, 9 nurses of a hospital, and 11 workers of a plane assembly plant, to try out the system, modified according to each user's work and answer a questionnaire. Figure 7 suggests that the majority of the workers, regardless of the workplace, endorsed the system. Some plant workers stated that they required more specific feedback. Figure 8 suggests that most workers find the system sufficiently easy to report the accidents/incidents.

3 PROSPECTS OF EXPANDING THE SYSTEM TO PUBLIC FACILITIES

In public facilities such as stations and parks, fatal or serious accidents like slip and falls occur sometimes. Even if we, fortunately, rarely meet with such accidents, we often feel unsafe. Because

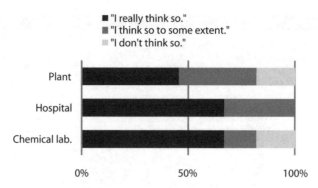

Figure 7.   Answers to the question "Do you think the feedback can prevent accidents?"

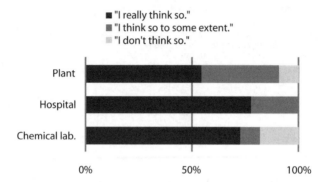

Figure 8.   Answers to the question "Can you easily report accidents or incidents through this system.

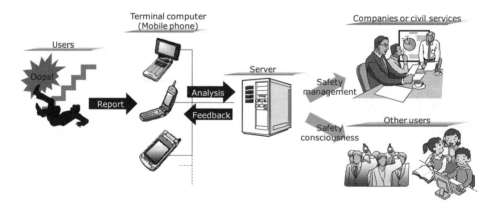

Figure 9.   Outline of the user-involvement accident/incident sharing system for public facilities.

we currently do not know how to share valuable experiences with other people, we soon forget them. Thus, we aim to develop the above-described user involvement accident/incident reporting system for use in public facilities, to enable users to easily and quickly report hazardous experiences, however trivial, and to share them with the community. Specifically, we plan to allow users to access the system from their cell phones, which is the most popular terminal for people to use anytime and anywhere. An outline of the expanded system is shown in Figure 9.

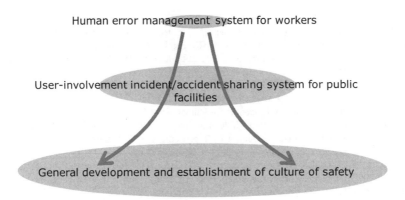

Figure 10.   Expansion of human-factor methodolog.

If this system is in wide use, it will be possible to easily gather ample data on various situations in which accidents and incidents occur. Feedback based on the above-described statistical method will also be useful to users in preventing fatal and serious accidents. Furthermore, this universal system will provide opportunities to communicate and exchange ideas on safety between service providers and recipients. It is expected to cultivate a culture of safety.

On the other hand, there is a risk that the reliability of the reported data may be reduced by opening the system up to the public. When access to the system is restricted within the intranet of the workplace, uncertain and imprecise data is less likely to be incorporated. However, if the system is opened up to the internet, this risk must be considered. A large amount of unreliable data may distort the multiple regression model and return incorrect feedback. One measure against this problem is to place "traps" in the checklist used to report accidents/incidents and automatically remove the data considered unreliable.

4   CONCLUSION

In this paper, we have explained the key idea behind the network-based accident/incident reporting system developed in our previous study (Nakanishi and Okada, 2005) and discussed the prospects of expanding the system to public facilities.

So far, sharing information on incidents as well as accidents has been an issue only in safety-critical industries. However, introduction of this concept to the public is expected to enable the development of a culture of safety. Further, it will not only improve people's safety but also our sense of ease surrounding hazards.

The time is now ripe to introduce and apply human-factor methodology to everyday situations, and if we are able to develop a human-error management system for workers and a user-involvement incident/accident sharing system for public facilities, a general culture of safety can be established (Figure 10). Mobile communication network is expected to be a key technology for their success.

REFERENCES

Heinrich, H.W., 1931, *Industrial Accident Prevention*, First edition, McGraw-Hill, New York.
Nakanishi, M. and Okada, Y., 2005, *Development of a network-based incident report system for open facilities*. Proceedings of the International Conference on CAES (Computer-Aided Ergonomics, Human Factors and Safety) in Košice, Slovak Republic, on CD-ROM.
Swain, A.D. 1967, Some limitations in using the simple multiplicative model in behavior quantification, In W.B. Askren (Ed.), Symposium on reliability of human performance in work, pp. 251–254, Aerospace Medical Research Laboratory.
Koseki, H. 2004, Hatto-Hiyari analyses, In Yukimachi, T. (Ed.), Human Factors for Human Error Prevention, pp. 255–262, Techno-Systems (In Japanese).

Morishita, M., Sakamoto, M., and Kashiwabara, M., 1997, An Anticipative safety Activity with Notice of Human Factors, Journal of the Society for Industrial Plant Human Factors of Japan, 2(1), pp. 71–76.

Tsukada, T., 1997, Activities of INSS/Institute of Social Research –"Hatto-Hiyari" (Near Incidents) for Measure Aimed at Preventing Human Error in Nuclear Power Plant–, Journal of the Society for Industrial Plant Human Factors of Japan, 2(1), pp. 77–83.

Yukimachi, T. and Nishikawa, K., 1997, A Study of Causality Analysis for Human Error, Journal of the Society for Industrial Plant Human Factors of Japan, 2(2), pp. 152–159.

Kawano, R., 1999, An Idea Generation Procedure for Human Error Reduction Techniques: Principles of Error Proof Technique Journal of the Society for Industrial Plant Human Factors of Japan, 4(2), pp. 121–130.

Hanaoka, S., et al., 1999, Human Error Prevention by Improvement of Work Practices and Environment Journal of the Society for Industrial Plant Human Factors of Japan, 4(2), pp. 131–138.

NASDA, 2000, NASDA-HDBK-10.

Nishikawa, K., 2002, Human Factors Extracted For Accident Study, Journal of the Society for Industrial Plant Human Factors of Japan, 7(1), pp. 50–53.

Takeda, M., 2002, Activity for Environment and Safety in the Iwakuni – Ohtake Works, Journal of the Society for Industrial Plant Human Factors of Japan, 7(1), pp. 54–64.

Ergonomic Trends from the East – Kumashiro (ed)
© *2010 Taylor & Francis Group, London, ISBN 978-0-415-88178-4*

Information-sharing on factors that induce human error in aviation

Akira Ishibashi, Daisuke Karikawa, Makoto Takahashi &
Toshio Wakabayashi
Department of Management Science and Technology,
Graduate School of Engineering, Tohoku University Sendai, Japan

1 INTRODUCTION

Accident statistics of Japanese land-sea-and-air transportation have indicated that the number of accidents and fatalities have gradually decreased in each transportation sector. However, various accidents and troubles have successively occurred since January 2005. These accidents and troubles have given negative effects on public trust in the safety of the public transportation system.

Case Example 1 (Aviation):

In January, 2005, the incident happened with an aircraft of a Japanese major airline erroneously started to takeoff roll without the takeoff clearance from air traffic controller.

In March, an aircraft of the same airline was operated without required operation of emergency exit erroneously. Moreover, an aircraft moved onto runway without controller's instruction, which forced another aircraft approaching the runway for landing to go around. Although each trouble did not result in serious accident, they made national headlines.

Case Example 2 (Ship):

In the marine transportation, the accident occurred in which a ferry crashed into a breakwater in May. 14 passengers were injured in the accident. In June, additionally, a sightseeing ship run aground and 26 passengers were injured.

Case Example 3 (Automobile):

In April, a rollover accident of a route bus happened with three fatalities and thirty injures. Furthermore a collision accident in which an express train and a motor truck stuck on a railroad crossing occurred.

Case Example 4 (Railroad):

In March, an operator of a manually-operated railroad crossing erroneously opened the crossing bars although a train was approaching. The accident killed two and injured two others. Additionally a severe railroad accident happened in April. In the accident, a rapid-service train derailed and crashed into an apartment building along railroad. The direct cause of derailment has been presumed to be late brake operation by the operator and excessive speed on a curve.

Table 1.   Transportation Accidents Statistics in Japan 2007.

	Fatalities	Injuries	Accidents
Aviation	10	25	23
Rail road	38	722	793
Automobile	5,744	1,034,515	833,019
Ship	87	899	2,579
Railroad crossing	128	295	355

These case examples has been analyzed from human factors perspectives and the related companies has developed remedial actions and safety improvement program, and released them to the public. The reports have indicated the following accident-inducing factors.

- Insufficiency of interactive communication between the headquarter and fields
- Insufficiency of management effort to spread safety-first policy to all parts of an organization
- Insufficiency of inter-section communication
- Insufficiency of interactive communication between the president and frontline workers

The Ministry of Land, Infrastructure, Transport and Tourism (MLIT) of Japan has established the exploratory committee for accident prevention caused by human errors in public transportation. About two months later, the committee published an interim report which emphasizes the necessity of safety management system (SMS) by the companies and the government. In response to the report, the MLIT has decided to introduce the SMS into Japanese public transportation systems.

2   SAFETY MANAGEMENT SYSTEM IN PUBLIC TRANSPORTATION

In order to improve safety in the public transportation, it is strongly required to construct the organizational safety promotion system and to carry out company-wide error prevention practices. The SMS obligates transportation companies to build the organizational safety management framework. At the same time, it also requires the government to make evaluation system to each company's safety activities.

As legislation toward the SMS, a laws in order to revise four basic laws for air, rail, road, and maritime in a lump was passed in the Diet and came into effect from October 1, 2006. Furthermore, the basic principle for the concrete practices based on the law has been approved in a Cabinet meeting, which provides that, as a countermeasure against frequent occurrences of accidents and troubles caused by human errors, the government authority should enhance countermeasures to prevent transport accidents by the construction of safety management system in companies, or by the promotion of safety-first attitude in the business operations. Airline companies, for example, are required following things by the SMS for enhancing proactive safety management system.

1) Construction of safety management system in each airline
   Each airline is required to establish safety management regulation, to assign, a person in charge and to report to the government authority.
2) Disclosure of safety information
   Each airline is required to gather safety information by incident reporting system etc. The safety information must be reported to the government authority and also must be opened to the public.
3) Evaluation of safety management system and administrative advice by the government authority.
   The authority regularly checks the safety management system of airlines and, if necessary, gives administrative advice.

3   RISK MANAGEMENT IN AVIATION

The SMS requires improvement of safety activities at a field level. However, for a field, not only framework for safety management specified in the SMS but also practical guidelines and case examples are necessary in order to drive forward concrete safety practices. In the aviation area, such practical knowledge and experience of organizational safety management have already been stored numerously because airlines have already performed it in the past. In addition, the safety activities have been promoted beyond organizational and national boundaries. Therefore, the safety activities in aviation are standardized internationally and the effectiveness has been widely demonstrated in the world wide airlines. In this chapter, the safety practices in aviation are briefly illustrated as pioneering case examples of concrete practices of the SMS.

3.1 *Crew Resource Management Training (Training for preventing decision making errors of experts)*

Although considerable experience and expertise are important source of experts' skills, some past accidents has demonstrated experts' tendency toward over-reliance on them without utilizing various resources such as other persons, information, systems and facilities, and so on.

For example, in the collision accident by two Boeing 747s at Tenerife on March 27th 1977, the experienced caption of KLM's B747 has erroneously started to takeoff run without takeoff clearance from the control tower based on his own decision without utilizing other resources efficiently under excessive time pressure originated from strict rules about crew's working hours. The Crew Resource Management (CRM) training was developed based on many lessons from the tragic accident in the history of aviation.

The target of the CRM training are to realize the method for achieving appropriate decision making with utilizing all available resources and also to acquire such decision making style. Modern worldwide airline pilots are required to take CRM training regularly. Although the target of the initial CRM is only pilots on a flight deck, it has been broaden to cabin attendants, dispatchers, aircraft mechanics in the later phases. In some countries, the target is further extended to all employers of an airline as Corporate Recourse Management.

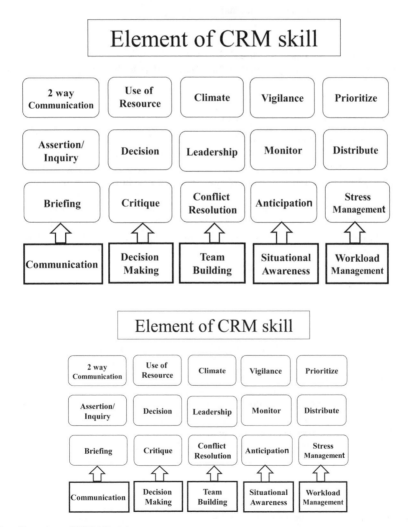

Figure 1. Overview of CRM Training.

45

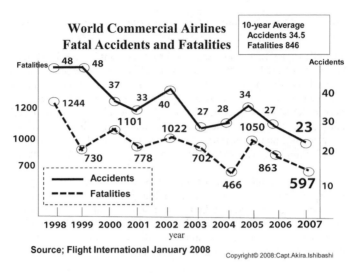

Figure 2.   Worldwide Aviation Fatal Accident Statistic 1998–2007.

Figure one illustrates the overview of CRM training. The ICAO's training manual recommends that the each item of CRM training should be modified to adapt to the specific culture and tradition of each airline.

3.2   *Aviation Safety Reporting System (ASRS)*

The Aviation Safety Reporting System (ASRS), a national safety reporting system which was established in US in 1976, is a typical example of proactive safety management method. The purpose of this reporting system is to gather and to analyze potential safety risks in daily operations. This reporting system has rapidly spread to other countries. In many cases, it has been operated by national budget.

The numerous collected data is stored huge databases and, in many cases, anyone can access the data freely. The important characters of this reporting system are anonymous and non-punitive for protection of reporters. The operation office function of this system is often commissioned to 3rd party research institutes in many cases in order to separate from government power of supervision and of punishment. In addition to such public safety reporting system, incident report activities at each field level has been widely adopted in aviation industry. This activities aim at quick resolution of potential safety problems at a field level.

3.3   *Toward improving industrial safety*

These safety practices in aviation can be basically applied to other industrial sectors as concrete safety activities required by the SMS. Their effectiveness in aviation can be shown by 23 years safety record of Japanese airlines and by small number of worldwide fatal accidents in statistics. Figure two indicates fatal accidents and fatalities of world commercial airlines. The number of fatal aviation accident has been steadily decreased in this decade. However, we should realize the fact that it took about 30 years to appear the effect of safety management activities such as the CRM and the ASRS as further decrease of the accident rate. The authors believe that this fact has indicates the importance of constant and persistent safety activities.

4   CONCLUDING REMARKS

The concept of proactive safety management has been actually implemented into the public trans-portation sectors in Japan by the SMS. Practical knowledge and experience of organizational safety management stored in aviation can and should be utilized to construct effective organizational safety management system in other domains. The authors are now developing a proactive safety management model towards practical use in various industrial sectors.

REFERENCES

The Ministry of Land, Infrastructure, Transport and Tourism (MLIT) of Japan, 2008, *The Transportation Safety Management System*, MLIT home page, http://www.mlit.go.jp/common/000013441.jpg.

ICAO Doc 9859 AN/460, *Safety Management Manual (SMM)*, Approved by the Secretary General published under his authority 2006.

ATEC Japan 2007, *Transportation Safety Management Seminar*, Proceedings of the Seminar pp. 1–17.

Mr. Y Tani *Japanese Transportation SMS*, Technical Department JCAB, pp. 1–28.

Capt. Miguel Ramos *SMS –The Perspective of ICAO*, Safety Management Programme ICAO.

Mr. David Learmount *Annual Safety Review*, Flight International 8–14 January 2008 pp. 30–38 London.

Capt. Akira Ishibashi 2006, *Why Accidents are repeated? Analysis of Human Factors, (Written in Japanese) JISHA The second edition 2006*.

Capt. Akira Ishibashi, *The leader ship to achieve "Zero-Risk"*, (Written in Japanese) jiyuukokuminsya publisher 2003.

NASA ASRS office, *Aviation Safety Reporting System Program Overview*, updated through Fourth Quarter 1999.

Capt. Akira Ishibashi, *Approach to Human Factors in Aviation field*, Academic Journal of AEM (The Japan Society of Applied Electromagnetics and Mechanics) Vol.15 No.1 9–15, 2007.

Capt. Akira Ishibashi, *A Study on the Analysis Method of Safety Reporting System, VTA Analysis Method* 7th SICE-SI Conference in Sapporo 2006, CD-ROM 600–601.

Capt. Akira Ishibashi, *CRM Training in Aviation Industry*, Lecture Proceedings at National Institute of Public Health 2007.

Chapter 3   Cutting edge of studies in safety ergonomics in Asia

Ergonomic Trends from the East – Kumashiro (ed)
© 2010 Taylor & Francis Group, London, ISBN 978-0-415-88178-4

A revisit to the human error assessment of the trip events in nuclear power plants

Yong-Hee Lee

I&C-Human Factors Research Division, Korea Atomic Energy Research Institute,
Daedeok-dero, Yuseong-gu, Daejeon, Korea

1 INTRODUCTION

Accidents in high reliability systems such as Nuclear Power Plants (NPPs) give rise to not only a loss of properties and life, but also social problems. They need active CounterMeasures (CMs) through technical efforts because the scale of a loss is huge. Most frequently used technique in NPPs is an event investigation analysis based on INPO's Human Performance Enhancement System (HPES), and the Korean Human Performance Enhancement System (K-HPES) in Korea, respectively. Event databases include their own events and information from various sources such as the IAEA, the regulatory bodies, and also from INPO and WANO.

There are a number of uncontrollable and hard to handle event sets because the nature of these events with a Human Error (HE) may often be threatened or very intensive. It is strongly required that systemic studies should be performed to grasp the whole picture of a current situation for hazard factors in NPPs. Many analysis methods associated with a HE in NPPs may be different from an explanation of the occurring events and the establishment of CMs. In the interest of them, we should consider these events with all the information about a system's condition and movement.

An Event happens causally, but a loss from a HE happens incidentally. A Human Error Analysis (HEA) as a technique to analyze events' causes is a HE causal analysis. The causes of the events can be searched by a causal backward reasoning analysis; on the other hand, possible barriers can be searched by a hazard analysis for elucidating the plausible causes for the events. A HE is a human-related event resulting in a kind of system loss such as a HE mechanism. CMs for the prevention of an accident are a set of plausible paths of similar events not a set of paths for an event sequence.

The objective of this paper is to propose an approach which can provide CMs against hazard factors in NPPs by using systematic procedures in substance. The procedures are; first, reconstructing the event to grasp the hazard factors, secondly, deducing the plausible paths of the event using the 4-M method (Man, Machine Media and Management) and the Industrial Accident Dynamics (IAD) diagram, and finally, deriving the CMs for the barriers to the paths. Then, case studies for a HE in NPPs using the proposed approach are performed for verification purposes. Our case studies clearly show that the proposed procedure can efficiently elucidate the hazard factors, and can effectively establish the CMs for safety purposes in NPPs.

2 A BRIEF REVIEW ON METHODS AND APPROACHES FOR HUMAN ERROR STUDY

2.1 *HE analysis methods*

Studies on HE analysis are classified into three types; quantitative approaches, qualitative approaches, and managerial approaches, according to an access method. Quantitative approaches use methodologies for the data management of HEs aiming at the computation of error probabilities. These techniques are grouped by two types of generations.

Table 1. A comparison for the person approach and the system approach.

Person Approach	System Approach
Wrong judgment	Structural design error
Insufficiency of period checkup and maintenance	Valve fail-open failure
	Insufficiency of safety culture
Deficiency of system comprehension	Insufficient human factors V&V
Unsatisfactory subject selection for period tests	Failure in H/W, S/W
Unsatisfactory training	Fault signal delivery in W/S
Insufficient procedure management	Insufficient transmission of accident cases
Deficiency of supervisor's control	Unsatisfactory procedure
Stress	Protecting cover
	Safety sign
	Unsafe of valve manipulation
	Function loss of local control panel

The first generation methods for HE are accident investigation and progression analysis, confusion matrix, operator action tree, socio-technical assessment of human reliability, expert estimation, etc. The second generation methods are cognitive reliability and error analysis method, a technique for HE analysis, generic error modeling system, Rasmussen's model, cognitive event tree system, cognitive environment simulator, etc. Quantitative analyses are insufficient to connect between a status analysis and a cause analysis, and hence they are general analysis methods rather than concrete analysis methods in application.

Qualitative analyses grasp cognitive behavioral characteristics through studies on theoretical and experimental HE as a psychological side. This approach focuses on theoretical characteristics of a cognitive act, and proposes only a high-level alternative plan like design concept of an error countermeasure. Therefore this approach has a limit for application to HE field in NPPs.

Lastly, for the managerial approaches having a report type, what is most frequently used technique in NPPs, is HE management system collecting and managing human error cases and analyzing human errors to make practical application of analyzing results. Theses are, for instance, incidents reporting system (IRS) of IAEA and OECD/NEA, HPES, K-HPES, and Japan human performance enhancement system (J-HPES), etc. A result of HE analysis from this approach works toward diminution of HEs through the improvements of the system, design, and work procedures.

However, these approaches are still not sufficient for a practical report system, and open to variation due to subjective judgments by analyzers' temperaments. A more effective approach to HE investigations in practice is strongly demanding [5].

2.2 *Approaches to HE investigation*

(1) Person vs. System Approach:

The HE problem may be viewed in two ways, the person approach and the system approach according to the Reason's suggestion [6]. Each has its model of error causation and each model gives rise to quite different philosophies of error management. The person approach focuses on the errors of individuals, blaming them for forgetfulness, inattention, or moral weakness. The system approach concentrates on the conditions under which individuals work and tries to build defenses to avert errors or mitigate their effects. Table 1 shows an example comparison for the person approach and the system approach.

(2) Socio-technical Approach:

Figure 1 shows a flow of loss prevention in socio-technical system. In order to evaluate the reliability of a socio-technical system, it is important to evaluate the system with a number of paths that correspond to a realistic sequence of events that could occur during the system's operation. The events in NPP are reconstructed and possible paths of the events are also conducted in this approach. There are many interfaces among departments or individuals that may not be clear and a communication error may intervene in high reliability organizations. Therefore, all possible interfaces should be performed carefully in an analysis for safety.

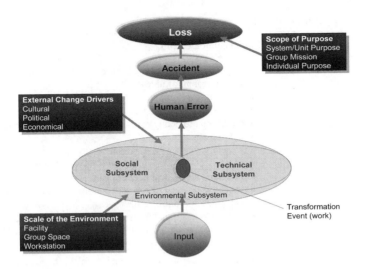

Figure 1. Socio-technical approach for barriers.

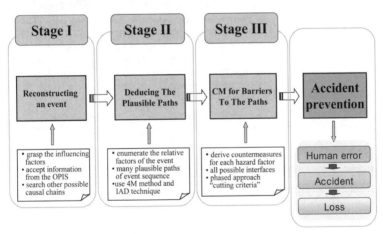

Figure 2. Overview of the proposed steps for establishing CMs.

(3) Approach for CMs vs. Causes:

CMs cannot be directly obtained from the causes. There are big differences between the approach for the CMs and the approach for the cause of an event. The dependency and the potentiality of the hazards in NPPs are defined by enumerating the relative factors of the events. The procedure in this research can search more hazardous factors as well as causal factors in each case than the preceding HEA so it may be considered as a better method for generating more effective CMs.

3 METHODS APPLIED AND THE RESULTS

Three stages are accomplished in this research. First of all, the event in a NPP is restructured by accepting the former information, and other possible causal chains are searched for each node of an event structure. Secondly, plausible paths of the events are deducted using the 4M method and the IAD diagram [1][2]. The dependency and the potentiality of the hazards are studied by enumerating the factors associated with the phases of the events according to a matrix type analysis at this stage. Finally, the CMs are derived for each hazard factor by means of stage I and stage II. This research

Table 2. Influencing elements list – Case 1.

Events	Influencing Elements
low-pressure turbine #1, stopping valve #2 shut	• faulty operation of pressure discharge valve of low-pressure turbine #1 and stopping valve #2 • unsatisfactory training • unsatisfactory procedure • not set up protecting cover to preventing the unintentional manipulation • no safety sign • unsafety of valve manipulation (fail-open failure) • function loss of local control panel • insufficient layout of pressure control switch for turbine stopping valve in TPC control panel • leading to manipulation error by not setting up protecting cover to prevent unintentional manipulation/no safety sign in TPC panel
rupture disk burst in MSR "A"	• pressure rising on MSR "A", "B" • MSR "A" 224psig "B" 213psig forming by the closing of low-pressure turbine #1 and stopping valve #1 on MSR "A" • insufficient analysis of pressure discharge process
low-pressure turbine #1, stopping valve #1 shut, high-pressure turbine control valve #1 shut	• burning out fuse damage of terminal box cable of low-pressure turbine #1, stopping valve #1 by breakdown • insufficient checkup of hydrogen leak in generator winding • insufficient safety confirmation of connecting valve • unsatisfactory auto operating condition in reactor power cutback system
reactor auto stop by P-8	• emergency manual stop for system protection

Table 3. Influencing elements list — Case 2.

Events	Influencing Elements
diminution of generator power voltage from 224 kV to 143 kV	• opening a circuit breaker of switchyard on manual operating mode in exciter operating mode (PCB 74400 and PCB 7411) • insufficient investigation about test condition • unsatisfactory procedure management • unsatisfactory training
reactor auto stop	• low-pressure in RCP mother line • design error • inappropriate basic value of auto/manual mode in controller • **failure in H/W** S/W and human **FMEA based design** • insufficient human factors V&V

proposes a systemic approach for an accident prevention against the hazard factors in NPPs. Figure 2 shows the stages that represent the procedure for an accident prevention establishment.

3.1 *Reconstructing an HE event*

An event in a NPP is restructured by accepting the information from the existing operational performance information system for NPPs (OPIS) [3], and other possible causal chains are searched for each node of an event structure. The additional influential elements are constructed by using an event based diagram or a cause & effect diagram after elucidating the influential elements from OPIS. Tables 2, 3 show the reconstructed influencing elements lists for the two different events.

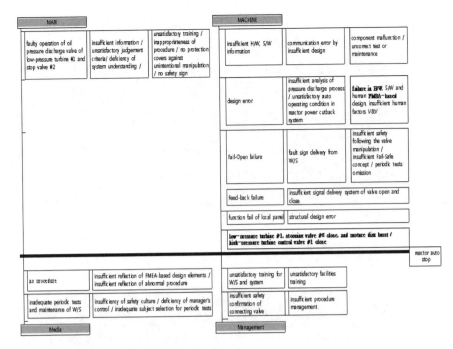

Figure 3. Hazard Factors for the 4-M Method – Case 1.

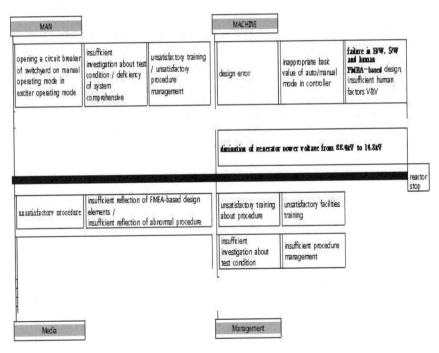

Figure 4. Hazard Factors for the 4-M Method – Case 2.

3.2 Deduction of the plausible paths from an event

Second stage classifies the hazard factors by the events. This stage clearly represents the relationships between the influential elements and each event. Also, the possible paths of these events are elucidated. This can help in deriving the CMs according to the events in a NPP.

55

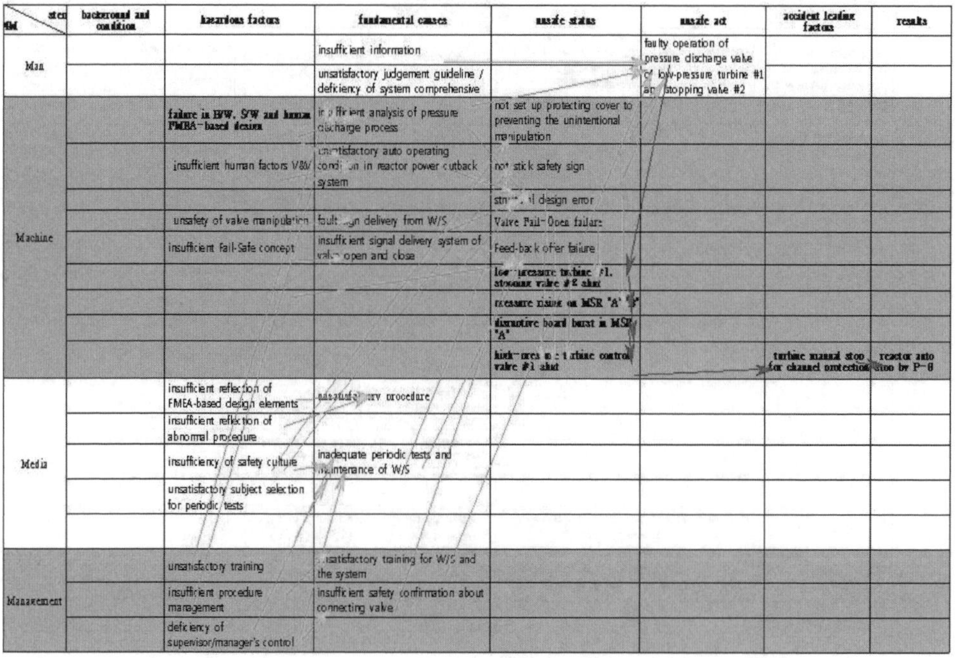

step 4M	background and condition	hazardous factors	fundamental causes	unsafe status	unsafe act	accident leading factors	results
Man			insufficient information		faulty operation of pressure discharge valve of low-pressure turbine #1 an stopping valve #2		
			unsatisfactory judgement guideline / deficiency of system comprehensive				
Machine		failure in HW, SW and human FMEA-based design	insufficient analysis of pressure discharge process	not set up protecting cover to preventing the unintentional manipulation			
		insufficient human factors V&V	unsatisfactory auto operating condition in reactor power cutback system	not stick safety sign			
		unsafety of valve manipulation	fault on delivery from W/S	structural design error			
		insufficient fail-Safe concept	insufficient signal delivery system of valve open and close	Valve Fail-Open failure			
				Feed-back offer failure			
				low-pressure turbine #1, stopping valve #2 shut			
				pressure raise on MSR 'A'			
				disruptive board burst in MSR 'A'			
				high area on turbine control valve #1 shut		turbine manual stop for channel protection	reactor auto stop by P-8
Media		insufficient reflection of FMEA-based design elements	unsatisfactory procedure				
		insufficient reflection of abnormal procedure					
		insufficiency of safety culture	inadequate periodic tests and maintenance of W/S				
		unsatisfactory subject selection for periodic tests					
Management		unsatisfactory training	unsatisfactory training for W/S and the system				
		insufficient procedure management	insufficient safety confirmation about connecting valve				
		deficiency of supervisor/manager's control					

Figure 5. Hazard Factors using the IAD Diagram – Case 1.

The dependency and the potentiality of the hazards in a NPP are defined by enumerating the relative factors of the events using the 4M method and the IAD diagram as shown in Figures 3, 4 and 5, respectively.

The IAD diagram uses seven stages (Background, Background + Initiating, Initiating, Intermediated, Immediate factors, Near-accident, and Accident) for the dependency and a visualization of the error. We used 7 stages to analyze the background and condition, hazardous factors, fundamental factors, unsafe status, unsafe act, accident leading factors, and results easily. Also, four factors (Machine; material and object of work; Human; Environment, Others) applicable to each factor were analyzed with a pattern of the 4M (Man, Machine, Media, and Management) method.

3.3 Deriving CMs

This stage derives the CMs for each hazard factor as shown in Tables 4. In order to evaluate the reliability of the high reliability organized systems, it is important to evaluate a system with a number of paths that correspond to a realistic sequence of the events that could occur during a system's operation.

Accident prevention can be established in these phases against the hazards in a NPP, composed of the HEs, the accidents, and the loss steps. There are many interfaces among departments or individuals which may not be clear and a communication error may intervene in high reliability organizations. Therefore, all possible interfaces should be performed carefully in an analysis from a safety point of view.

4 CONCLUSIONS AND DISCUSSIONS

This research suggests an effective systematic approach that strategically focus to the CMs for an accident prevention. It firstly derives the CMs rather than the causes of an event by utilizing the influential elements list, the 4-M method, the IAD diagram, and proposes a safeguard by representing the possible paths to a loss in a NPP. The procedure proposed in this paper can search more hazardous factors rather than the causal factors in each case comparing to the preceding

Table 4. CMs for each Hazard Factor – Case 1.

Hazard factors	Results of hazard	Causes	Countermeasures
• wrong judgement	• fault operation of pressure discharge valve of low-pressure turbine #1 and stopping valve #2	• deficiency of control system comprehensive • structural design error • unsatisfactory facilities training	• procedure improvement • technical training about MCR control system • setting up protecting cover to preventing unintentional manipulation • safety sign
• pressure rising MSR "A", "B"	• rupture disk burst	• stopping valve close in low-pressure turbine #1 • A design that MSR rupture disk can burst in the high pressure	• safety countermeasures for human error prevention • human factors review for Fail-open failure conditions • procedure supplementation for possible items of Fail-open failure • plans for modifications of pressure discharging process
• insufficiency of period checkup and maintenance	• high-pressure turbine control valve #1 shut	• control valve(GV)#! and low-pressure turbine#1 shut abnormally by steam bursting • high-pressure turbine control valve#1 shut	• checkup and maintenance of hydrogen leak in generator winding • human factors review for the conditions of Fail-open failure • procedure supplementation about possible items of Fail-open failure
• structural design error	• valve Fail-open failure	• insufficient human factors V&V • failure in H/W, S/W and human FMEA-based design	• human factors V&V implementation for remote control system • H/W; S/W, and human FMEA-based design • confirmation of normal status and construction of systemic procedure about abnormal management in installation of remote control system
• valve Fail-open failure	• function fail of local control panel	• fault signal delivery in W/S	• training about signal delivery of remote control system • adding notices about signal delivery of remote control system to procedures
• inadequate periodic tests and maintenance of W/S and network	• networking error between Workstation and local control panel	• unsatisfactory subject selection for period tests	• adding systems related to fail-open valve to periodic tests
• insufficiency of safety culture	• faulty judgement selection (shutdown rebooting)	• insufficient transmission of accident cases	• safety culture establishment about work culture and reporting customs

HEA of the reported NPP trips, so it can be a better method to generate more effective CMs. This approach might be more effective for the achievement of higher level of operational safety in NPPs.

For a further research, it is necessary to develop a computerized tool which searches for the enumeration of all possible paths within the event structure, and screening them by applying a cost benefit analysis. Also, the criteria for making decisions on screening paths and stopping the enumerations must be formalized for an application of the method in this research.

REFERENCES

S. H. Hwang, D. H. Kim, Y. H. Lee, A Case Study for a Human Error Analysis in Nuclear Power Plants, In Proceedings of the Conference on Ergonomics Society of Korea, 2007.

J. K. Park, T. I. Jang, J. W. Lee, J. C. Park, H. C. Lee, Y. H. Lee, A Case Study for the Human Error Analysis in Nuclear Power Plants Using a Work Domain Model, In Proc. of the Ergonomics Society of Korea, 2007.

Korea Institute of Nuclear Safety (KINS), http://opis.kins.re.kr/

Y. H. Lee, A Discussion for a more Effective Approach to Human Error Studies in Industries, Proc. Joint Conf. ESK & JES, Osaka, Japan, 2006.

Y. H. Lee, et al., Human Error Cases in Nuclear Power Plants: 2002~2007 in Korean, KAERI, 2007.

J. Reason, Human Error: Models and Management, Br. Med. J, 2000.

Ergonomic Trends from the East – Kumashiro (ed)
© 2010 Taylor & Francis Group, London, ISBN 978-0-415-88178-4

Exploring nuclear power operation issues of interacting with digital human-system interfaces

Sheau-Farn Max Liang
Department of Industrial Engineering and Management, National Taipei University of Technology, Taiwan, ROC

Sheue-Ling Hwang
Department of Industrial Engineering and Engineering Management, National Tsing Hua University, Taiwan, ROC

Po-Yi Chen
Department of Industrial Engineering and Management, National Taipei University of Technology, Taiwan, ROC

Yi Jhen Yang & Tzu Yi Yeh Liu
Department of Industrial Engineering and Engineering Management, National Tsing Hua University, Taiwan, ROC

Chang-Fu Chuang
Atomic Energy Council, Taiwan, ROC

ABSTRACT: Digital Human-System Interfaces (HSIs) have been applied as the main course of operations in the Main Control Room (MCR) of the Fourth Nuclear Power Plant (FNPP) in Taiwan. Potential safety and performance issues of interacting with the digital HSIs were explored through focus group interviews with ten operators at the FNPP. Nine major issues were revealed from the interviews. These issues were then associated with 18 high-level HSI design review principles. The findings of this study provided guidance of further improvements on the HSI design and operator training for ensuring the nuclear safety.

Keywords: Nuclear Power Plant, Human-System Interface, Nuclear Safety

1   INTRODUCTION

Nuclear Power Plant (NPP) has been operated for 30 years in Taiwan. The Fourth Nuclear Power Plant (FNPP) with two 1350-MWe Advanced Boiling Water Reactor (ABWR) units is currently under construction and planned to be commercially operated in 2009 and 2010. Different from the Human-System Interfaces (HSIs) in previous three NPPs, digital HSIs, such as touch-screen Video Display Units (VDUs), are the major course for operators to retrieve information and control the system in the Main Control Room (MCR). The principal location for safety related control actions is the MCR (IAEA, 2000; IAEA, 2002). Figure 1 indicates the layout of the FNPP MCR. A Wide Display Panel (WDP) is on the front wall of the MCR, and the Main Control Console (MCC) is located between the WDP and the Shift Supervisor Console (SSC). A schematic display of equipment and flow lines is on the middle area of the WDP to provide general information of operation. System-level alarm indicators are displayed on the top of the WDP. The Plant Status Display (PSD) is on the left and a Large Variable Display (LVD) is on the right. 19 touch-screen VDUs and some hard control switches and buttons are located on the panel under the WDP. 13 vertical and 10 horizontal touch-screen VDUs are on the MCC, and three VDUs are on the SSC. These three VDUs are

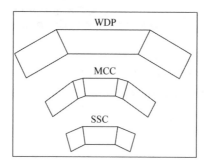

Figure 1. Layout of the Main Control Room (WDP: Wide Display Panel; MCC: Main Control Console; SSC: Shift Supervisor Console).

only with monitor function, but the 42 VDUs on the WDP and MCC are with monitor and control functions. 13 of them are designed for safety systems, other 29 are for non-safety systems.

The operation will surely be benefited from digital HSIs. However, new technology may also bring in new problems. Related design principles on HSIs are reviewed in the next section. Focus group interviews with the operators at the FNPP for exploring potential safety and performance issues of operations with the digital HSIs are described followed the review. Results of the interviews and discussions are presented in the last.

2 HUMAN-SYSTEM INTERFACES

18 high-level principles for the HSI design are proposed by the U.S. Nuclear Regulatory Commission and grouped into four categories: general, primary task design, secondary task control, and task support (O'Hara, Brown, Lewis and Persensky, 2002). General principles emphasize that the HSI design should (1) support personnel safety, be compatible with their (2) cognitive and (3) physiological capabilities, be (4) simple, and be (5) consistent with procedures and training. Operators' primary tasks of process monitoring, decision-making, and control should be supported by the HSI design in (6) situation awareness. The HSI design should be compatible with (7) tasks and (8) user models, have well (9) organization of HSI elements and (10) logical/explicit structure, and consider (11) timeliness of tasks, (12) the compatibility between controls and displays, and provide (13) feedback. Operators' secondary tasks, such as navigating through displays, manipulating windows, and accessing data, should be supported by the HSI design to minimize operators' (14) cognitive workload and (15) the number of steps to accomplish an action. Finally, the HSI design should be (16) flexible so tasks can be accomplished in more than one way, provide (17) user guidance and support and (18) error tolerance and control (O'Hara, Brown, Lewis and Persensky, 2002). These high-level HSI design review principles were used as the scheme to classify the potential safety and performance issues revealed from the focus group interviews with the operators at the FNPP.

3 FOCUS GROUP INTERVIEWS

Ten operators from the FNPP were recruited for the focus group interviews. During the interviews, the operators were encouraged to discuss any operation issues with regard to the HSI design in the MCR. Potential issues with associated high-level HSI design review principles are listed as below:

- The vertical bar indicator on the WDP which represents the reactor water level is a composite value and difficult for operators to interpret.
  Principle (2), (4), (6), (7), (8), (10), (14), (15)
- The resolution of projected screen on the LVD is insufficient and makes operators difficult to read.

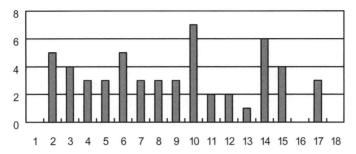

Figure 2. Frequency of Issues on Each HSI Design Principles.

Principle (3), (5), (10), (14)
- The decimal point in the digit display on the WDP for showing the pressure of drywell is too small to read.
  Principle (3), (6), (10)
- The display format of the electric power capacity on the WDP (by values) is different from the format on the procedure documents (by percentage).
  Principle (2), (5), (7), (10), (14), (15)
- In abnormal situation, alarm system produces considerable sound and flash that interfere the operation.
  Principle (3), (4), (6), (9), (11), (14), (17)
- Thousands of alarms displayed in acronyms make operators difficult to memorize their meanings and distinguish each other.
  Principle (2), (4), (10), (14), (17)
- System response lag sometimes happens when operators interact with the touch-screen VDUs.
  Principle (2), (5), (6), (7), (8), (10), (11), (12), (13), (14), (15)
- There is a glare when operators read screens of the horizontal VDUs at a certain view angle.
  Principle (3), (9)
- It is sometimes difficult to distinguish enabled buttons from disabled buttons on the screen of the VDUs.
  Principle (2), (6), (8), (9), (10), (12), (15), (17)

The frequency of issues on each HSI design principle is presented in Figure 2. As indicated, most issues are related to the (10) logical/explicit structure, (14) cognitive workload, (2) cognitive compatibility, and (6) situation awareness.

The findings of this study provided guidance of further improvements for the HSI design and operator training to minimize potential human errors, and ultimately, to enhance the effectiveness and efficiency of operation and ensure safety.

ACKNOWLEDGEMENTS

This research was funded by the National Science Council and the Atomic Energy Council of Taiwan (NSC 97-NU-7-007-005).

REFERENCES

IAEA (2000). *Safety of Nuclear Power Plants: Design*, Safety Standards Series No. NS-R-1, International Atomic Energy Agency, Vienna.
IAEA (2002). *Instrumentation and Control Systems Important to Safety in Nuclear Power Plants: Design*, Safety Standards Series No. NS-G-1.3, International Atomic Energy Agency, Vienna.
O'Hara, J. M., Brown, W.S., Lewis, P.M. and Persensky, J. J. (2002). *Human-System Interface Design Review Guidelines*, NUREG-0700, Rev. 2, U.S. Nuclear Regulatory Commission, Washington, DC.

Chapter 4   Common qualification system for work conditions and ergonomics among the East Asia

Ergonomic Trends from the East – Kumashiro (ed)
© 2010 Taylor & Francis Group, London, ISBN 978-0-415-88178-4

Introduction of WMSDs prevention law and system in Korea

Kwan S. Lee
Hongik University, Seoul, Korea

Dong-Kyung Lee
Korea Occupational Safety and Health Agency, Inchon, Korea

1 INTRODUCTION

The presence of musculoskeletal burden tasks and work-related musculoskeletal disorders (WMSDs) of industrial workers had not been well-known until 2003 in Korea. But in July, 2003, the Korean occupational safety & health law which was changed to add the proprietor duty in preventing work-related MSDs of workers to the business owner's responsibility became in effect. Since then WMSDs have become a big issue in Korea. The manufacturing industry especially, auto and shipping industry, complied with this law early in 2003 since they have been blamed for causing overexertion in most time. But in these days it is spreading out to non-manufacturing industries gradually. In this study, Korean WMSD laws and recent problems related to this law in Korea are introduced and changes are proposed to resolve these problems.

After the addition of the proprietor duty to the occupational safety & health, a guideline was introduced to help in setting up the work-related musculoskeletal disorders (WMSDs) prevention system in an organization. This system was expected to decrease the number of patients suffering from musculoskeletal disorders at work.

The number of occurrence of WMSDs in 2004 was 4,112. A decrease of 93.1% compared to that in the previous year. In 2003, the number of occurrence of WMSDs was 4,532. In 2004, 1,569 workplaces reported to have WMSDs. This is 19.0% decrease from the previous year as shown in Table 1. However, it is not very clear that this system has been very effective or not since the number of workers with WMSDs started to increase in 2006. The increase was partially caused by the change of the rule in classifying the WMSDs. The rule was changed to include lumbago due to accident as WMSDs in 2006.

2 OCCUPATIONAL SAFETY AND HEALTH LAW

The Korean government added Clause 1-5, Article 24 (health measures) of the Occupational Safety and Health Law to impose the obligation of preventing musculoskeletal disorders to business owners.

The Occupational Safety and Health Law

> ① In the course of doing their business, business owners must take the necessary measures for the prevention of the following health problems:
> 1.~4.: "no change"
> 5. Health problems caused by repetitive operations or operations placing excessive burden on the human body

Then, regulations on the criteria for occupational health were revised to add Article 9 (prevention of health problems due to the burden placed on the musculoskeletal system). It defined the specific

Table 1. Status of occurrence of musculoskeletal disorders per year (unit: place, person).

Workplaces Recording Musculoskeletal Disorders (rate of increase/decrease)				Patients with Musculoskeletal Disorders (rate of increase/decrease)			
2003	2004	2005	2006	2003	2004	2005	2006
1,938	1,569 (Δ19.0%)	1,383 (Δ11.8%)	4,606 (233.0%)	4,532	4,112 (Δ9.3%)	2,901 (Δ29.5%)	6,233 (114.9%)

※ Change in the criteria for statistical classification in 2006: Lumbago due to accident was added as MSDs (3,612 persons)

Table 2. The ratio of burden jobs to whole jobs.

	Employer			Employee		
Task	No.	Average rate (%)	Standard Deviation	No.	Average rate (%)	Standard Deviation
Production	255	25.3	22.6	169	28.4	26.2
Administrative position	137	8.2	12.7	65	7.8	12.6

responsibilities of business owners and the responsibilities include examining the harmful elements, improving the working environment, taking medical steps, notifying the harmfulness, and establishing and implementing programs to prevent musculoskeletal disorders. The scope of MSD hazardous tasks were to be determined by the Minister of Labor and thus a total of 11 operations were defined by the notification of the scope of MSD hazardous tasks as shown below;.

1. Concentrated use of keyboard of mouse (for 4 hours)
2. Repetitive operation with the use of the neck, shoulder, elbow, wrist, or hand (for 2 hours)
3. Operation with the elbow positioned away from the trunk of the body (for 2 hours)
4. Operation with the neck or waist bent or twisted (for 2 hours)
5. Operation involving squatting or with the knees bent down (for 2 hours)
6. Operation involving the lifting of objects weighing more than 1 kg with the fingers of one hand (for 2 hours)
7. Operation involving the lifting of objects weighing more than 4.5 kg with one hand (for 2 hours)
8. Operation involving the lifting of an object weighing more than 25 kg (10 times)
9. Operation involving the lifting of an object weighing more than 10 kg (25 times)
10. Operation involving the lifting of an object weighing more than 4.5 kg (for 2 hours)
11. Operation that applies impact on the hands or knees (for 2 hours, 10 times per hour)

These 11 burden jobs are the same as the burden jobs shown in Washington State standards in U.S.

3 PROBLEMS

It was found that definitions of terms such as "MSD hazardous tasks" and "operations placing a burden on the musculoskeletal system" were not very clear and thus personnel in industry had problems in selecting the burden jobs at work places. Table 2 shows that there are some discrepancy in determining the burden jobs between employers and employees. The discrepancy was bigger for production jobs. However, the discrepancy may not be very significant and this was confirmed by Lee (2005) who reported that 92% of jobs turned out to be burden jobs among jobs which were designated as burden jobs by companies. The discrepancy might be caused by the different interpretation of the burden jobs as reported by Ki (2007).

Table 3. The result of Investigation of ergonomic factors and survey (units: numbers, (%)).

Categories	ergonomic investigation	MSD symptom survey					
		No		Done		계	
employers	no	137	(40.3)	1	(0.3)	138	(40.6)
	done	25	(7.4)	177	(52.1)	202	(59.4)
	total	162	(47.6)	178	(52.4)	340	(100.0)
employees	no	116	(46.4)	1	(0.4)	117	(46.8)
	done	22	(8.8)	111	(44.4)	133	(53.2)
	total	138	(55.2)	112	(44.8)	250	(100.0)

It was also found that a large proportion of companies have not complied with this law yet. This is partially due to a loose inspection system by Ministry of Labor. Mostly the government officials visit large companies for inspection.

Further, there were very opposite demand on the investigation cycle between employers/ representatives of employers and employees/representatives of employees. As expected the employers' side requested a longer cycle time than 3 years which is the current cycle time and the employees' side requested a shorter cycle than 3 years. It was also found that the law should clarify the qualification of investigator and tools which should be used to assess jobs to determine the musculoskeletal stress level. People also requested the definition and scope of improvement required in case a job is found as a burden jobs. There were complaints about the ambiguity of the sign or symptoms of the WMSD's.

4 CONCLUSIONS

It is very complicated to prevent MSDs. There is no one special solution for it. The present MSDs regulations in Korea must be modified to the effective system for MSD prevention as we have discussed so far. However, to make this MSD prevention campaign successful, the most important step is to make stakeholders feel the need and benefit of self-regulation of the MSDs by stakeholders. The second is to make regulations minimum and thus every enterprise can comply. This regulation should be complied throughout Korea as the Korean government intended to reduce the number of occurrences of WMSDs.

REFERENCES

Ki, Do-hyung, 2007, KOSHA OSHRI 2007-124-1055, 2007
Lee, Chang Min, 2005, A Study on the Status of Burden Jobs of Musculoskeletal Disorders in Korea, *A Study Report*, KOSHA, 2005

Chapter 5   Cognitive ergonomics 1

Ergonomic Trends from the East – Kumashiro (ed)
© *2010 Taylor & Francis Group, London, ISBN 978-0-415-88178-4*

Relation between the change of pupil diameter and the mental workload

Kimihiro Yamanaka & Mitsuyuki Kawakami
Faculty of System Design, Tokyo Metropolitan University, Hino-shi Tokyo, Japan

ABSTRACT: This study aims to develop a physiological information system for monitoring human reliability in man-machine systems. The mental state of a worker, such as autonomic nervous activity, is evaluated using pupil diameter. In this study, two kinds of experiments are carried out. The former experiment is to investigate the relation between the change of pupil diameter and time pressure. On the other hand, it is investigated that the relation between the change of pupil diameter and resource pressure in the latter experiment. As well, in this experiment the Event-Related Potentials (ERP) obtained from electroencephalogram (EEG) as a quantitative index are measured to evaluate mental workload. It was found that (1) measurement of pupil diameter is an effective indicator of autonomic nervous activity, (2) based on the results of electroencephalogram, the amplitude of event-related potentials are dependent on mental workload.

1 INTRODUCTION

Human error is closely related to the safety and structural health of a man-machine system. The reliability of a mechanical element is improved by advancements in science and the technology. However, there remains the problem of human reliability. Human error is caused by an interaction between subjective uncertainties and various kinds of factors such as mechanical equipment and the work environment (Rassmussen J, 1983., Reason J, 1990., Hollnagel E, 1993., Rassmussen J, 1986.). Human performance is an important factor in man-machine systems. Therefore, in order to prevent human error we have to consider many factors such as the mental state of the worker, and the design of the mechanical system as well as environment conditions (Fitts P.M and Jones R.E, 1947.).

All existing systems are operated by workers. In an automated system, the machine can work without an operator. However, error detection and system maintenance remain the operator's roles. Therefore, all mechanical equipments are considered man-machine systems. Recently, because man-machine systems have become not only large-scale but also complex, the reliability of the system has become dependent on the operator, thus making it important to deal with human error in such systems. However, since human error is dependent on events composed of various factors, little research had been performed regarding the mechanism of human error (Rassmussen J, et. al., 1981., Malone T. B, et. al., 1980.).

From this point of view, the present study attempts to use an ergonomic approach to measure workers' mental state and mental workload. In the experiment, it aims to clarify the relation between the change of pupil diameter and autonomic nervous activity. As well, the event related potentials (ERP) obtained from an electroencephalogram (EEG) as a quantitative index are measured to evaluate the mental workload.

2 EXPERIMENTAL SETUP

A schematic of the experimental device is shown in Figure 1. The electrocardiogram (ECG) was measured using multi-telemeter system (Nihon Kohden, Web-5500). An electroencephalogram was also measured with the disc electrode attached based on a ten-twenty electrode system (Jasper H, 1958) such a Fz, Cz, and Pz. The sampling frequency was 1000 Hz. The analogue outputs of the

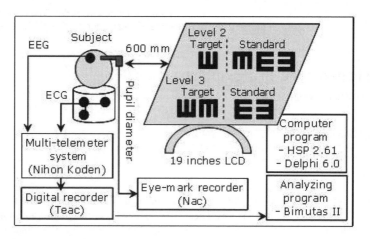

Figure 1. Diagram of experimental devices.

multi-telemeter system were transferred to an A/D converter recorder (Teac, DR-m50). On the other hand, the pupil diameter was measured using an eye-mark recorder (Nac, EMR-8B). The sampling frequency was 60 Hz. The out put data is recorded by a personal computer. A 19-inch liquid crystal display (Dell, E197FP) was situated 600 mm from the eye of the subject. The brightness and the illuminance were, respectively, 76.2 *cd* and 514 *lx* in front of the liquid crystal display. Two kinds of buttons were held in each of the subject's hands as shown in Figure 1.

In this study, two kinds of experiments were carried out. In the first experiment, the subject was instructed to perform a time pressure task. A one digit number from 0 to 9 with a size of 1.0° was shown on the liquid crystal display at regular intervals. The subject was required to register on a numerical keypad the one digit number shown on the liquid crystal display within a specific time limit. The three levels of time pressure task were 1.5, 1.0, and 0.5 sec. In the second experiment, the resource pressure of the information processing is used as the experimental task. This experiment was divided into levels 1, 2, and 3. The task of level 1 involved a simple auditory stimulus which was programmed by HOT soup processor 2.61 in line with an oddball task (Sutton S, et. al., 1965., Dunean J and Donchin E, 1977.). The auditory stimulus had a pure sound of 1000 db as a target sound and a pure sound of 2000 db as a standard sound. The ratio of target sound to standard sound was 0.20. The recurrence interval of the auditory stimuli was 1500 ± 500 ms. The subject was required to count the number of occurrences of the target sound. In level 2, the visual stimulus was added to the level 1 task. A visual stimulus with the size of 1.0° had a specific shape in Snallen's chart as a target object while a similar object was provided as a non-target. The visual stimulus appeared on the centre of the liquid crystal display at randomly distributed coordinates. The recurrence interval of visual auditory stimulus was 1700 ms. The position and time interval of the visual stimulus were programmed using Delphi 6.0 in line with the experimental paradigm. The subject was required to push the button on the right side as soon as possible when recognizing the target object. In level 3, two target objects were provided. The subject was required to push the button on the right side as soon as possible when recognizing target object 1. On the other hand, the subject was required to push the button on the left side when target object 2 appeared.

In this experiment, 5 healthy university students (mean age, 22.4 years; range, 21–24 years) were selected as subjects in order to examine the relation between changes in pupil diameter and autonomic nervous activity.

3   RESULTS AND DISCUSSIONS

Figure 2 shows the results of the experiment involving the time pressure task. The relation between time pressure and heart rate variability such as LF/HF is shown in Figure 2. It is well known that heart rate variability composed by LF (low frequency component), MF (mid frequency component),

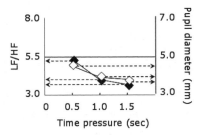

Figure 2. Relation between time pressure and mental workload.

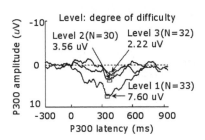

Figure 3. Ensemble average of P300.

and HF (high frequency component). The value of LF is frequency band from 0.04 to 0.15 Hz, and shows both the sympathetic nerve activity and the parasympathetic nerve activity. On the other hand, the value of HF is frequency band from 0.15 to 0.4 Hz, and shows parasympathetic nerve activity. For that reason, the sympathetic nerve activity is shown by the value of LF/HF (Brentson G, 1977., Grossman P and Svebak S, 1987., Lacey B. C and Lacey J. I, 1978., Ohsuga M, et. al., 2001., Sayers B, 1973.). For comparison, the value of LF/HF at rest is 2.44. This figure reveals two findings. First, it is seen that the value of LF/HF in time pressure 0.5 sec which is more than double that at other time pressures. Secondly, a typical tendency is revealed in which increases in time pressure are accompanied by increases in LF/HF value. These results show that the activity of the sympathetic nervous system dominates the parasympathetic nervous system under time pressure.

As well as, Figure 2 shows the relation between time pressure and pupil diameter. For comparison, pupil diameter at rest is 3.71 mm. In this figure, it is seen that pupil diameter under a time pressure of 0.5 sec is bigger than that under other time pressures. Also, a typical tendency is seen in which increases in time pressure are accompanied by increases in pupil diameter. These results indicate a common tendency between changes in pupil diameter and changes in LF/HF value. In other words, it is seen that the change in the pupil diameter is an effective indicator of autonomic nervous activity.

Figure 3 shows an ensemble average of P300 potentials on Pz as obtained by the signal averaging method. In this figure, N indicates the number of averaging. As pointed out in the reference (Amari S and Toyama K, 2000., Proverbio M.A, et. al., 2002.), the P300 amplitude reflects the degree of concentration for information processing. On the other hand, there is a limitation of the resources for information processing. When the amount of the information processing increases, one must share this task from one's limited resources for information processing (Courchesne C. K, et. al., 1975., Picton T. W and Hillyard S. A, 1974., Squires N. K, et. al., 1975.). From these points of view, this figure shows a great difference of P300 amplitude between the three tasks. In this figure, it is seen that P300 amplitude is dependent on task level, because the more difficult the task becomes, the smaller the P300 amplitude becomes. However, no relation is seen between task level and P300 latency.

The relation between P300 amplitude and heart rate variability such as LF/HF is represented in Figure 4. It is seen that the larger the P300 amplitude, the larger the LF/HF value. As well as, it is found that with increases in P300 come increases in pupil diameter, as shown in Figure 4. These results demonstrate that as the quantity of information to be processed increases, the LF/HF value

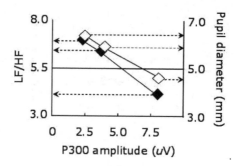

Figure 4. Relation between P300 and mental workload.

increases as does pupil diameter. Thus, it is possible to evaluate the degree of concentration on a task in terms of changes in pupil diameter.

4 CONCLUSION

This study describes the development of a physiological information system for monitoring human reliability in man-machine systems. Pupil diameter was used to evaluate the metal state of a subject under work stress. The obtained results can be summarized as follows.

1) It was found that measurement of pupil diameter is an effective indicator of autonomic nervous activity.
2) Based on the results of electroencephalogram, the amplitude of event-related potentials is dependent on mental workload.

ACKNOWLEDGMENT

The authors wish to acknowledge Kazuya Murata of Graduate School of Tokyo Metropolitan University for his experimental contributions. The authors also express great thanks to the Grand-in Aid for Scientific Research Found of the Ministry of Education, Science Sports and Culture of Japan (Grant No. 20510161) for this financial support.

REFERENCES

Amari S and Toyama K, 2000, Dictionary of Brain Science, *Japan, Asakura Publiser*, (In Japanese).
Brentson G. G, Bigger J. T, Eckberg D. L, Paul Grossman P, Kaufmann G, Nagaraja G, et. al., 1977, Heart rate variability Origins, methods, and interpretive caveats, *Psychophysiol.*, 34(6), pp.623–648.
Courchesne C. K, Hillyard S. A and Galambos R, 1975, Stimulus novelty, task relevance, and the visual evoked potential in man, *Electroencephalogr. Clin. Newrophysiol.*, 39, pp.131–142.
Dunean-Johnson C.C and Donchin E, 1977, On quantifying Surprize; The variation in event-related potentials with subjective probability, *Psychi-Physiology*, 14, pp.456–467.
Fitts P.M and Jones R.E, 1947, Psychological aspevts of instrument display, *U.S. Air Forces Air Material Command, Engineering Division*, No. TSEAA-694-12A.
Grossman P and Svebak S, 1987, Respiratory sinus arrhythmia as an index of parasympathetic cardiac control during active coping, *Psychophysiol.*, 24(2), pp.228–235.
Hollnagel E, 1933, Human Reliability Analysis Context and Control, *Academic Press*, p.145.
Jasper H, 1958, Ten-twenty electrode system of the International Federation, *Electroencephalogr. Clin. Neurophysiol.*, 10, pp.371–375.
Lacey B. C and Lacey J. I, 1978, Two way communication between theheart and the brain, *Am. Psychol.*, Feb, pp.99–113.

Malone T. B, et. al., 1980, Human Factors Evauation of Control Room Design and Operator Performance at Three Mile Island -2, *NUREG/CR-1270*, vol.1.

Ohsuga M, Shimono F and Genno H, 2001, Assessment of phasic work stress using autonomic indices, *Int. J. of Psychophysiol.*, 40, pp.211–220.

Picton T. W and Hillyard S. A, 1974, Human auditory evoked potentials, *II. Effects of attention. Electroencephalogr. Clin. Neurophysiol.*, 63, pp.191–199.

Proverbio M.A, Esposito P and Zani A, 2002, Early Involvement of the Temporal Area in Antinational Selection of Grating Orientation: an ERP Study, *Journal of Cognitive Brain Resarch*, 13, pp.139–151.

Rassmussen J, et. al., 1981, Classification System for Reporting Event Involving Human Malfunction, Riso-M-2240.

Rassmussen J, 1983, Skills, Rules, Knowledge Signals, Signs and Symbols And Other Distinctions in Human Performance Models, *IEEE Trans. On SMC*, SMC-13, 3, pp.257–267.

Rassmussen J, 1986, Information Processing and Human Computer Interaction, *North Holland*, p.135.

Reason J, 1990, Human Error, *Cambridge Univ. Press*, p.53.

Sayers B, 1973, Analysis of heart rate variability, *Ergonomics*, 16, pp.17–32.

Squires N. K, Squires K. C and Hillyard S. A, 1975, Two varieties of long-latency positive waves evoked by unpredictable auditory stimuli in man, *Electroencephalogr. Clin.Neurophysiol.*, 38, pp.387–401.

Sutton S, Baren M, Zubin J and John R, 1965, Evoked-potential correlates of stimulus uncertainly, 150, *Science*, pp.1187–1188.

Ergonomic Trends from the East – Kumashiro (ed)
© *2010 Taylor & Francis Group, London, ISBN 978-0-415-88178-4*

Electroencephalographic study of visual fatigue

Bin-Wei Hsu
Department of Industrial Engineering and Engineering Management,
National Tsing Hua University, Taiwan, ROC
Department of Industrial Engineering and Management, Chinmin College, Taiwan, ROC

Mao-Jiun J. Wang
Department of Industrial Engineering and Engineering Management,
National Tsing Hua University, Taiwan, ROC

1 INTRODUCTION

For visual fatigue measures, Critical Fusion Frequency (CFF) has been considered to reflect neuron impulse transmission from retinal ganglion cells to the primary visual cortex and is used as an indicator of visual fatigue (Murata *et al.*, 1991). Gunnarsson and Soderberg (1983) indicated NPA (Near Point Accommodation) can reflect the degree of visual fatigue. In addition, Subjective Rating (SR) is also a commonly used index. The advantage of SR is easy using and having high surface validity. But it must be used with other indices because lacking diagnostic abilities to accurately pinpoint the cause of visual fatigue. SR has obvious correlation with fatigue, CFF, and operation time (Shieh *et al.*, 1996). Chi and Lin (1998) stated that the sensitivity of accommodation, CFF or SR of fatigue appear to be higher when task time was over 60 minutes. Therefore it is recommended to use these indices in lengthened sessions.

Iwasaki *et al.* (1988) stated electroencephalography (EEG)-VEP (visual evoked potential) and VEP latency was associated with VDT work fatigue, especially in short term task. Therefore, it is postulated that the brainwaves would be a sensitive visual fatigue index. Except for the VEPs, EEG signal could be subjected by Fast Fourier Transformation (FFT) and expressed it as power δ, θ, α and β. These bands would be suitable to indicate fatigue, wakefulness, sleepiness and alertness et al. Moreover, Eoh *et al.* (2005) used EEG ratio indices to measure the fatigue level. Further, Cheng *et al.* (2007) pointed out that the main fatigue induced from VDT task was in the occipital region by the ratio index $(\alpha + \theta)/\beta$ increasing. Others showed EEG power were very similar for θ power and self-rated fatigue (Lai *et al.*, 2001) and the combination of α and θ activity gives more consistent EEG results than if they were evaluated separately.

The research aims to find the applicability of the EEG power in visual fatigue evaluation by finding the relationships between EEG and other visual fatigue indices. The goals of this research were: (1) To determine the relationship between the EEG power and visual fatigue and to clarify which EEG power indices (region/ rhythm/ basic index and ratio index) could best reflect the degree of visual fatigue? (2) To understand the correlations among the EEG power, CFF, NPA, and subjective rating.

2 METHOD

2.1 *Experimental setting*

Twenty participants were male and with mean age of 19.15 years ranging from 18 to 22. They were all with normal or corrected-to-normal vision, and at least two years of experience playing video games. The environment simulated a living room with a 16:9 PDP TV. The viewing distance was 2 to 2.5 m, which could be adjusted according to the subjects' needs. The task was a car racing video game – PS2's Gran Turismo 4 (GT4) used on the Sony PlayStation 2 hardware, and was set at A-spec mode with test course. During the task, subjects drove on a monotonous ellipse roadway

session	BV	V1	V2	V3	V4	V5	V6	V7	V8	V9	V10	V11	EV
minutes	0th min	10th min	20th min	30th min	40th min	50th min	60th min	70th min	80th min	90th min	100th min	110th min	120th min

Figure 1. The 12 sessions and 13 measurements of the task.

with no barrier and competition, and chose the fastest and safest path by controlling the car's speed and direction. The roadway length of each lap is 10.34 km. The finishing line was set at the end point of the 50th lap. No subjects arrived the finishing line until the task time out. From the beginning to end, the task took 120 minutes, and was divided into 12 sessions for data recording.

The subjects were informed to avoid alcohol and caffeine in one day before the experiments, and asked to sign informed consent form in conformity with the law on biomedical research on human volunteers. Training was performed, but no further training was given if the subject could achieve a bronze medal of the task. On the test day, the task started at 14:00 PM after the pre-process of the experiment. The task was divided into 12 sessions for data recording in every 10 minutes.

2.2 Data recording and experimental design

NPA, CFF, EEG and SR were recorded during the experiment. NPA was measured via an Accommodo-polyrecorder, and was calculated by the average of two trials. CFF was measured via a Lafayette flicker, and was calculated by the average of two ascending trials.

The EEG signals were collected via a BRAIN-QUICK VIDEO-EEG system. The electrodes were pasted at four locations (O1, O2, P3, P4) following the international 10–20 systems. Raw data were amplified and sampled at 500 Hz with the 0.05~50 Hz filter passbands. FFT was applied for all artifact-free trials to represent the frequency bands of δ, θ, α, and β activity, respectively. The δ band was not included in this study, since it happens in deep sleep and usually overlaps with artefacts. The EEG indices were classified into two types—the basic index, and the ratio index. The basic indices mean the relative power of θ, α, and β bands. The ratio indices are the calculated data - θ/α, β/α, $(\alpha + \theta)/\beta$. After each single session, rating scale with 6 descriptive items was taken to collect SR of visual discomfort. Each item was rated on a five point scale for assessing severity of discomfort.

As for the experimental design, the dependent variables included: (1) EEG basic index and ratio index for the 4 location of cerebrum, (2) CFF, (3) NPA, (4) SR; The independent variables included: (1) "Experimental session", including BV(before the task),V1(the 10th min),....and EV(ending task)(see figure one), (2) "Location"– (O1,O2,P3,P4). ANOVA, Duncan post hot analysis, and Pearson's r correlation analysis were carried out with a probability of error $\alpha < 0.05$ to make advance analysis.

3 RESULTS AND DISCUSSIONS

3.1 The evaluation of visual fatigue of this task

One-way ANOVA were carried out to test time effects of visual fatigue of this task. The average of CFF, NPA, and SR were dependent variables and "Session" was independent variable. The ANOVA results revealed significant difference in the three fatigue indices between the different sessions (CFF: $F(12,143) = 9.312$, $p < 0.01$; NPA:$F(12,143) = 22.952$, $p < 0.01$; SR:$F(12,143) = 80.679$, $p < 0.01$), it implied visual fatigue significantly increased following the operating time increase. In the trend plot and Duncan analysis of CFF for the "Session", the first significant difference was occurred on the 10th minutes, the second apparent difference appeared on the 50th~60th minutes (see figure two and table one), and no significant difference tills the end of the task. Comparing with CFF, the occurrence time of significant difference of NPA shifted 10 minutes backward around the 60th~70th minutes (see figure three and table one). And then, the apparent difference appeared in the 110th~120th minutes. As to SR, the change trend was similar to the NPA's, but it did not have significant difference in the 10th~20th minutes (see figure four and table one).

From above results, it implies that the occurrence time of significant difference in each index would different. CFF seemed to be more sensitive then NPA and SR in 60 minutes, but showed

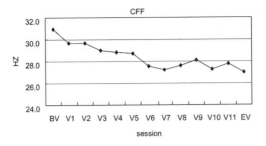

Figure 2.   CFF change trend in different session.

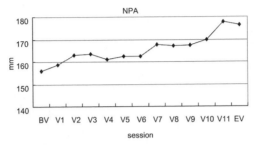

Figure 3.   NPA change trend in different session.

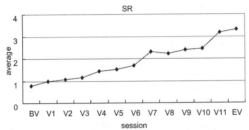

Figure 4.   SR change trend in different session.

Table 1.   The summarized ANOVA results for EEG indices.

Index	Location	Session	Interaction
α	0.0086**	0.0092**	0.2432
β	0.0039**	0.0068**	0.8546
θ	0.0078**	0.8562	0.4890
θ/α	0.0298*	0.5478	0.1933
β/α	0.0342*	0.0028**	0.0698
$(\alpha + \theta)/\beta$	0.0063**	0.0034**	0.0231

* Significant at $\alpha = 0.05$, ** Significant at $\alpha = 0.01$.

an opposite trend in the second hour. It is worthy discussing if the tendency also appeared in EEG indices.

3.2   *The ANOVA results of EEG Basic and ratio indices*

Two-way ANOVA were conducted to examine the 4 locations and 13 sessions effects on EEG indices. The ANOVA results are summarized in Table 1. All EEG indices showed significant difference in location. And the EEG indices α, β, β/α, and $(\alpha + \theta)/\beta$ were found to have significant difference in different session. It also implied these indices value significantly changed following the operating time and increase of visual fatigue. (See figure five and table two). No interaction effect was

79

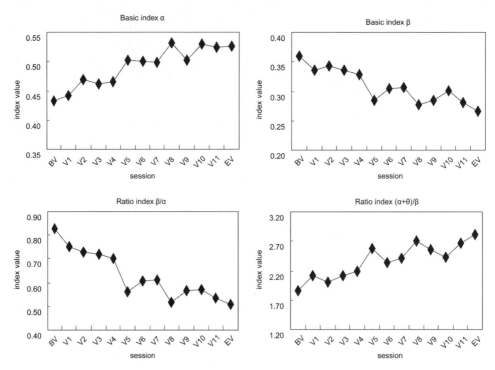

Figure 5. The change trend of EEG indices.

Table 2. Duncan results of the visual fatigue indices at $\alpha = 0.05$.

Indices	Visual fatigue (slight <==> severity) Sessions(13 measurements)												
CFF	BV	V1	V2	V3	V4	V5	V9	V11	V8	V6	V10	V7	EV
	A												
		B	B	B	B	B							
				C	C	C	C	C	C	C	C	C	C
NPA	BV	V1	V4	V6	V5	V2	V3	V8	V9	V7	V10	EV	V11
	A	A											
		B	B	B	B	B	B						
			C	C	C	C	C	C	C	C	C		
												D	D
SR	BV	V1	V2	V3	V4	V5	V6	V8	V7	V9	V10	V11	EV
	A	A	A	A	A	A	A						
								B	B	B	B		
												C	C

significant. Then the conclusion would be draw preliminary, EEG power were appropriate indices to measure visual fatigue. The 16 valid EEG indices found out by the combination of 4 appropriate indices and 4 locations were as follows: O1, O2, P3, P4, O1β, O2β, P3β, P4β, O1 (β/α), O2 (β/α), P3 (β/α), P4 (β/α), O1$(\alpha+\theta)/\beta$, O2$(\alpha+\theta)/\beta$, P3$(\alpha+\theta)/\beta$, P4$(\alpha+\theta)/\beta$.

In the Duncan post hoc analysis for the factor of location, 4 location were separated into two groups—occipital region (O1, O2), and parietal region (P3, P4) for the basic index α. Except for the basic indices α, the location of the other five indices were separated into three groups–(O1, O2), (P3) and (P4). (at $\alpha < 0.05$). It indicated that no difference between left and right hemisphere in occipital region. It also implied for the cost reduction, just choosing one electrode site between

Table 3. Duncan analysis results in EEG indices at $\alpha = 0.05$.

EEG indices	Visual fatigue (slight<==>severity) 12 Sessions(13 measurements)												
α	BV	V1	V3	V4	V2	V7	V6	V9	V5	V11	EV	V10	V8
	A	A	A	A									
			B	B	B								
						C	C	C	C				
								D	D	D	D	D	D
β	BV	V2	V3	V1	V4	V7	V6	V10	V5	V9	V11	V8	EV
	A	A	A	A	A								
						B	B	B	B	B	B	B	B
β/α	BV	V1	V2	V3	V4	V7	V6	V10	V9	V5	V11	V8	EV
	A	A											
		B	B	B	B								
						C	C	C	C	C	C	C	C
$(\alpha+\theta)/\beta$	BV	V2	V1	V3	V4	V6	V7	V10	V9	V5	V11	V8	EV
	A	A											
		B	B	B	B								
						C	C	C					
									D	D	D	D	D

Table 4. Correlation analysis results between EEG and CFF, NPA, SR.

EEG indices	CFF var. 0~60	NPA var. 0~120	0~60	SR 0~120	0~60	0~120
O1α	*↑	*↑	*↑	*↑	*↑	*↑
O2α	*↑	*↑	*↑		*↑	*↑
P3α			*↓		*↓	
P4α			*↓		*↓	
O1β	*↓					
O2β	*↓					
P3β						
P4β						
O1 (β/α)	**↓		*↓	*↓	*↓	*↓
O2 (β/α)	*↓		*↓		*↓	*↓
P3 (β/α)			*↓		*↓	
P4 (β/α)			*↓		*↓	
O1$(\alpha+\theta)/\beta$	**↑	*↑	**↑	*↑	*↑	*↑
O2$(\alpha+\theta)/\beta$	**↑	*↑	***↑	*↑	*↑	*↑
P3$(\alpha+\theta)/\beta$						
P4$(\alpha+\theta)/\beta$						

* Significant at $\alpha = 0.05$, ** Significant at $\alpha = 0.01$.
↑ positive correlation ↓negative correlation

O1 and O2 shall be adequate. In the Duncan post hoc analysis for the factor of session, basic indices α and β showed significant difference until about the 50th minutes (see table three). But the ratio indices β/α and $(\alpha+\theta)/\beta$ appeared the significant difference in about 20 minutes later. It indicated if the task time < 50 minutes for visual fatigue measurement, the ratio indices seemed more suitable for use. Moreover, both the ratio and basic indices would be appropriate if the task time > 50 minutes. Comparing with other indices, it also shown the occurrence time of EEG indices abrupt increase was about 10 minutes earlier than the moment of subjective rating and NPA intense variation, but it was almost simultaneous with the variation of CFF.

3.3 Correlation between EEG indices and other visual fatigue indices

A correlation analysis was conducted between the 16 valid indices and the variation of CFF ($CFF_{var.}$), NPA ($NPA_{var.}$), average of SR. The variation equations of CFF, NPA are presented in the following:

$CFF_{var.} = (CFF_{initial}-CFF_{after})/CFF_{initial}$; $NPA_{var.} = (NPA_{after}-NPA_{initial})/NPA_{initial}$. The analysis was divided into two periods- 0∼60 and 0∼120 minutes for a more exhaustive analysis because the significant difference occurred during the 50th∼60th minutes of the task, and the results were listed in Table 4.

No significant correlation was found between $P3\beta$, $P4\beta$, $P3(\alpha+\theta)/\beta$, $P4(\alpha+\theta)/\beta$ and the three visual fatigue indices in two periods. But, $O1\alpha$, $O1(\alpha+\theta)/\beta$, and $O2(\alpha+\theta)/\beta$ showed remarkable correlations throughout the two periods. $O1\alpha$, $O2\alpha$, $O1(\beta/\alpha)$, $O2(\beta/\alpha)$, $O1(\alpha+\theta)/\beta$, $O2(\alpha+\theta)/\beta$ also showed the significant correlations in the 0∼60 minutes task period. From the results, it can be easily point out which indices were suitable in visual fatigue measurement. To simplify the numerous indices above, it should be adequate to sample the indices from O1 if these indices were good in O1 and O2 simultaneously. Overall, it can be summarized $O1\alpha$ and $O1(\alpha+\theta)/\beta$ were appropriate visual fatigue indices in 0∼120 minutes task period. And, $O1(\beta/\alpha)$ was also a suitable index in 0∼60 minutes. It was also found that the $CFF_{var.}$ was significantly related with EEG in occipital region. In addition, the results also showed the basic index α and the ratio index β/α had an obvious correlation between the $NPA_{var.}$ and the EEG indices in the 0∼60 minutes task period, and the correlation results of NPA were similar with SR's.

4 CONCLUSION

As for the EEG indices and visual fatigue measurement, EEG powers sampling from the occipital lobe (O1, O2) were valid visual fatigue indices. Choose one electrode site of occipital lobe should be adequate because of no difference between left and right hemisphere. If the task time <50 minutes, the ratio indices seemed to be more suitable for use. If the task time >50 minutes, both the ratio indices and the basic indices would be appropriate. As for the recommend indices for visual fatigue measure, the best indices of the 120 minutes task were $O1\alpha$ and $O1(\alpha+\theta)/\beta$. Besides the two indices, $O1(\beta/\alpha)$ was suitable to evaluate visual fatigue in the 60 minutes task.

Regarding to the relationship between EEG indices and other visual fatigue indices, the moment of the abrupt increase of EEG indices was about 10 minutes earlier than that of SR and NPA, but it was almost simultaneous with CFF. It revealed that CFF and the brainwaves from occipital lobe had many similarities, and the change trend of NPA was similar with SR which was significantly correlated with α-wave variation. It is worthy for further evaluation.

REFERENCES

Cheng, S. Y., Lee, H.Y., Shu, C. M. and Hsu, H. T., 2007, Electroencephalographic study of mental fatigue in VDT task. In *Journal of Medical and Biological Engineering*. Vol.27, 3rd, pp.124–131.

Chi, C. F. and Lin, F. T., 1998, A comparison of seven visual fatigue assessment techniques in three data-acquisition VDT tasks. In *Human Factors*. Vol.40, 4th, pp. 577–590.

Eoh, H. J., Chung, M. K. and Kim S. H., 2005, Electroencephalographic study of drowsiness in simulated driving with sleep deprivation. In *International Journal of Industrial Ergonomics*, Vol.35, pp.307–320.

Gunnarsson, E. and Soderberg, I., 1983, Eye strain resulting from VDT work at the Swedish telecommunications administration, In *Applied Ergonomics*, Vol.14, pp.61–69.

Iwasaki, T. and Kurimoto, S., 1988, Eye-strain and changes in accommodation of the eye and in visual evoked potential following quantified visual load. In *Ergonomics*, Vol.31, pp.1743–1751.

Lal, S.K.L. and Craig, A., 2001, A critical review of the psychophysiology of driver fatigue. In *Biological Psychology*, Vol.55, pp.173–194.

Murata, K., Araki, S., Kawakami, N., Saito, Y. and Hino, E., 1991, Central Nervous System Effects and Visual Fatigue in VDT Workers. In *International Archives of Occupational and Environmental Health*. Vol.63, pp.109–113.

Shieh, K. K. and Chen, M. T., 1996, CFF and subjective visual fatigue as a function of VDT task characteristics. In *The 4th Pan Pacific Conference on Occupational Ergonomics*, Ergonomics Society of Taiwan, pp. 143–146.

Ergonomic Trends from the East – Kumashiro (ed)
© *2010 Taylor & Francis Group, London, ISBN 978-0-415-88178-4*

Pressure-pain tolerance at different hand locations as wearing gloves under various finger skin temperature conditions

Yuh-Chuan Shih[1] & Yo-May Wang

Department of Logistics Management, National Defense University, Beitou District, Taipei City, Taiwan

ABSTRACT: This study examines the effects of gender, gloved condition, and the location of hand on the Pressure-Pain Tolerance (PPTo) at different Finger Skin Temperature (FST). The experimental design is a combination of nested-factorial and split-plot. Ten men and ten women were recruited and nested within the gender. The gloved level includes bare hand, wearing one or two layers of latex gloves. The FST levels are 15, 20, and 25°C, and FST is considered as the whole plot. The hand location includes fingers, metacarpal, and palm; four test points are selected from each location and nested within the location. This experiment was performed in a water immersion. The ANOVA result indicates that men PPTo (7.3 kg) are greater than those for women (5 kg). PPTo at 25°C is the least. Additionally, the more layers of gloves are worn, the greater PPTo are observed. As to the location effect, PPTo on palm and metacarpal is the highest and the least, respectively.

1 INTRODUCTION

The use of powered hand tools seems to be taken for granted, however, in many situations unpowered hand tools are still commonly used. Furthermore, manual working in cold environment has become more and more popular nowadays, for example, engaging in manual materials handling in freezing foods distribution centers, processing or packing freezing foods, etc. Wearing gloves in a cold environment could not only protect hands from contact with extreme hot/cold materials/objects, but also insulate the hand and decrease the speed of heat dissipation. Unfortunately, the usage of gloves usually entails an impairment in performance, such as increasing manipulating time (Bensel, 1993), impairing hand sensitivity (Phillips *et al.*, 1997; Shih *et al.*, 2001), reducing the range of hand movement (Bellingar and Slocum, 1993), and decreasing grip strength (Kovacs *et al.*, 2002; Shih, 2007).

It has been reported that wearing gloves in an immersion test delays the finger skin temperature (FST) decrease and reduces feelings of pain (Suizu *et al.*, 2004; Suizu and Harada, 2005). The local reaction to cold is a decrease in blood flow and thus in heat dissipation (Edwards and Burton, 1960), and this lowers local skin temperature. The influence of skin temperature is mostly studied at a local level. Daanen *et al.* (1993) found impaired finger dexterity as FST fell below 14 °C. Schieffer *et al.* (1984) found a slight reduction and a strong decrease in manual dexterity at 20–22°C and 15–16°C on FST, respectively.

Pain is a warning sign of impending tissue damage, and as such should not be ignored. Existence of a high pressure on the surface of the hand, arising from grasping and guiding a tool handle, could cause a sense of discomfort or pain under sustained loading. The sensation of pain in the hand due to high sustained external pressure has been cited by Fraser (1980) as a limiting factor in the performance of work with hand held tools. Sensitivity of the hand to pressure has been investigated using an algometer to establish the pressure-discomfort and pressure-pain threshold (Muralidhar and Bishu, 2000; Fransson-Hall and Kilbom, 1993). These studies concluded that the thenar area, the skin fold between the thumb and index finger, and region around os pisiform have lower pressure-discomfort and pressure-pain threshold in relation to the rest of the hand surface. The

---

[1] Corresponding author: E-mail: river.amy@msa.hinet.net

Table 1. Anthropometric data of subjects (SD: standard deviation).

Items	Age (yr.)	Height (cm)	Weight (kg)	Hand length (cm)*	Palm breadth (cm) *
Male	21.7(3.2)	173.3(3.5)	70.4(5.8)	19.0(0.5)	8.2(0.3)
Female	26.7(5.9)	158.5(1.1)	49.9(4.4)	16.9(0.5)	6.8(0.2)

* left hand and tested in this study

tips of digits IV and V, and the zone near the fourth metacarpal were also found to exhibit lower pressure-pain threshold. Muralidhar and Bishu (2000) indicated that wearing gloves could increase the pressure-discomfort threshold. Wearing gloves to operate hand tools in a cold environment is not avoided, but the FST effect on pressure-pain tolerance (PPTo) seems to be less discussed.

Therefore, present study examines the sensitivity of distinct hand locations to maximal applied pressure under different gloved conditions in different immersing temperatures. The response is PPTo, the maximal applied pressure which the subject can not tolerate any more.

2 METHODS

2.1 Subjects

Ten male and ten female subjects were recruited, and they were healthy and without any musculoskeletal disorders. All were right-handed and anthropometric data are presented in Table 1.

2.2 Materials and apparatus

A water tank made by Firstek Co. (Model: B102) was used. It maintains the temperature constantly at a desired level by an electronic thermo-sensor and a heater with a resolution of $0.3°C$. The range is from the ambient temperature plus $5°C$ to $80°C$. A submersible cooler made by Firstek Co. (Model: HC-101) was also used. It cools the water temperature range from the ambient to $-20°C$. At $20°C$, the cooling efficiency is $750 Kcal/hr$. This cooler and the former water tank were used together to regulate the water temperature. In addition, a digital thermometer and hygrometer was used to monitor the ambient temperature and humidity at the same time (TECPEL Co.; Model: DTM301). The temperature range measured is from $-10°C$ to $+50°C$, and relative humidity measured ranges from 20% to 99%. A digital 4-channel thermometer made by TECPEL Co. (Model: DTM319) was used to record FST. The sampling rate was 6 Hz, and it was connected to a personal computer with an RS-232. A fabricated algometer equipped with a load cell (made by Rightronic Brand, Model U3S1, capacity: 10 kg) was used to test the PPTo. The diameter of its round iron rod is 0.9 cm (area is about $0.64 cm^2$). The load cell further connects with a 12-bit A/D convert card. The water tank was placed beneath the rod during formal experiment for the sake of controlling the FST and measuring the applied pressure at the same time. Finally, surgical gloves made of latex (Hau-Hsin Co., Model: 1010) were used, and there are six sizes available.

2.3 Experimental design

An experimental design combining nested-factorial and split-plot was employed. The factors included gender, region (finger, metacarpal, and palm), location (four locations, see Figure 1, were selected in each region and nested within the region factor), HST (whole plot and including 15, 20, and $25°C$), and glove (sub-plot and including bare-hand, wearing one or two layers of latex gloves). Each treatment was replicated twice. The dependent variable is PPTo (in kg/cm^2). The level of significance (α) was set at 0.05.

2.4 Procedures

All participants were well informed of the goals and procedures first. Secondly, the test locations on left hand were marked. This hand was placed with the palm facing upwards, and the subject was

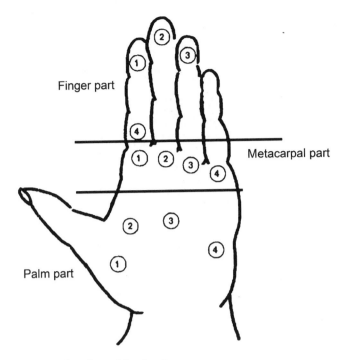

Figure 1. The selected locations for each hand region.

asked to grasp the handwheel of the algometer with the right hand, and gently lowered the contact rod onto the designated locations randomly. The entire test was carried out under the subjects' own control. The subjects were asked to gently increase the pressure on their palm until the pressure on the location turned into a feeling of pain untolerated, and experimenter placed the keyboard to mark the applied pressure at once. This applied pressure is defined as PPTo. Prior to the formal measurement, many trials were given to familiarize subjects with the whole procedure.

The mean ambient temperature (standard deviation, SD) was 19.7°C (1.17), and mean relative humidity (SD) was 63.2% (7.28). During formal measurement, subjects were first asked to immerse their hand into the water tank until FST reached the designated temperature, then operating the algometer to measure the PPTo. In order to standardize limb submersion across subjects, the subjects were instructed that they should place their left hands into the stimulus water up to about 1/3 of the forearm above the wrist. Before immersion, two channels of the 4-channel digital thermometer were applied to monitor the FST on the ventral side of the distal phalanges of the little finger of the left hand.

3 RESULTS AND DISCUSSION

The ANOVA results shown in Table 2 indicate that all main effects are significant ($p < 0.05$), as well as the interactions of Sex × FST, Sex × Region, and FST × Glove. For sex effect, from Figures 2 and 3, male subjects were found to have higher PPTo than female subjects did. Past studies also reported that male has greater PPTo (Fransson-Hall and Kiblom, 1993). Lower FST leads to higher PPTo, see Figures 2 and 4. It perhaps results from more numbness due to lower FST which leads to less sensitive to pain perception. Additionally, wearing thicker gloves was able to tolerate higher PPTo (Figure 4). Gloves are able to absorb some of the applied pressure, and it resulted in higher PPTo. This result is supported by the finding of Muralidhar and Bishu (2000), who indicated that wearing gloves could increase the pressure-discomfort threshold by 25-65%. As to the region effect, it can be observed from Figure 3 that the metacarpal is the most sensitive and the palm is the least.

Table 2. The ANOVA results for PPto.

Sources of variations	d.f.	S.S.	M.S.	F-value	p-value
Sex	1	8831.9	8831.9	2244.74	0.0000
subject (Sex)	18	10454.6	580.8	147.62	0.0000
FST	2	687.9	343.9	87.41	0.0000
Glove	2	230.5	115.3	29.29	0.0000
Region	2	1420.6	710.3	180.53	0.0000
Location (Region)	9	1647.3	183.0	46.52	0.0000
Sex*FST	2	78.2	39.1	9.94	0.0000
Sex*Glove	2	22.8	11.4	2.89	0.0556
Sex*Region	2	193.8	96.9	24.63	0.0000
FST*Glove	4	100.0	25.0	6.36	0.0000
FST*Region	4	12.4	3.1	0.79	0.5341
Glove*Region	4	30.9	7.7	1.96	0.0974
Sex*FST*Glove	4	9.2	2.3	0.58	0.6748
Sex*FST*Region	4	24.2	6.0	1.53	0.1893
Sex*Glove*Region	4	1.9	0.5	0.12	0.9740
FST*Glove*Region	8	8.1	1.0	0.26	0.9790
Sex*FST*Glove*Region	8	22.0	2.8	0.70	0.6920
Error	4239	16678.3	3.9		
Total	4319	40454.6			

For more detailed locations from Figure 5, the L3 and L4 of palm region still had the greatest PPTo, and almost all of the locations of the metacarpal region had the least PPTo.

4 CONCLUSIONS

From aforementioned results, males have greater PPTo, and wearing gloves and lower FST are able to augment PPTo. Next, in general, the metacarpal is the most sensitive and the palm is the least to PPTo.

ACKNOWLEDGEMENTS

This paper presents the results from a project sponsored by National Science Council. The project number was NSC95- 2221- E123-002-MY2.

REFERENCES

Bellingar, T.A. and Slocum, A. C., 1993. Effect of protective gloves on hand movement: an exploratory study. *Applied Ergonomics*, 24(4): 1055–1062.

Bensel, C. K., 1993. The effects of various thickness of chemical protective gloves on manual dexterity. *Ergonomics*, 36(6): 687–696.

Daanen, H.A.M., Wammes, L.J.A., and Vrijkotte, T.G.M., 1993. Windchill and dexterity. *Report IZF A-7, TNO Institute for Perception*, Soesterberg, NL.

Edwards, M. and Burton, A.C., 1960. Correlation of heat output and blood flow in the finger, especially in cold-induced vasodilatation. *Journal of Applied Physiology*, 15(2): 201–208.

Fransson-Hall, C. and Kilbom, Å., 1993. Sensitivity of the hand to surfacr pressure. *Applied Ergonomics*, 24: 181–189.

Fraser, T.M., Ergonomic principles in the design of hand tools. 1980. *Occupational Safety and Health Series* no. 44 (International Labour Office, Genevé).

Kovacs, K., Splittstoesser, R., Maronitis, A., and Marras, W.S., 2002. Grip force and muscle activity differences due to glove type. *AIHA Journal*, 63: 269–274.

Muralidhar, A. and Bishu, A.A., 2000. Safety performance of gloves using the pressure tolerance of the hand. *Ergonomics*, 43(5): 561–572.

Phillips, A.M., Birch, N.C., and Ribbans, W.J., 1997. Protective gloves for use in high-risk patients: how much do they affect the dexterity of the surgeon? *Annuals of the Royal College of Surgeons England*, 79: 124–127.

Schieffer, R.E., Kok, R., Lewis, M.I., and Meese, G.B., 1984. Finger skin temperature and manual dexterity; some inter-group difference. *Applied Ergonomics*, 15(2): 135–141.

Shih, R.H., Vasarhelyi, E.M., Dubrowski, A., and Carnahan, H., 2001. The effect of latex gloves on the kinetics of grasping. *International Journal of Industrial Ergonomics*, 28: 265–273.

Shih, Y.C., 2007. Glove and Gender Effects on Muscular Fatigue Evaluated by Endurance and Maximal Voluntary Contraction Measures. *Human Factors*, 49(1): 110–119.

Suizu, K. and Harada, N., 2005. Effects of waterproof covering on hand immersion tests using water at 10°C, 12°C and 15°C for diagnosis of hand-arm vibration syndrome. *International Archives Occupation Environment Health*, 78: 311–318.

Suizu, K., Inoue, M., Fujimura, T., Morita, H., Inagaki, J., Kan, H., and Harada, N., 2004. Influence of waterproof covering on finger skin temperature and hand pain during immersion test for diagnosing hand-arm vibration syndrome. *Industrial Health*, 42: 79–82.

Daily recordings on task performance of the two mental tasks and menstrual associated symptoms

Keiko Kasamatsu
Tokyo Metropolitan University, Asahigaoka, Hino, Tokyo, Japan

Risa Araki & Saori Yoneda
Kanazawa Institute of Technology, Yatsukaho, Hakusan, Ishikawa, Japan

Hiroyuki Izumi & Masaharu Kumashiro
University of Occupational and Environmental Health, Iseigaoka, Yahatanishi, Kitakyusyu, Fukuoka, Japan

ABSTRACT: There is a biological difference between men and women. There are phases of menstrual cycle and menopausal disorder of the peculiarity to the women in the character differences. The menstruation is a phase in which large majority is experienced in the female worker. It is necessary to support to live a comfortable life and work for female. The menstrual cycle variation on the task performance and the menstrual associated symptoms were examined in this research. The menstrual associated symptoms and task performance of the two kind of mental tasks were recorded every day to grasp physical and mental conditions. The future goal of this research was to develop the self-health management system for female that understand the task performance and the menstrual associated symptoms.

1 INTRODUCTION

In the working environment in recent years, the chance to use the information and communications related devices such as PC has increased. The information terminal is used in not only the office work but also all offices. Therefore, a present work form is not past physical load work but mental load work, and the influence of the mental load on health has strengthened. As the employment situation of Japan, the female worker increases and the scope of female occupations has expanded. Consideration and the approach concerning the employment among women on the employer side have changed greatly. The female workers are being used as important manpower. The equality of employment has come to be secured by the Equal Employment Opportunity Law etc. However, the matter that has been installed up to now for the women's protection was abolished, and women came also to do labor equal with men. This contains the possibility of forcing the large encumbrance on women than before. Because the women have a periodic, physiological change of menstruation, and there are some women influenced the body and mentally according to the menstrual cycle. Therefore, it is an important problem to clarify the size of the labor load of the women in occupational health and safety, and the idea of productivity.

There are some biological differences between men and women. There are phases of menstrual cycle and menopausal disorder of the peculiarity to the women in the character differences. The menstruation is a phase in which large majority is experienced in the female workers. Legal maintenance and social support of the menstruation and the memopausal disorder are insufficient yet. Various mental and physical symptoms appear according to this menstrual cycle. Previous researches indicated that some women suffer from Premenstrual Syndrome (PMS) and Premenstrual Dysphoric Disorder (PMDD) (Dalton and Holton, 1999) and as many as 85 percent of women of reproductive age report having premenstrual symptoms at some time during their lives (Grady-Weliky, TA, 2003). It is necessary to support to live a comfortable life and work for female.

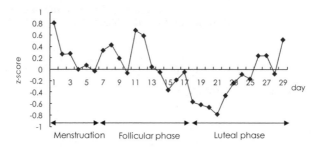

Figure 1.   The score of modified MDQ.

The menstrual cycle variation on the task performance and the menstrual associated symptoms were examined in this research. The menstrual associated symptoms and task performance of the two kind of mental tasks were recorded every day to grasp physical and mental conditions and develop self health management system for female. The results of menstrual associated symptoms and task performance show chapter 2 and 3 separately. These indices were measured every day at same period.

2   SURVEY OF THE MENSTRUAL CYCLE AND MENSTRUAL ASSOCIATED SYMPTOMS

2.1   *The menstrual cycle*

The menstrual cycle was recognized by the basal body temperature. The basal body temperature was taken and recorded every morning for 2∼3 month to confirm biphasic temperature period. The menstruation, follicular phase and luteal phase were understood.

2.2   *The menstrual associated symptoms*

The MDQ (Menstrual Distress Questionnaire) was developed to create a questionnaire about menstrual cycle complaints by Moos (Moos, 1968) and the questionnaire used in this research was modified the MDQ. The modified MDQ with 47 items in Japanese was administered to investigate physical and mental conditions every day in this survey. These items were classified into eight factors: Pain, Concentration, Behavioral change, Autonomic reactions, Water retention, Negative affect, Arousal, and Control. The subjects filled out the modified MDQ at the end of a day.

2.3   *Subjects*

The criteria for the subjects were single women with a regular menstrual cycle, who did not take hormones, oral contraceptive pills, or other drugs, who were not pregnant, and gynecological disorder. The subjects to a survey were eighteen women. The eighteen subjects to a survey were reported as with a regular menstrual cycle by themselves. However, eight subjects of them had not regular cycle or many omissions of recording in data. Therefore, they were removed from the analysis result.

2.4   *Results of the survey*

The average days of menstrual cycle on the ten subjects (Subject A∼J) were 30.1 ± 5.74 days. The average body basal temperature of high temperature period on them was $36.19 \pm 0.158°C$, low temperature period was $36.45 \pm 0.200°C$.

MDQ score is estimated as not absolute index but variation. Therefore, the menstrual cycle variation of MDQ score was examined by standardized score (z-score) for all subjects. As the result, the MDQ score was high on the menstruation, the beginning of follicular phase, and premenstruation. (Figure 1).

3 EXPERIMENTAL METHODS

3.1 *Experimental tasks*

The two experimental tasks were used a mental arithmetic task (Task A) as a task with thinking and a graphic identification task (Task B) as a task without thinking. These tasks were performed every day for the menstrual cycle. These experimental tasks were created using Visual Studio as experimental system.

Task A was a mental arithmetic task involving the addition of two digits. The expression of addition of two figures appeared on computer screen. The answer was inputted from ten-key and the next question was presented when the enter key was inputted. The number of questions was composed of 30. The answer of each question was given feedback to subjects. The response time and true-false of each question were recorded by experimental system.

Task B was a graphic identification task. The subject recognized the mark of "○", "□" and "△" which were presented on experimental screen randomly, and inputted the appropriate key. The judgement by the subject was inputted using ten-key. The mark of "○" was related "4" of ten-key, "□" was "5", "△" was "6". The next mark was presented when the ten-key was inputted. The number of questions was composed of 30. The answer of each question was given feedback to subjects. The response time and true-false of each question were recorded by experimental system.

3.2 *Measurement indices*

The accuracy rate and response time were measured as the task performance every day. The score of MDQ was calculated to investigate the menstrual associated symptoms every day. The basal body temperature was recorded every morning for experimental period. These indices were examined by standardized score (z-score) because these were estimated as not absolute index but variation.

The data of beginning of menstruation was used on the basal body temperature, MDQ score, and the results of experimental task. Learning level was related on the experimental task. Therefore, the subjects practiced the task enough, and the data for a first week was not used.

3.3 *Subjects*

The ten female subjects(Subject A~J) had the biphasic temperature period and the number of days in the menstrual cycle was determined to be within the normal range such as chapter 2.3.

3.4 *Experimental procedure*

Experimental periods were 2~3 cycles. The subjects measured the basal body temperature every morning, filled in the questionnaire and performed the experimental task every day. The experimental task was installed subject's PC to perform at their home. The order of two tasks was Task A, next Task B. The subjects were enlightened that the experiment did not reattempted, and every question was thought and judged steadily.

4 EXPERIMENTAL RESULTS

The response time and accuracy rate on two experimental tasks, basal body temperature, and the score of MDQ were examined menstrual cycle variation as daily data from menstruation beginning. The data of them was converted to standardized score (z-score). The score of MDQ was showed on Figure 1. The accuracy rate and response time of Task A were showed on Figure 2 and 3, them of Task B were showed on Figure 4 and 5.

The accuracy rate of Task A was higher than 0 point for first three days of menstruation, after that, it decreased. It rose toward ovulation, and decreased gradually at luteal phase. On the other hand, the response time was low point on 5th and 6th day of menstruation, and it was a comparatively steady pace though the task pace was a slow at the menstruation and follicular phase. It was recognized that the task pace moved up and down at the luteal phase, and stability was low. It tended to do task accurately and slowly in the first half of menstruation.

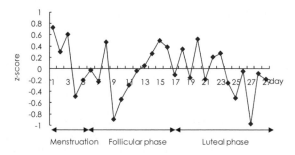

Figure 2.　Accuracy rate on Task A.

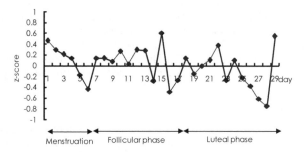

Figure 3.　Response time on Task A.

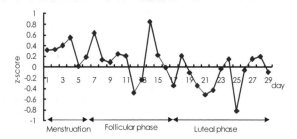

Figure 4.　Accuracy rate on Task B.

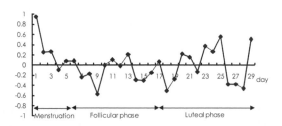

Figure 5.　Response time on Task B.

The accuracy rate of Task B was higher than 0 point on menstruation and follicular phase, lower than it on luteal phase. The response time was long on menstruation, that is, the task pace was slow on menstruation. The task performance had worsened comparatively at the luteal phase. It caused that the accuracy rate of 14th days was high and response time was fast, and the accuracy rate of 20th and 21th day was low and response time was late.

5 DISCUSSION

It tended to be high on the score of MDQ and accuracy rate, and be late on response time on menstruation totally. The accuracy rate was low and response time was late when there was no discomfort, that is, the score of MDQ was low, on Task B.

Kasamatsu et al. (2007) reported that it was shown on the menstruation of a graphic identification task that the number of accuracy rate had decreased though the response time was long. The similar result was recognized as for this research. It was confirmed that the task performance was tended to influence by the menstruation on a graphic identification task, simple reaction task without thinking. It is necessary to introduce the system to prevent the error beforehand because it has misgivings about receiving the influence at the menstrual cycle in monotonous mental work. The menstrual cycle may influence the task performance on the mental task with thinking, the other factors also influence the task performance on it. The mental task with thinking is a cooperation working among perception, cognition, thinking, and motion. Therefore, a further examination is necessary to clarify these relations.

6 CONCLUSIONS

The menstrual cycle variation on the task performance and the menstrual associated symptoms were examined in this research. It was recognized that the task performance was influenced by the menstrual cycle on monotonous task like the graphic identification task, and decreased. The task performance on the mental task with thinking tended to be influenced due to symptom concerning pain of menstruation, however, it was indicated that the influence at the menstrual cycle was small.

A part of this study was supported by Grant-in-Aid for Young Scientists (B) (18790407), Grant-in Aid for Scientific Research from the Ministry of Education, Culture, Sports, Science and Technology in Japan.

REFERENCES

Dalton, K., Holton, W., 1999, *Once a month -Understanding and treating PMS-*, 6th ed., (Hunter House Inc., Alameda), pp.44–80 and pp.159–171.

Grady-Weliky, TA., 2003, Premenstrual dysphoric disorder, *The New England Journal of Medicine*, Vol.348, No.5, pp.433–438.

Moos, R.H., 1968, The development of a menstrual distress questionnaire, *Psychosomatic Medicine*, Vol.30, No.6, pp.853–867.

Kasamatsu, K., Furuta, S., Izumi, H., Kumashiro, M., 2007, The daily recordings on the task performance of three mental tasks, Proceedings of the 14th Annual Meeting & Conference, Ergonomics Society of Taiwan, CD-ROM.

Ergonomic Trends from the East – Kumashiro (ed)
© 2010 Taylor & Francis Group, London, ISBN 978-0-415-88178-4

Stimulus-response compatibility effect for hand and foot controls with visual signals on eccentric and central positions in horizontal orientation

Ken W.L. Chan & Alan H.S. Chan

Department of Manufacturing Engineering and Engineering Management, City University of Hong Kong, Kowloon Tong, Hong Kong

ABSTRACT: Most of the studies on spatial Stimulus Response (S-R) compatibility are limited to the use of hand controls, and there are insufficient ergonomics guidelines on the use of foot controls with due consideration on the possible importance of spatial compatibility. If some controls of a process can be given over to the feet, there would be an obvious advantage in freeing the hands for other tasks. To have the feet responsible for manipulating controls, it is of great importance to understand how well the feet interact and cooperate with the hands in response execution. This study examined the performance of thirty two Chinese right-handed/footed participants for four-choice reaction tasks with compatible and incompatible mappings of visual signals and mixed controls of hand and foot on a horizontal array. A significant interaction between visual signal positions and hand and foot control positions was demonstrated, revealing the existence of spatial S-R compatibility effect similar to that found with the sole use of hand controls. The results were translated into practical recommendations for design of control-display configurations of hand and foot controls.

1 INTRODUCTION

Generally, foot controls are not as widely employed in industrial applications as hand controls. For applications equipped with foot controls like automobiles and airplanes, almost all the primary controls are given over to hands. If some of the controls can be assigned to the feet, there would be an obvious advantage of leaving the hands free for other tasks, which demand higher precision and dexterity. Spatial S-R compatibility is an important consideration in human-machine interface, which refers to the situation that the selection of a response is directly related to the position of stimulus (Sanders and McCormick, 1993). When the relationship between stimuli and responses is direct and natural, it is described as compatible; while the relationship is indirect and unnatural, it is described as incompatible (Proctor and Vu, 2006). It is a common finding that compatible pairings will lead to faster reaction times (RTs) and lower error rates than that of incompatible pairings. At present, there are no specific research studies in examining the spatial compatibility relationship regarding the simultaneous manipulation of hand and foot controls in response to the displays. Given the potential importance and usefulness of employing foot controls in industrial applications and the significance of spatial compatibility relationship in control tasks, the spatial S-R compatibility effect for hand and foot controls with a simple configuration of horizontal visual signals on human response performance was thus explored in this study.

2 METHOD

2.1 *Subjects*

Thirty two Chinese students of City University of Hong Kong (20 males and 12 females) of ages 20 to 26 participated in this study. They were all right-handers /footers, and had normal or corrected-to-normal vision.

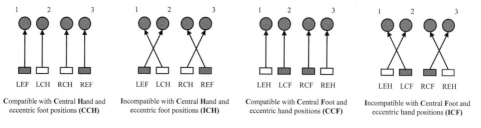

1 2 3	1 2 3	1 2 3	1 2 3
LEF LCH RCH REF	LEF LCH RCH REF	LEH LCF RCF REH	LEH LCF RCF REH
Compatible with Central Hand and eccentric foot positions (CCH)	Incompatible with Central Hand and eccentric foot positions (ICH)	Compatible with Central Foot and eccentric hand positions (CCF)	Incompatible with Central Foot and eccentric hand positions (ICF)

Figure 1. The experimental setup of this study. Two-hand buttons or foot pedals were arranged on either eccentric or central positions according to different spatial mapping conditions.

2.2 *Apparatus and Stimuli*

Four 12-volt yellow LED lights served as visual signals and a 5-volt green LED at the center position (between LEDs 2 and 3) served as a warning signal and fixation point, which were horizontally arrayed on a table. For visual stimulus presentation, either one of the four LEDs was presented in a trial. The two buttons and two-foot pedals were interfaced with the computer software and hardware for enabling subjects to respond to the corresponding visual signals, where they were placed on the table and floor respectively.

2.3 *Precudure and Design*

In each trial, one of the four horizontally arrayed LEDs (1, 2, 3 and 4) was presented to the subjects. Four response keys of which two buttons or two pedal keys on either eccentric or central positions were provided for inputting responses (Figure 1). With the given signals and input positions, four spatial stimulus-response (S-R) mapping conditions were tested as follows: Compatible with Central Hand and eccentric foot positions (CCH), Incompatible with Central Hand and eccentric foot positions (ICH), Compatible with Central Foot and eccentric hand positions (CCF), and Incompatible with Central Foot and eccentric hand positions (ICF). To prevent subjects from being confused by the complicated spatial mapping conditions and to make the first run simple, four LEDs (1, 2, 3 and 4), in which LEDs 1 or 2 constituted a 'left stimulus', and LEDs 3 or 4 constituted a 'right stimulus', were responded ONLY by left side key (for LEDs 1 or 2) and right side key (for LEDs 3 or 4) of either central and eccentric key positions respectively. In the CCH mapping condition, the four LEDs and response keys of central hand and eccentric foot were arranged congruously in horizontal dimension so that subjects would respond by pressing the left eccentric foot key to LED 1, left central hand key to LED 2, right central hand key to LED 3, and right eccentric foot key to LED 4. In the ICH mapping condition, S-R mapping was incompatible to the condition of CCH and subjects would respond by pressing left central hand key to LED 1, left eccentric foot key to LED 2, right eccentric foot key to LED 3, and right central hand key to LED 4. In the CCF mapping condition, congruous S-R mapping was found for the LEDs and response keys of central foot and eccentric hand and subjects would respond by pressing left eccentric hand key to LED 1, left central foot key to LED 2, right central foot key to LED 3, and right eccentric hand key to LED 4. In the ICF mapping condition, relatively, S-R incompatible was incompatible to the CCF mapping condition and subjects would respond by pressing left central foot key to LED 1, left eccentric hand key to LED 2, right eccentric hand key to LED 3, and right central foot key to LED 4. The 32 subjects in four groups were tested in different sequences with the four blocks of S-R mapping conditions in a counterbalanced order. Each block contained eight practice trials and twenty-four testing trials.

During the test, subjects sat at a distance of about 500 mm directly in front of the LEDs. They were asked to position their hands and feet on the two buttons and pedals respectively with a posture convenient for pressing the devices. Each trial started with the display of a green LED for serving as a warning signal and fixation point. After a delay of 1 to 4 seconds, one of the four LEDs was lighted up randomly. In response to the signal, subjects then pressed or trod on the appropriate key according to the compatibility conditions being tested. In all trials, subjects were asked to react as fast and accurately as they could. No feedback on the accuracy was given.

Table 1. Mean RTs of different stimulus positions and response key positions. The shortest RTs for individual S-R mappings are bold-faced.

Stimulus position	Response key position	Mean RTs (ms)	Average RTs for the signal position (ms)
LED 1 (Left-eccentric stimulus)	Left-Central Foot (LCF)	726	625
	Left-Central Hand (LCH)	671	
	Left-Eccentric Foot (LEF)	592	
	Left-Eccentric Hand (LEH)	512	
LED 2 (Left-central stimulus)	Left-Central Foot (LCF)	611	614
	Left-Central Hand (LCH)	485	
	Left-Eccentric Foot (LEF)	754	
	Left-Eccentric Hand (LEH)	623	
LED 3 (Right-central stimulus)	Right-Central Foot (RCF)	583	585
	Right-Central Hand (RCH)	473	
	Right-Eccentric Foot (REF)	720	
	Right-Eccentric Hand (REH)	573	
LED 4 (Right-eccentric stimulus)	Right-Central Foot (RCF)	699	607
	Right-Central Hand (RCH)	672	
	Right-Eccentric Foot (REF)	583	
	Right-Eccentric Hand (REH)	476	

3 RESULTS

A total of 3,072 (32 subjects x 4 conditions x 24 trials) responses were collected in this study. Overall, 153 (4.98%) incorrect responses were received. The mean and standard deviation of reaction times (RTs) were 608 ms and 209 ms, respectively. The mean RTs of correct responses at different stimulus positions (LEDs 1, 2, 3 and 4) and response key positions (LCF, LCH, LEF, LEH, RCF, RCH, REF and REH) are summarized in Table 1. Amongst the sixteen mean RTs, the shortest value was 473 ms obtained from LED 3 (right-central stimulus) with right-central hand key position while the longest time was 754 ms obtained from LED 2 (left-central stimulus) with left-eccentric foot key position.

The order of mean RTs across the four S-R mapping conditions was CCH (532 ms), CCF (546 ms), ICF (655 ms) and ICH (704 ms). The results showed that RT was influenced by the relative positions of visual stimulus and response key. This was evidenced by the obvious difference of 172 ms and 109 ms obtained from the compatible and incompatible S-R mappings of positioning of 'central hand and eccentric foot' and 'central foot and eccentric hand' respectively. Further examination of RTs was performed with an analysis of variance (ANOVA). The results showed that factors of signal position effect [$F(3,93) = 5.64$, $p < 0.01$], response key effect [$F(7,217) = 28.28$, $p < 0.0001$], and S-R mapping effect [$F(3,93) = 75.57$, $p < 0.0001$] were significant.

The order of mean RTs across the four visual signals was LED 3 (585 ms), LED 4 (607 ms), LED 2 (614 ms) and LED 1 (625 ms). A post hoc pairwise comparison (LSD) classified signal positions into two subsets, revealing that the responses to the right LED were faster than to left LED and that LED 1 and LED 3 was significantly different from each other ($p < 0.05$). The result clearly confirmed the right-side advantage for the right-handed/footed subjects. For the significant factor of key position, the order of the mean RTs across the eight response key positions was REH (525 ms), LEH (568 ms), RCH (573 ms), LCH (578 ms), RCF (641 ms), REF (652 ms), LCF (669 ms) and LEF (673 ms). The pairwise comparison (LSD) classified the RTs of the response key positions into three subsets in which the result clearly showed obvious right-hand advantage for right-handed subjects in treading the response keys; however there was no difference in treading between the left and right pedal keys.

The significant S-R mapping effect revealed a salient spatial stimulus-response (S-R) compatibility effect in this test. Responses were fastest in the compatible condition in which hand and foot were in the central and eccentric positions respectively and slowest in the incompatible condition in which hand and foot were respectively in the central and eccentric positions. The pairwise comparison (LSD) classified the reaction times of different S-R mapping conditions into three subsets. This indicated that the compatible and incompatible mappings were significantly different from each other and that spatial correspondence between positions of display and control yielded faster reaction times than spatial non-correspondence which was consistent with the results of other spatial compatibility studies. The significant difference between 'ICF' and 'ICH' suggested that if the hand and foot effectors are classified in horizontal array, hands would have a faster reaction than feet when they were in 'eccentric' positions.

4 DISCUSSION

In the horizontal array of hand-and-foot visual displays and controls, subjects responded faster with their hands. However, there is no difference found between the hand and foot in terms of error percentage, providing that the feet could perform as accurately as hands. Response using the dominant right hand was significantly faster than that of the non-dominant left hand. This is consistent with the finding of Annett (2002) that right-handers showed significant right-hand and right-foot performance advantage. Response preference was also affected by the positions of the four visual signals. Subjects responded faster and more accurate to the LED 3 (right central signal) than to the LED 1 (left eccentric signal). Wang, Zhou, Zhou, and Chen (2007) showed that the left hemisphere of the right-handed people plays the dominant role in recognizing the global properties of an environment. The right visual field, which controlled by the left hemisphere, was found to be faster and more accurate at identifying 'global' differences than the left visual field for right handers. The significant factor effect of S-R mapping revealed that the existence of spatial S-R compatibility effect, which is similar to that found with the sole use of hand controls (Proctor and Vu, 2006; Chan, Chan and Yu, 2007). Responses were faster and more accurate to compatible than incompatible mappings. An RT improvement of 140 ms was found with the correspondence S-R mappings (CCH and CCF conditions) when compared to the non-correspondence S-R mappings (ICH and ICF conditions). Reponses were faster and more accurate to the mapping of ICF than to the ICH, which is probably due to the end-state comfort effect (e.g. Short and Cauraugh, 1999) of positioning the hand and foot respectively in the eccentric and central locations. An end-state comfort effect is attained when subjects have completed a movement in a comfortable posture.

5 CONCLUSION

The relative positions of visual signals and keys of hand and foot should be spatially compatible for better system performance in terms of response times and response errors.

The horizontal array of hand and foot controls respectively in the eccentric and central positions leads to the best human performance.

Responses with right hand are faster than that with left hand for right-handed subjects.

Right-handed subjects respond faster to the right visual field than to the left.

ACKNOWLEDGMENT

The work described in this paper was fully supported by a grant from the Research Grants Council of the Hong Kong Special Administrative Region, China (Project No. CityU 110306).

REFERENCES

Annett, M. (2002). Handedness and brain asymmetry: the right shift theory. New York: Taylor and Francis.
Chan, A. H. S., Chan, K. W. L., and Yu, R. F. (2007). Auditory stimulus-response compatibility and control-display design, Theoretical Issues in Ergonomics Science, accepted for publication.
Proctor, R. W. and Vu, K.-P. L. (2006). Stimulus-Response Compatibility Principles. Taylor & Francis Group: Boca Raton, Florida.

Sanders, M. S. and McCormick, E. J. (Eds.) (1993). Human Factors in Engineering and Design. (7th ed.). Singapore: McGraw-Hill.

Short, M. W. and Cauraugh, J. H. (1999). Precision hypothesis and the end-state comfort effect. Acta Psychologica, 100(3), 243–252.

Wang, B., Zhou, T. G., Zhou, Y. and Chen, L. (2007). Global topological dominance in the left hemisphere. PNAS, 104, 21014–21019.

Chapter 6   Safety ergonomics

Ergonomic Trends from the East – Kumashiro (ed)
© 2010 Taylor & Francis Group, London, ISBN 978-0-415-88178-4

Hazard and risk analysis for decommissioning safety assessment of nuclear facilities

Kwan-Seong Jeong
*Division of Decommissioning Technology Development, Korea Atomic Energy Research Institute,
Daedeok-daero, Yuseong-gu, Daejeon, Republic of Korea*

Hyeon-Kyo Lim
*Department of Safety Engineering, Chungbuk National University, Sungbong-ro,
Heungduk-gu, Cheongju, Chungbuk, Republic of Korea*

1 INTRODUCTION

The radiological and non-radiological hazards arise during decommissioning activities. The non-radiological or industrial hazards to which workers are subjected during the decommissioning and dismantling process may be greater than those experienced during the operational lifetime of the facility.

The hazards associated with decommissioning are important not only because they may be a direct cause of harm to workers but also because their occurrence may, indirectly, result in increased radiological hazard.

According to the reports of decommissioning for nuclear facilities, in addition to radiation exposure and contamination, industry safety (DOE/EH-0578) occurs such as falls, heavy equipment hazards, structural hazards (sharp metal and debris). And substantial increase in numbers/amount of scaffold use, confined use, welding/grinding atmospheres, dusts/vapors nuisance atmospheres and heavy lifts (loading and rigging).

To protect the workers and prevent accidents during decommissioning activities, the reduction of risks should be taken actions by systematically analyzing and quantitatively assessing the causes and characteristics of potential hazards. The safety assessment should consider the consequences from occurrences and recommend appropriate measures and controls to minimize risks. The extent and level of detail of safety assessments should correspond to the types of hazards and their potential consequences.

2 A SYSTEMATIC PROCEDURE OF DECOMMISSIONING SAFETY ASSESSMENT

The safety assessment should be developed in a systematic manner which proportionate to the hazard potential of the facility and the possible consequences of the decommissioning activities under evaluation. Safety assessments for decommissioning should be based on the framework defined in Figure 1.

3 IDENTIFICATION OF HAZARDS IN DECOMMISSIONING ACTIVITY

A number of industrial hazards are associated with the decommissioning of any radiological facility. Many of these hazards are routine to the non-nuclear industry and their mitigation consists of standard industrial safety practices. According to the reports of a decommissioning for nuclear facilities (DOE/EH-0578), in addition to radiation exposure and contamination, industry safety occurs such as falls, heavy equipment hazards, structural hazards (sharp metal and debris). And

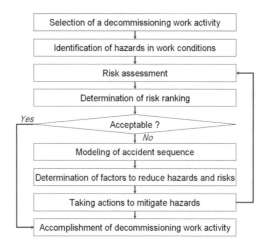

```
┌──────────────────────────────────────────────┐
│  Selection of a decommissioning work activity │
└──────────────────────────────────────────────┘
                      │
┌──────────────────────────────────────────────┐
│   Identification of hazards in work conditions │
└──────────────────────────────────────────────┘
                      │
┌──────────────────────────────────────────────┐
│               Risk assessment                 │◄───┐
└──────────────────────────────────────────────┘    │
                      │                              │
┌──────────────────────────────────────────────┐    │
│          Determination of risk ranking        │    │
└──────────────────────────────────────────────┘    │
                      │                              │
 Yes    ╱─────────────────────────────────╲          │
 ◄──────        Acceptable ?                ─────────│
        ╲─────────────────────────────────╱          │
                      │ No                            │
┌──────────────────────────────────────────────┐    │
│          Modeling of accident sequence         │    │
└──────────────────────────────────────────────┘    │
                      │                              │
┌──────────────────────────────────────────────┐    │
│ Determination of factors to reduce hazards and risks │
└──────────────────────────────────────────────┘    │
                      │                              │
┌──────────────────────────────────────────────┐    │
│         Taking actions to mitigate hazards     │────┘
└──────────────────────────────────────────────┘
                      │
┌──────────────────────────────────────────────┐
│  Accomplishment of decommissioning work activity │
└──────────────────────────────────────────────┘
```

Figure 1. A procedure of decommissioning safety assessment.

a substantial increase in the numbers/amount of scaffold use, confined use, welding/grinding atmospheres, dusts/vapors nuisance atmospheres and heavy lifts (loading and rigging).

The principal hazards anticipated during a decommissioning include physical hazards and potential exposures to activated material during a dismantling, surveying, moving and packaging potentially activated components. Workers need to be protected by eliminating or reducing the radiological and non-radiological hazards that may arise during routine decommissioning activities and as well as during accidents (IAEA, 2005).

Overall radiological risks can be lower during a decommissioning than during a regular operation (ISTC, 2008). However, the nature of decommissioning activities can mean that there is an enhanced risk of an exposure for some workers during a decommissioning. Remote handling and robotics technologies can greatly mitigate these risks, but when there are unavailable, a worker's exposure must be carefully managed. Similarly, the ingestion and inhalation of radionuclides from a surface contamination present a genuine risk that must be clearly addressed by standard worker protection measures. The potential for a criticality and breach of a containment are usually of less concern, but in some scenarios-such as the case where fissile material remains in process equipment-the possibility must be recognized and field activities planned accordingly. Containment systems can be particularly problematic. Those used during operation may no longer be working, and even if they are, there is no assurance that they can match the increased and varying demands of decommissioning activities. Radiological protection against these hazards is provided by a number of technical and managerial measures, including an isolation and removal of radioactive material, a spill prevention and dust/aerosol suppression techniques, bulk shielding of workers, discrete individual shielding through personnel protective clothing etc., training, air filtering, wastewater treatment, and appropriate waste-disposal techniques. Non-radiological hazards include fire (the most common risk due to the presence of flames in cutting technologies coupled with the accumulation of potentially combustible wastes), explosions (originating in dusts produced), toxic material (particularly in aged facilities where material no longer allowable [e.g., asbestos] may be present) and electrical and physical hazards (e.g., noise, confined space risks, impact trauma from falling objects, etc).

Hazard identification should begin by identifying all the potential radiological/non-radiological hazards which harm could be realized. A radiological hazard is a worker's exposure. And non-radiological hazards include industrial safety practices such as fire, explosions, falling, collision, etc closer to construction safety than operational safety. Radiological and non-radiological hazards for a decommissioning safety of nuclear facilities are illustrated in Figure 2. Radiological hazard is a worker radiation exposure and non-radiological hazards include a fall, upset/rollover, falling objects, collapse/destruction, crushing/winding, electric shock, fire/explosion and other toxic & hazardous environments.

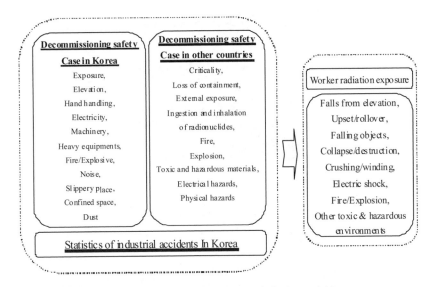

Figure 2. Radiological & non-radiological hazards in decommissioning activities.

Workers need to be protected by eliminating or reducing the radiological and non-radiological hazards that may arise during routine decommissioning activities and as well as during accidents (IAEA).

Hazard identification should begin by identifying all potential radiological/non-radiological hazards which harm could be realized. A radiological hazard is a worker exposure. And non-radiological hazards include industrial safety practices such as fire, explosions, falling, collision, etc closer to construction safety than operational safety.

4 QUANTITIATIVE RISK RANKING METHOD

An approach to quantitative risk ranking consists of tow ways. Risk ranking approach to a radiological hazard is results of exposure assessment within work scenario and pathway. And risk ranking approach to a non-radiological hazard is quantitative risk ranking by using a risk matrix.

The risk matrix is determined by the consequences and likelihood. A risk matrix method has been used for many years in industry and by the US military to rank different risks in order of importance. The merit of risk matrix can be classified by quantitative values. A risk assessment matrix allows assessment of the risk from each identified hazard. It help operator prioritize potential risks and determine which risks need documented controls.

4.1 *Risk ranking of a radiological hazard*

A radiological risk matrix uses the detailed level of worker exposure. The detailed level of worker exposure are classified by making reference to the radiation safety management manual of nuclear power plant operations, the decommissioning project of KRR-2 and UCF in Korea, ICRP and INES (IAEA and OECD/NEA, 2001).

Overall radiological risks can be lower during decommissioning than during operation. The nature of decommissioning activities can mean that there is an enhanced risk of exposure for some workers. Remote handling and robotics technologies can greatly mitigate these risks, but when these are un available, worker exposure must be carefully managed. The established dose limits must be fulfilled and applicable dose constraints should restrict the projected individual doses.

4.2 *Risk ranking of a non-radiological hazard*

A non-radiological risk matrix consists of consequences' level and likelihood's level of a non-radiological hazard.

Table 1. Levels of worker exposure.

Level of Exposure (Unit:mSv)	
R1	< 0.1
R2	0.1 ~ 0.5
R3	0.5 ~ 2
R4	2 ~ 20
R5	20 ~ 50
R6	50 ~ 250
R7	250 >

Table 2. Levels of consequence and likelihood.

Level	Injury Period	Level	Description
1	No Injury	1	< 10%
2	1 week ~ 1 month	2	< 25%
3	1 month ~ 3 month	3	25% ~ 50%
4	3 month ~ 1 year	4	50% ~ 75%
5	> 1year or one death	5	> 75%

Table 3. Risk calculation of a non-radiological hazard.

Level of consequence	Injury period		Level of likelihood	Likelihood
1	No injury		1	< 10%
2	1week ~ 1month	X	2	10% ~ 25%
3	1month ~ 3month		3	25% ~ 50 %
4	3month ~ 1year		4	50% ~ 75 %
5	>1year or death		5	75 % >

Likelihood \ Consequence	1	2	3	4	5
5	5	10	15	20	25
4	4	8	12	16	20
3	3	6	9	12	15
2	2	4	6	8	10
1	1	2	3	4	5

In general, according to standard MIL-STD-882D (US DOE, MIL-STD-882D), the categorization of a consequence and likelihood depends on the type of activity or specifics of the processes involved. This is a basis to constitute a plane matrix with cells each representing a certain risk category. Sometimes, especially for simple risk assessments, there may be 3 × 3 cells matrix, 5 × 5 cells matrix, and for process plants risk assessment of a larger structure, a 7 × 4 cells matrix used. This work advocates a 5 × 5 cells risk matrix – meaning that there are 5 different levels of likelihood and 5 different levels of severity of consequences. The detailed levels of consequence are classified by making reference to the accident injury rate of the Korea Occupational Safety & Health Agency (KOSHA). And the detailed levels of consequence and likelihood are classified by the occurrence rates shown in Table 2.

Risk calculation can be performed by consequences and likelihood. Risk of the non-radiological hazard calculation is assessed as follows.

- 16~25: Stop accomplishing a decommissioning work activity and after elimination and reduction of hazard accomplish one.
- 6~15: Accomplish a decommissioning work activity under actions and control
- 1~5: Accomplish a decommissioning work activity as it is

4.3 *Risk ranking method of combining the radiological risk and the non-radiological risk*

In case that both radiological hazard and non-radiological hazard exist, using risk priority, by comparing the risk of scenarios and works, the precedence of risk reduction is determined.

Table 4. Risk ranking of the radiological and non-radiological risk.

Hazard Type	Level	Priority
Radiological	R7	Priority 1
	R6	
	R5	
	R4	
Non-radiological	25	Priority 2
	20	
	16	
Radiological	R3	Priority 3
	R2	
Non-radiological	15	Priority 4
	13	
Radiological	R1	Priority 5
Non-radiological	6, 8, 9, 10	Priority 6
Non-radiological	1, 2, 3, 4, 5	Priority 7

In case that only non-radiological hazard exists, by comparing the risk of scenarios and works and comparing each maximum value of a non-radiological risk ranking score (risk matrix), the precedence of risk reduction are determined.

5 CASE STUDY

The proposed methodology has been successfully applied to KRR-2. KRR-2 (Korean Research Reactor-2) was the second nuclear reactor, a TRIGA MARK III type, with an open pool and a movable core, and its power was 2MWt. Its first criticality was reached in 1972 and it had been operated for 55,000 hours until decision to decommissioning in 1995. The main purpose of KRR-2 was the production of radioisotopes and neutron utilization research. KRR-2 was consisted of 12 laboratories, 10 lead hot cells and 2 concrete hot cells, which used for experiments with radioisotopes (KAERI/TR-1654/2000).

It has been applied as a case study examining hazards, removal of the reactor core, removal of the rotary specimen rack, and removal of the horizontal and vertical column, at three decommissioning workplaces in KRR-2. Having completed the case study, hazards and risks in decommissioning workplaces were assessed in 3 steps, that is, hazard identification, hazard evaluation, and hazard control. First, hazard identification involved indentifying the hazards associated with the activity of each process and type of potential hazardous events. A simple way of identifying hazards for a particular work activity is to divide the work activity into steps of carrying out the work and analyze the steps individually for the presence of hazards. It is also important to differentiate between hazards and incidents which are events caused by inadequate control of hazards. Second, hazard evaluation is the process of estimating the risks for the hazards. This was used as a base for prioritizing actions to control these hazards and minimize risks. Third, hazard controls were selected to reduce or confine the risk. These systematic procedures and results seemed promising in supporting the decision-making process for decommissioning activities towards the better working environments.

5.1 Removal of the rotary specimen rack in KRR-2 reactor

5.1.1 Physical characteristics from removal of the rotary specimen rack
The Rotary Specimen Rack (RSR) assembly consists of an annular aluminum rack, and outer ring-shaped, seal–welded aluminum housing. The rack which can be rotated inside the housing; supports 41 evenly spaced aluminum tubes, open at the top and all but number 1 closed at the bottom, serving as receptables for specimen containers. The RSR housing encircles the core shroud which is suspended from the reactor core bridge, approximately 7m into the tank. The RSR employs graphite blocks to maintain symmetrical positioning and two stainless-steel pins mounted on the buoyancy chambers. These run in guide slots on the core shroud and prevent the housing rotating and limit

107

Items	KRR-2 (TRIGA Mark-III type)
Reactor Type	Open pool, Movable core
Total Operating Time (Hours)	55,000
Total Generating Power (MWh)	69,000
Max. Neutron Flux (n/cm^2·sec)	7×10^{13}

Fuel
  Contents of U (w/o)
  Enrichment (w/o)
  Cladding
  Chemical composition
	
	8.5
	70
	304SS
	Er-U-ZrH$_{0.6}$

Moderator/ Coolant	H$_2$O
Reflector	H$_2$O
Control rod	B$_4$C

Figure 3.   Characteristics of KRR-2.

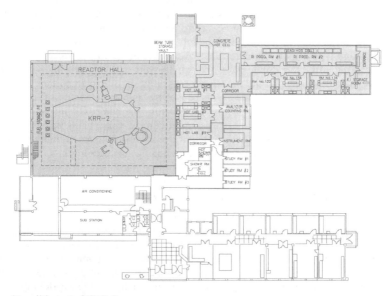

Figure 4.   Overall layout of KRR-2.

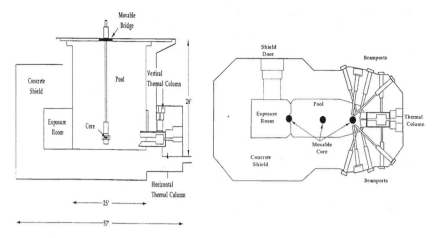

Figure 5. The reactor's cutting view of KRR-2.

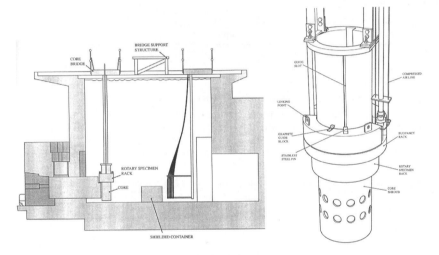

Figure 6. The RSR's cutting view of KRR-2.

the upper and lower positions of the RSR. The buoyancy chambers consist of two aluminum tanks, open at the bottom, and attached to the top of the RSR by 8 stainless-steel bolts (4 per chamber). Compressed air, supplied from the core bridge via a flexible line, displaces the water which, when vented, refilled. This system controls the vertical movement of the facility. Tools and equipments used during the decommissioning activities are conventional engineering tools, long reach tool to disconnect guide pins, long reach tool to connect cables to RSR lifting points, hydraulic shear tool, shear tool deployment system, and shielded container (KAERI/TR-1654/2000).

5.1.2 *Major tasks in the work package of 'removal of the rotary specimen rack'*
The tasks for 'removal of rotary specimen rack' to be accomplished are 'removal of RSR from KRR-2 reactor tank' and 'transfer to volume reduction facility and remove the RSR'.

5.1.3 *Major hazardous events from removal of the rotary specimen rack*
The tasks for 'removal of rotary specimen rack' and 'transfer to volume reduction facility and remove the RSR' are expected to give rise to the following several hazardous events.

• Working above the reactor, someone or something falling into the reactor core.
• Movement within the reactor hall of the rotary specimen rack removal equipment.

Table 5. A radiological risk assessment of hazardous event from removal of the RSR.

Hazardous event	Hazard type	Risk level	Description
Radiation dose uptake to the workforce	Worker radiation exposure	R4	3.482 mSv

Table 6. A non-radiological risk assessment of hazardous event from removal of the RSR.

Hazardous event	Hazard	Consequence	Likelihood	Risk level
Working above the reactor, someone or something falling into the reactor core	Falling from elevation	5	5	25
Movement within the reactor hall of the rotary specimen rack removal equipment	Falling objects	5	5	25
Transfer and handling of the container in and out of the core	Collapse/ destruction	2	5	10
Injury to personnel from improper lifting techniques	Falling from elevation	3	5	15

- Transfer and handle the container in and out of the core.
- Injury to personnel from improper lifting techniques.
- Radiation dose uptake to the workforce.

The predicted dose uptake for the removal of the tasks was expected to be 3.842 mSv.

5.1.4 *Risk assessment of hazardous events from removal of the rotary specimen rack*

Table 5 shows the radiological risk assessment's result of hazardous events from removal of the RSR. And Table 6 presents the non-radiological risk assessment's result of hazardous events from removal of the RSR.

5.1.5 *Measures and controls to reduce the hazards*

Measures and controls for reducing the hazards of the tasks associated with 'removal of rotary specimen rack' are below.

- Health physics personnel must be in attendance during the entire operation.
- Only personnel required to be working on the decommissioning activities must be allowed in the controlled area.
- The appropriate protective clothing must be worn i.e., overalls, hard hats, safety shoes, rubber gloves.
- All personnel working in the controlled area must be issued with a film badge and personal alarm dosemeter.
- All personnel working over the reactor must be required to wear a safety harness.
- Manual handling assessments and appropriate lifting tackle must be provided.
- All electrical isolations must be carried out by an approved person. An isolation certificate must be provided prior to work commencing.
- All operations must be controlled by way of a Permit to Work.

5.2 *A composite risk assessment of the three decommissioning tasks in KRR-2*

Table 7 summarizes the composite risk assessment's results of three decommissioning work packages in KRR-2. Through the proposed composite risk assessment method, risk levels and priorities

Table 7.   Results of a composite risk assessment of the three decommissioning tasks in KRR-2

Task	Hazardous event	Hazard	Hazard category	Risk level	Priority
Removal of reactor core and associated equipment	Radiation dose uptake to the workforce	Worker radiation exposure	Radiological	R3	3
	Working above the reactor, someone or something falling into the core	Falling from elevation	Non-radiological	25	2
	Handling the core	Crushing/winding		9	6
	Damage to the core	Other toxic & hazardous environments		9	6
	Injury to personnel from improper lifting techniques	Falling objects		25	2
Removal of horizontal and vertical thermal column	Radiation dose uptake to the workforce	Worker radiation exposure	Radiological	R4	1
	Injury to personnel from improper lifting techniques	Falling from elevation	Non-radiological	15	4
	Possible airborne contamination	Other toxic & hazardous environments		10	6
Removal of RSR from KRR-2 reactor tank	Radiation dose uptake to the workforce	Worker radiation exposure	Radiological	R4	1
	Working above the reactor, someone or something falling into the reactor core.	Falling from elevation	Non-radiological	25	2
	Movement within the reactor hall of the rotary specimen rack removal equipment.	Falling objects		25	2
	Transfer and handling of the container in and out of the core.	Collapse/ destruction		10	6
	Injury to personnel from improper lifting techniques	Falling from elevation		15	4

on the tasks of each work package are calculated and decided by hazardous event, hazard category, risk level, and priority.

REFERENCES

KAERI/TR-1654/2000, 2000, K.J. Jung, U.S. Jung, K.H. Jung, S.K. Park, D.G. Lee, H.R. Kim, J.K. Kim, S.H. Yang, and B.J. Lee, Decommissioning Plan–Decommissioning Project for KRR1 & 2
IAEA Safety Reports Series No. 45, 2005, Standard Format and Content for Safety related Decommissioning documents
IAEA and OECD/NEA, 2001, The International Nuclear Event Scale (INES) User's Manual
ITRC, 2008, Decontamination and Decommissioning of Radiologically Contaminated Facilities
US DOE, MIL-STD-882D, US Department of Defense, System Safety Program Requirements
US DOE/EH-0578, Statistical Evaluation of DOE D&D Occurrences

Ergonomic Trends from the East – Kumashiro (ed)
© 2010 Taylor & Francis Group, London, ISBN 978-0-415-88178-4

A study of work improvements for seafarers

Shuji Hisamune, Hiroaki Fukui, Suguru Tamura & Nobuo Kimura
Department of Economics, Takasaki City University of Economics (Institute for Nautical Training, Yokohama, Japan, Department of Fishery Science Hokkaido University, Hakodate, Japan)

1 PURPOSE

The incidence of ship crew work accidents is higher than that of other industries. The disaster incidence for ship crew members in 2004 fiscal year was 10 per 1000 workers, which was 2.5 times higher than the incidence in other industries. The death rate from crew accidents in 2004 was 0.4 per 1000 workers, which was 4 times higher than the rate in other industries. Thus it is very important to improve the working conditions on ships by involving the crew in raising the level of work efficiency and preventing work accidents. This work is ship-specific, because the inboard equipment varies greatly according to the type and size of ship and the area. Agreements about IMO labor standards are integrated into the International Labour Organization (ILO) agreement. In 2001 the ILO adopted guidelines for Occupational Safety and Health Management Systems (ILO-OSH 2001), which aim at providing the crew, carries out the safety measures to improve safety and health at work. The WISE (Work Improvements in Small Enterprises) training method, which is a participatory action-oriented training tool, has been developed with the support of the ILO and proposes model education WIB (Work Improvements on Board) to implement an action-oriented, systematic approach to improving safety on board ships.

2 METHOD

The work improvement activity was practiced with the cooperation of the Institute for Nautical Training. The lecture was three hours long and included lesson the meaning of work improvement, the selection of a good improvement case, and the use of a checklist, and each group presented three good and the weak points. After this program, the crew improved weak point.

Date: August 30, 2005 Taisei Maru 5800 gross ton Institute for Nautical Training

The participant: 34 crews.

In the occupational category, they were 19 engine people, and 15 deck people. They were 15 officer and 19 members. The average age was $38.8 +/- 11.5$ years. All participants classified each item into one of three categories: "It was good (Improvement was unnecessary)", "Improvement is necessary", "Improvement is not possible to implement".

3 RESULT

3.1 *Extraction of problems using a checklist*

(1) The entire evaluation

All members evaluate the weak point by the checklist. The item in need of improvement is flow by all members (34 people). For the item "Clearly mark movement of people and materials", the rating "Improvement is necessary" was 29.4%. For the item "Set up clear separation to or fences to prevent workers may approach a hazardous situation", the rating "Improvement is necessary" was 23.5%. For the item "Improve artificial lighting or provide spot lighting", the rating "Improvement is necessary" was 23.5%.

Table 1. The list of improvement activity of the deck department.

Improvement idea	Improvement easy?	Corrective strategy	Time	Cost
Peace for the mooring cable is stumbling.	Ivestigation	–	–	–
The stand of windlass operation is high from the floor side.	Difficulty	The middle stand is installed.	–	–
There is danger of the fall in the stand of crane operation.	Difficulty	The lid is applied.	–	–
There is a part without the protection cover of the lifeboat.	Difficulty	The protection cover is set up.	–	–
There is a ledge in the power simply box etc.	Easy	The warning color of the attention rousing is	–	–
There is no stopper of the bottle can tank lid.	Investigation	It only has to put up the stopper.	–	–
The door catcher in the mesa room is loose.	Easy	Adjustment	–	–
Height on the floor side on both sides was difference.	Easy	The stand of the height adjustment is set up.	–	–
The head is thrown at the gangway motor and the wire.	Investigation	The limitation bar is applied.	–	–
There are a ledge and danger of knocking against on the passage	Easy	The warning color was put.	0.05 h	¥0
Stumbling prevention in hatch cover rail.	Easy	The warning color was put	0.3 h	¥0
Danger of throwing the head is high in the trainee with a high	Easy	The cushioning material is applied.	0.1 h	¥0
There is danger of the fall at lifeboat safey rope.	Easy	The position in which the end of the safety rope is taken is charged.	1 h	¥0

(2) Deck department

The item in need of improvement is flow in the deck department (15 people). For the item "Clearly mark movement of people and materials", the rating "Improvement is necessary" was 40.0%. For the item "Make different switches and controls within easy distinguishable from each other by changing position, shapes and colors", the rating "Improvement is necessary" was 40.0%. For the item "Set up clear separation to or fences to prevent workers may approach hazardous situations", the rating "Improvement is necessary" was 33.3%.

(3) Engine part

The item in need of improvement is shown in flow in the engine part (19 people).

For the item "Clearly mark movement of people and materials", the rating "Improvement is necessary" was 26.4%. For the item "Improve artificial lighting or provide spot lighting", the rating "Improvement is necessary" was 26.4%. For the item "Clear aisles and passageways from obstacles or stumbling hazards", the rating "Improvement is necessary" was 21.1%. For the item "Remove or relocate sharp, dangerous or hot objects so that the workers' hands, feet or heads may not be hurt inadvertently during work", the rating "Improvement is necessary" was 21.1%. For the item "Improve artificial lighting or provide spot lighting", the rating "Improvement is necessary" was 21.1%. In the improvement item to need improvement, the deck department and the engine part was different.

3.2 *Results of improvement activities*

The safety was improved in each case. The improvements took three weeks to complete, which had the following advantages:

• The improvement is easy. 2. The cost is low. 3. The improvement is effective.
  The result is shown.
  Deck improvement result

Table 2. The list of improvement activity of the engine department.

Improvement idea	Improvement easy?	Corrective strategy	Time	Cost
There are a lot of dangerous places in a tall crew in the engine room.	Investigation	A dangerous part is specified for the crew with a high height.	–	–
The arrangement and order in the store a little more.	Easy	It is noted that the crew does not do in a disorderly manner.	–	–
The name is put in the storage place.	–	The storage place is displayed one by one.	–	–
The stair is fall.	Easy	Iron pipe was welded in stair.	2.5 h	¥1,000
Rotation is danger in the upper part of the countershaft	Easy	The signboard was made.	0.5 h	¥100
The handrail for the fall prevention is increased.	Easy	The iron pipe was welded.	2.5 h	¥1,000
Guard is put on the coupling of the pump. (GSand ballast pump.	Investigation	The drive conformation of the pump can be facilitated.	–	–
The limited part illumination is needed.	–	It corresponds with the torch lamp of individual	–	–
It is necessary to display the escape route from the engine room the lower.	Investigation	The fluoresce signboard is necessary.	–	–
It is the wide between the handrail and the step, the danger of the fall is thought.	Investigation	The iron pipe was welded.	2.5 h	¥1,000
There is no stand which operates the value set up at a high position.	difficult.	The ship shakes, and there is danger of the fall when the step is put.	–	–
The finger enters the hole of the handrail and it is likely to injure.	Easy	The tampion is put the hole of the handrail.	1.0 h every 5 pieces	Scrap use

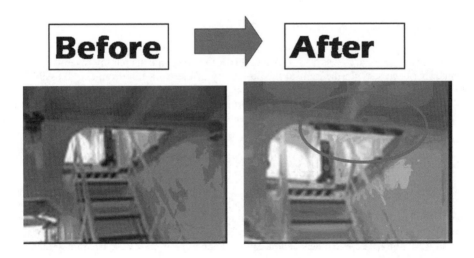

Caution above your head

Figure 1.   Case of Improvement (Deck).

Prevention of falls

Figure 2.   Case of Improvement (Engine).

Table 1 and Figure 1 show the improvement activity of the deck department. The improvements included the precautionary painting for the umbo and the installation of cushioning material (deck department).
Engine improvement result
Table 2 and Figure 2 show the improvement activity in the engine room. The example was closing the hole which welded the iron pipe. Cost and time are not shown in the table before the improvement activity. This activity recorded the papers.

3.3   *Evaluation of improvement activity*

In the item gauging the necessity of work improvement, the rating "Very necessary" was 37.5% and "Somewhat necessary" was 56.2%. About 90% of the respondents felt improvement was very or somewhat necessary. In the item of the possibility of the work improvement, "Possible" was 46.9%, and "Possible a little" was 31.2%. About 90% of the respondents felt improvement was very or somewhat possible. In free answer concerning the independent improvement activity, there were a lot of opinions of the uplift of their motivation for the improvement.

4   CONSIDERATION

Enactment of an international agreement concerning training of the crew, the requirement of a qualification certificate, and establishment of standards of duty (the STCW agreement) was adopted in the general convention of "International Maritime Organization (IMO)" in crew's international tendency for safety in July 1978. Our country ratified the agreement in May 1982, and it came into effect in April 1984. The agreements about labour standards for crew have been integrated in the ILO now.

It is expected to become an international standard so that crew health and safety can be defended as well as other IMO agreements. On land, ILO OSH 2001, which the ILO adopted in 2001, has been used as an indicator of an international standard of industrial safety.

The following effects were achieved by this work improvement. Crews could improve the work in the group. Crew understand the need for improvement easily. All members proposed three problems by using the checklist. A dangerous, harmful factor and the risk are specific, and are foreseen, and

Prevention not to catch your finger

Figure 3.   Case of Improvement (Engine).

evaluated. The application of this WIB can supplement a present regulation regarding safety and sanitation, and improve the ship's working environment based on the results of this study.

We will be necessary to act more in the future to enlarge the scope of work improvement activities. We propose the use of this system as the safety and health management system of the Ministry of Land, Infrastructure and Transport.

REFERENCES

Ministry of Land, Infrastructure and Transport, 2005 Crew work accident and disease (Article 111 of Crew Law).
HISAMUNE, S., AMAGAI, K., et al.,2006, A Study of Factors Relating to Work Accidents among Seamen, Industrial Health, Vol. 44, No. 1. pp. 144–149.
HISAMUNE, S., et al,. 2007, Empirical Study on Seamen of Work Improvement, Journal of Japan Navigation Society, No. 115, pp. 141–146.
ILO, 1998, Ergonomics Checkpoint, Institute of Science for Labour
KAWAKAMI, T., and KHAI, T., 1997,Sharing Positive Experiences in Making Changes in Work and Life in a Local District in Vietnam, Journal of Human Ergology, No. 26, pp. 129–140.
KOGI, K., 2006, Advances in Participatory Occupational Health Aimed at Good Practices in Small Enterprises and the Informal Sector, Industrial Health, No. 44, pp. 31–34.
TAKEYAMA, H., et al., 2006, A Case Study on Evaluations of Improvements Implemented by WISE Projects in the Philippines, Industrial Health, No.44, pp. 53–57.

Chapter 7   Ergonomics in occupational health 1

Ergonomic Trends from the East – Kumashiro (ed)
© 2010 Taylor & Francis Group, London, ISBN 978-0-415-88178-4

Development and practical use of software for work burden analysis

Seiko Taki, Yasuhiro Kajihara, Hideki Yukishima & Mitsuyuki Kawakami
Division of Management Systems Engineering, Faculty of System Design,
Tokyo Metropolitan University, Tokyo, Japan

1 INTRODUCTION

In conventional research, virtual reality was, primarily used to shorten the time required to transition from product design to production system design. The purpose of more recent research employing virtual reality has primarily been the shortening of time required to transition from production system design into a mass-production phase. The developed conventional systems improved manufacturers' competitive power with respect to time efficiency, yet the work burden of employees performing tasks associated with conventional virtual reality based systems has not been evaluated to date. As a worker will exhibit various signs of fatigue as a result of unsuitable work posture, systems that have been poorly designed in this respect will result in lower degrees of productivity. As such, the assessment of work burden in the design of a production system should be made a priority in order to minimize worker fatigue. Virtual Reality (VR) is a useful tool in resolving this problem, as VR allows us to obtain data pertaining to biomechanics, which is required for the evaluation of work posture. Results of the work posture evaluation are used to improve the work environment and process conditions, in order to mitigate the aforementioned risks.

To address the above, we have developed practical software that aims to reduce the physical burden of work, which results from employees necessarily maintaining poor posture in the execution of tasks. Workstations and a human model are drawn using Computer Aided Design (CAD) tools. These representations are shown in a virtual workspace. The motion of the human model is planned step by step through the manipulation of the model via mouse and keyboard. Data pertaining to work postures held by the human model during the course of one cycle of motion is recorded and the degree of work burden that is associated with the postures held is calculated continuously. The OWAS (Ovako Working posture Analyzing System) method is used to evaluate work burden. A work procedure guidance manual is then produced based upon the data. Furthermore, the data is used for employee training purposes. Thus, the design of the production system, the planning of operating procedures and the analysis of the burden associated with executing various work tasks are carried out continuously.

2 PROPOSED METHOD

This system allows one to define the motions of a worker within a virtual space during the conduct of some task, in order to analyze the resultant physical burden (work burden). Motions are indicated using a mouse and keyboard rather than a motion capture device. The design of the work area, the planning of work procedures, the development of a work guidance sheet explaining the work procedures and the evaluation of work burden are all performed continuously on a computer. The work burden is calculated based upon work posture using the OWAS method.

2.1 *Function of the system*

Practical software has been developed to analyze the work burden resulting for the execution of different tasks. In order to analyze the work burden, the motion of a worker is first planned using a

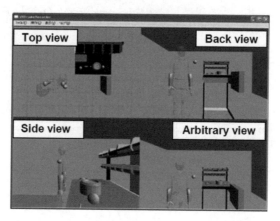

Figure 1.   Screen for planning operating procedure.

3D animation. The range of motion is analyzed, with the locus of analysis being the worker's joints. The range of motion is evaluated based upon principles of motion economy, and those motions that are less than desirable are ultimately identified.

Based upon the workers posture, the work burden is continuously estimated, from beginning to end, using the OWAS method. Workstations and work procedures are then modified in an effort to adjust the work burden. The planning and modification of work procedures is performed in an iterative manner until a final, optimal procedure is determined. This final plan serves as a standard operating procedure. The standard time is led from the standard operating procedure, calculated via the PTS method.

2.2   Planning and replaying the operating procedure

(i) Work planning of the operating procedure

In order to analyze the work burden of a given task, the motion of a worker is laid out in a 3D animation. Figure 1 depicts a screen of the operating procedure planning process. The screen is divided into four sub windows. The sub windows present a top, back and side view of the worker, as well as one additional, arbitrary point of view. The presentation of these four viewpoints lessens the frequency with which the operator must modify the viewpoint during the planning process. Motion data, that is, data pertaining to the posture of the human model and parts being assembled, is recorded and later used in replaying the work procedure. The operating procedure is executed with descriptive text and comments added to the recorded motion data.

(ii) Replay of the work operating procedure

The movement of the human model is replayed by loading the data that was recorded during the course of work planning. The working posture maintained by the model, within the virtual work area, is then evaluated against the developed work procedure manual. Figure 2 depicts a sample screen image that the operator might view when replaying the work procedure that reflects a given work procedure manual. The replay speed is controlled using control keys located on the lower left side of the window. The defined motion of the human model is assessed to determine whether the all tasks are carried out to completion. Furthermore, guidance in the form of comments can be incorporated to describe each motion.

(iii) Output of the work posture data

The data pertaining to the motion of the human body and hands during the course of the task execution is recorded for subsequent analysis of the associated work burden. The data pertaining to the location of the assembly components grasped by the worker's hands are also output, simultaneously. Locations are expressed in cm (centimetres), angles are expressed in ° (degrees) and elapsed time is expressed in seconds.

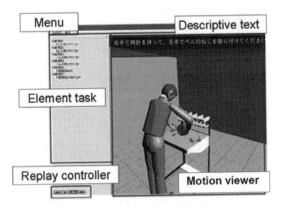

Figure 2.   Screen for replaying work procedure.

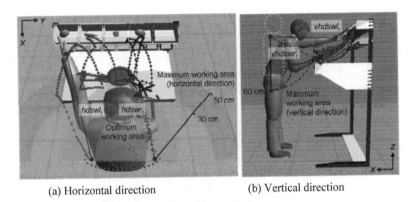

(a) Horizontal direction   (b) Vertical direction

Figure 3.   Locus chart of worker's hands and maximum working area.

2.3   *Analysis of work burden*

(i) Evaluation of range of motion

The range of motion is assessed with the locus of analysis as the worker's joints. The range is assessed based upon the principles of motion economy, and unsuitable motions are identified. For example, Figure 3 presents a locus chart of a worker's two wrists during the course of task execution. Diversification of distances from the shoulder to the wrist each hand is evaluated over the entire duration of the task, as shown in Figure 4. The horizontal and vertical distances ($hdswr_t$, $hdswl_t$, $vdswr_t$, $vdswl_t$) at time (t) are calculated from the shoulder position ($Sr_t(xsr_t,$ $ysr_t, zsr_t)$ or $Sl_t(xsl_t, ysl_t, zsl_t)$) and the wrist position ($Wr_t(xwr_t, ywr_t, zwr_t)$ or $Wl_t(xwl_t, ywl_t,$ $zwl_t)$) in the following Equation (1).

$$hdswr_t = \sqrt{\left|xsr_t - xwr_t\right|^2 + \left|ysr_t - ywr_t\right|^2}$$

$$hdswl_t = \sqrt{\left|xsl_t - xwl_t\right|^2 + \left|ysl_t - ywl_t\right|^2}$$

$$vdswr_t = \sqrt{\left|xsr_t - xwr_t\right|^2 + \left|zsr_t - zwr_t\right|^2}$$

$$vdswl_t = \sqrt{\left|xsl_t - xwl_t\right|^2 + \left|zsl_t - zwl_t\right|^2}$$

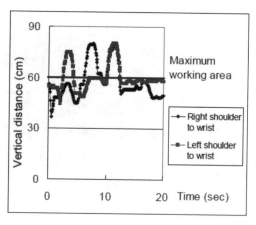

Figure 4. Vertical distances from shoulder to wrist.

Table 1. Posture code.

Back	(1) Upright
	(2) Leaning forced or backward
	(3) Twist or leaning side
	(4) Twist with lean to side
Arms	(1) Both arms are under shoulder
	(2) Either arm is above shoulder,
	(3) Both arm are above shoulder
Legs	(1) Sitting
	(2) Standing up
	(3) Standing on one leg
	(4) A half-sitting posture
	(5) Standing on one leg in a half-sitting posture
	(6) Kneel on one knee or both knees
	(7) Walk or move
Weight of force	(1) Less then 10 kg (w ≤ 10 kg)
	(2) From 10 to 20 kg (10 < w ≤ 20 kg)
	(3) Over 20 kg (w > 20 kg)

(ii) Evaluation of work burden

The work burden associated with a given task is evaluated, using the OWAS method, based upon the posture data that is recorded. The OWAS method first evaluates a given work posture from four aspects: the back, the upper limbs, the legs, and weight distribution. An appraisal of work burden is expressed as four digits (the posture code). Possible posture codes are presented in Table 1. In addition to the detail conveyed by a given posture classification, the range (or space) limitation is also provided in abbreviated form.

Moreover, by employing the OWAS method, the intensity of the work burden associated with a work posture, and thus the requirement for improvement in task operating procedures, is determined based upon posture codes, as shown in Table 2. The result of this determination is expressed as one of four levels of Action Category (AC).

3 APPLICATION EXAMPLE

In order to better understand the functionality of the developed software, consider the following practical example, in which two types of assembly guidance manuals have been designed pertaining

Table 2. Action category (AC) decided from posture codes.

back	arms	1			2			3			4			5			6			7			legs weight or force
		1	2	3	1	2	3	1	2	3	1	2	3	1	2	3	1	2	3	1	2	3	
1	1	1	1	1	1	1	1	1	1	1	2	2	2	2	2	2	1	1	1	1	1	1	
	2	1	1	1	1	1	1	1	1	1	2	2	2	2	2	2	1	1	1	1	1	1	
	3	1	1	1	1	1	1	1	1	1	2	2	3	2	2	3	1	1	1	1	1	2	
2	1	2	2	3	2	2	3	2	2	3	3	3	3	3	3	3	2	2	2	2	3	3	
	2	2	2	3	2	2	3	2	3	3	3	4	4	3	4	4	3	3	4	2	3	4	
	3	3	3	4	2	2	3	3	3	3	3	4	4	4	4	4	4	4	4	2	3	4	
3	1	1	1	1	1	1	1	1	1	2	3	3	3	4	4	4	1	1	1	1	1	1	
	2	2	2	3	1	1	1	1	2	4	4	4	4	4	4	4	3	3	3	1	1	1	
	3	2	2	3	1	1	1	2	3	3	4	4	4	4	4	4	4	4	4	1	1	1	
4	1	2	3	3	2	2	3	2	2	3	4	4	4	4	4	4	4	4	4	2	3	4	
	2	3	3	4	2	3	4	3	3	4	4	4	4	4	4	4	4	4	4	2	3	4	
	3	4	4	4	2	3	4	3	3	4	4	4	4	4	4	4	4	4	4	2	3	4	

AC1: This posture does not produce musculoskeletal burden. Improvement is unnecessary.
AC2: This posture is harmful to musculoskeletal system.
AC3: This posture is harmful to musculoskeletal system.
AC4: This posture is harmful to musculoskeletal system. This posture should be improved immediately.

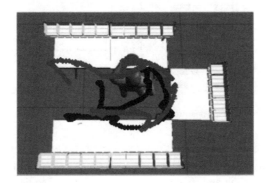

Figure 5. Locus chart of the both hands.

to an alarm clock. An operator carries out the clock assembly working within a space delimited by a U-shaped boundary lines. Different components were assigned different initial resting points in each case. In case 1, components are only located upon work desks. However, in case 2, the components are located on the work desks as well as the shelves.

As workers perform the designed working procedures, the associated work burden is assessed. The range of motion is evaluated, with the locus of analysis being the worker's wrists. Unsuitable motions are identified, and the resultant Action Category (AC) associated with each scenario is then identified based upon the exhibited work posture, specifically with respect to the worker's back, arms and legs.

3.1 Case 1: clock components located only upon the work desk

(i) Evaluation of range of motion

Figure 5 depicts a locus chart of the worker's hands during the course of the task. The variation in the distance from shoulder to wrist of each hand is evaluated over the duration of the task, as shown in Figure 6. These results indicate that there is no problem, as the worker assembled the components without straying outside of the maximum working area.

(ii) Evaluation of work burden

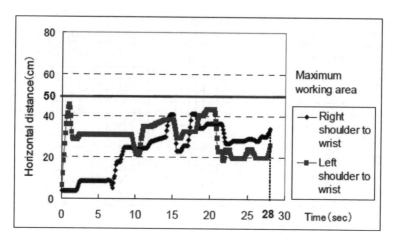

Figure 6.   Horizontal distance.

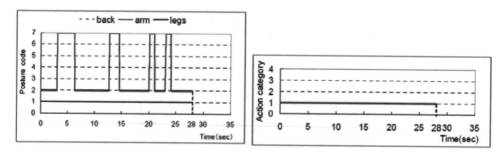

Figure 7.   Variation in work burden (posture codes and action category) over the duration of task execution for case 1.

Figure 6 depicts charts of the resulting posture codes associated with the worker's execution of the task, as well as the associated action category (AC). The determination of the action category (AC) for each work situation was made based upon the observed work posture code, with respect to the worker's back, arms and legs. These results would indicate that there is no problem here in terms of work burden.

3.2   *Case 2: clock components were located on both the work desk and shelves*

(i) Evaluation of range of motion

Figure 8 presents a locus chart of the worker's hands during the course of the task execution. Variation in the distance from shoulder to wrist of each hand is evaluated from the beginning to the end of the task, as shown in Figure 9. These results suggest that the positioning of the clock components can be improved, as the worker was required to stray from the bounded working area in order to complete the task.

(ii) Evaluation of work burden;

Figure 10 presents graphical representations of the posture codes and the action category (AC). The determination of the action category (AC) for each work scenario was made based upon the observed work posture code, with respect to the worker's back, arms and legs. This result suggests that the postures maintained by the working during the task execution are harmful to the musculoskeletal system, thus it needs improvement in the near future.

The aforementioned analysis greatly improves the design process as it allows one to carry out assembly training and to perform an examination of work burden, before the physical availability of the components for product assembly.

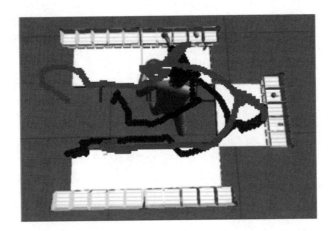

Figure 8.   Locus chart of the both hands.

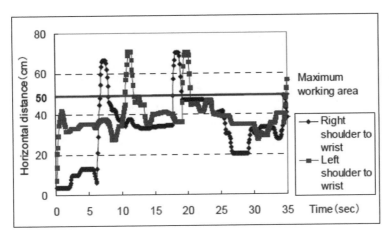

Figure 9.   Horizontal distance.

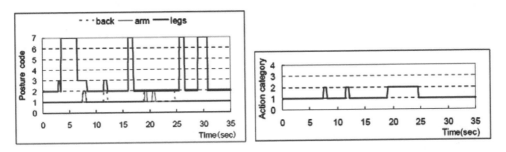

Figure 10.   Variation in work burden (posture codes and action category) during the course of task execution in case 2.

4   CONCLUSION

The results of this research can be summarized as follows.

(i) Software aiding the design of operating procedures was developed.
(ii) This system makes it possible to simultaneously carry out the evaluation of work burden associated with a given task, as well as the planning of work procedures.

(iii) This system makes it possible to design the layout of the work area, plan the task execution sequence, evaluate the associated work burden and develop the training manual for the task, without interruption, at any time or place.

REFERENCES

Kajihara Y. et. al., 2008, Analysis of effects of virtual reality based assembly training, J. Jpn. Ind. Manage. Assoc., 59, (2), pp. 162–172.
Watanabe Y., 2004, Hirou no kagaku, Kinpodo.
Louhevaara V. et. al., 1992, OWAS: a method for the evaluation of postural load during work, Instituteof occupational health.

Ergonomic evaluation of dual display type VDT workstation

Shin Saito
Mie Prefectural College of Nursing, Mie, Japan

Takahiro Nakatsukasa
Faculty of Engineering, Mie University, Mie, Japan

Toshio Matsuoka
Mie Prefecture Industrial Research Institute, Mie, Japan

Soichi Nakamura
Faculty of Engineering, Mie University, Mie, Japan

Norikazu Ohnishi
Mie Prefectural College of Nursing, Mie, Japan

Yosuke Sanbayashi
Tokyo Metropolitan College of Industrial Technology, Tokyo, Japan

Ryojun Ikeura & Kazuki Mizutani
Faculty of Engineering, Mie University, Mie, Japan

ABSTRACT: The aim of this study was to evaluate working conditions in subjects using multi-display Visual Display Terminal (VDT) workstations in terms of work performance, head movement and electromyogram (EMG) activity of the neck muscles. A comparative study using single and dual display type VDTs was undertaken to obtain ergonomic data applicable to the development of guidelines and recommendations for the use of VDTs with dual displays. Subjects comprised 10 healthy students who performed data search and entry tasks for 5 min. Single and dual display workstations were provided for each subject.

The work efficiency of subjects using the single display was poorer than that of those using the dual display. Head movement was greater with the dual display than with the single display. EMG activity did not differ significantly between the two groups for any of the muscles assessed. The dual display provided better work efficiency with a lower physiological workload than the single display.

1 INTRODUCTION

In recent years, dual display type VDT workstations have come into use in workplaces for applications such as movie editing, medical diagnostic imaging and securities transactions. Since most VDTs have a liquid crystal display, it is easy to build multi-display VDTs. Dual displays have the advantage of allowing the use of multiple non-overlapping open windows (Saito *et al.*, 2005).

No guidelines or recommendations have been defined for the use of dual display type VDTs since such displays have not yet gained widespread public use (Ministry of Health, Labour and Welfare, 2002).

The purpose of this study was to evaluate the working conditions of users operating a dual display VDT in terms of work efficiency, head movement and electromyogram (EMG) activity of the neck muscles. Single and dual displays were compared to obtain ergonomic data applicable to the establishment of guidelines and recommendations for dual display VDT workstations.

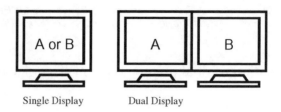

Single Display   Dual Display

Figure 1. Workstation devices.

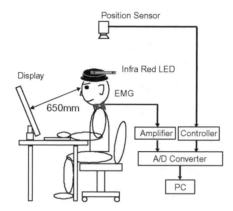

Figure 2. Experimental apparatus.

2 METHOD

Figure 1 shows the workstation devices used in this experiment. Ten healthy students were selected as a subject of this experiment. Each subject was provided with two kinds of VDT workstation device: a single and a dual display. The single display was a 15-inch monitor, while the dual display comprised two 15-inch displays. All displays were of the colour liquid crystal type. The resolution of the single and dual displays was 1024 (H) × 768 (V) dots. The personal computer used in this experiment had a G450 type video card for twin outputs (Matrox Graphics, Canada).

  The subject had to memorize 6 digits number from the word processing window, and the 6 digits number was reversely arranged. Next, the subject was required to input the 6 digits number which was reversely arranged in a specified place on the spreadsheet window.

  Figure 2 shows the experimental apparatus. Head movement was detected using a C5949 position sensor (Hamamatsu Photonics, Hamamatsu) and measured in terms of rotation angles. The rotation angle was integrated for 5 min under each experimental condition (single versus dual display). Surface EMG was recorded for the left and right sternocleidomastoid and trapezius muscles. The EMG signal was amplified using a BA1008 bio-amplifier (TEAC, Tokyo). Analog Recorder Pro measurement software (G1 System, Aichi) was used to record the data and analyse the head rotation angle and percent maximum voluntary contraction (%MVC).

3 RESULTS

Figure 3 shows the comparison of the average number of correct answers. It was the maximum, and in the number of correct answers, of single display, 16 and the minimum were 7, and 10.7 on the average. On the other hand, the maximum was 16, the minimum was 9 to the number of correct answers of dual display, and the average was 13.2. Significant difference was indicated in each experimental condition by paired t-test ($p < 0.05$).

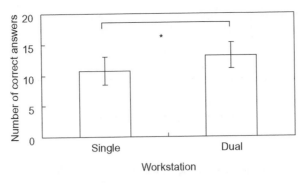

Figure 3. Average number of correct answers (n = 10, p < 0.05).

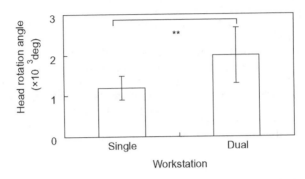

Figure 4. Average and SD of the multiplication head rotation angle (n = 10, p < 0.01).

Table 1. %MVC of sternocleidomastoid muscles (n = 10, NS: not significant).

Muscle	Left			Right		
Display	Single	Dual		Single	Dual	
Average	5.7	6.1	NS	8.1	7.2	NS
SD	2.4	2.6		6.3	4.7	

Figure 4 shows the average and SD of the multiplication head rotation angle in each experimental condition. The average head rotation angle of subjects using a single display was 20° and that of subjects using a dual display was 30°. The head rotation angle of subjects using a single display was significantly smaller than that of subjects using a dual display (p < 0.05, paired t-test).

Tables 1 and 2 show the average and SD of the %MVC of EMG activity for the left and right sternocleidomastoid and trapezius muscles in each group. The average EMG activity did not differ significantly between workstation types for either the sternocleidomastoid or the trapezius muscle.

4 DISCUSSION

Subjects using the dual display showed much better work efficiency compared with those using the single display. This was attributed to the large display area of the dual display compared with the single display. Moreover, when switching windows in the single display, we considered that having to operate the taskbar would affect work efficiency.

Table 2. %MVC of trapezius muscles (n = 10, NS: not significant).

Muscle Display	Left			Right		
	Single	Dual		Single	Dual	
Average	14.5	14.9	NS	16.4	19.1	NS
SD	5.7	5.1		7.9	8.2	

In general, movement of the head and neck in VDT work is minimal. Therefore, blood flow to the muscles of the neck of VDT operators is poor, rendering them susceptible to musculoskeletal disorders. Head movement was less for users of the single display compared with users of the dual display. This leads to an increase in static muscle work, which in turn places excessive load on the musculoskeletal systems of the neck and back (Schüldt et al., 1986). The head rotation angle was greater in subjects using the dual display than in those using the single display, as the latter has lesser lateral (left and right) extensions.

Using a dual display may decrease the static muscle work in the neck. Consequently, the use of a dual display might reduce the physiological load compared with the use of a single display. Blood flow to the muscles may be relatively high in users of dual displays compared with those operating single displays because of the increased muscle activity in the head and neck. Although the rotation angle of the head differed significantly between the users of the two display types, EMG activity did not differ significantly between the two groups. The physiological workload in the neck in users of dual displays is thus considered to be minimal.

5 CONCLUSION

Compared with the single display type VDT, the dual display type VDT has the advantages of better work efficiency and a lower physiological workload for the user. These basic data are applicable to the establishment of guidelines and recommendation for multi-display type VDT workstations.

REFERENCES

Ministry of Health, Labour and Welfare, 2002, Outline of new VDT work guideline, Report announced document, (in Japanese).

Saito Sh, Ohnishi N, Katoh Z, Miyao M, Ito K, Ikeura R, Mizutani K: An ergonomic study on multi screen type VDT work, Journal of Ergonomics in Occupational Safety and Health, 7(1), 17–22, 2005 (in Japanese).

Schüldt K, Ekholm J, Harms -Ringdahl K, Németh G, Arborelius U: Effects of changes sitting work posture on static neck and shoulder muscle activity, Ergonomics, 29, 1525–1537, 1986.

Evaluating fatigue among workers working long hours in Japan to provide in-person guidance

Shoko Kawanami & Seichi Horie

Department of Health Policy and Management, Institute of Industrial Ecological Sciences,
University of Occupational and Environmental Health, Kitakyushu, Japan

1 INTRODUCTION

The custom of long working hours in Japan has increased rates of myocardial infarction, arrhythmia, stroke, and depression (Spurgeon, 1997, Wada, 2002). A new policy requiring in-person guidance (hereafter simply "guidance") by a physician for workers who often put in long working hours went into effect in 2006. Under Industrial Safety and Health Law, in-person guidance must be provided to workers whose number of working hours beyond the normal 40 hours per week reaches 100 hours in any given month and who show signs of accumulated fatigue. At the site at which guidance is provided, the physician is required to confirm working conditions, levels of accumulated fatigue, and general mental and physical health status. To help ensure effective guidance, the Occupational Health Promotion Foundation was commissioned by the government to develop a textbook incorporating checklists and a manual for physicians to use at sites where guidance is provided (2006). For workers whose work hours or observable symptoms fall short of mandatory guidance under the above criteria, the law requires employers to provide a modified version of the guidance, whether provided by a physician or other personnel. However, no specific methods or indices for evaluating accumulated fatigue when screening for the need for such guidance are specified. We investigated how fatigue was evaluated in actual practice when screening workers for guidance.

2 METHODS

In an October 2006 survey targeting 652 physicians, all members of the Japan Society for Occupational Health (JSOH), and 255 registered physicians at Local Industrial Health Centers (LIHC), we reviewed answers provided in response to questionnaires designed to assess accumulated fatigue to determine what factors are considered when screening workers and implementing the required guidance.

3 RESULTS

Of the 396 respondents (response rate: 44%), we analyzed data for 279 physicians reporting experience in providing the type of guidance in question. At most workplaces at which the physicians had been assigned, the primary parameter for screening workers to whom guidance was provided was the number of working hours. In addition, while the law requires employers to provide guidance only in cases involving workers who request it, 89 of 279 (31.9%) physicians reported providing guidance for all workers who exceeded a certain worktime criteria, whether or not the workers actually requested it. This proportion was higher among JSOH (64/192; 33.3%) than among LIHC (25/87; 13.0%). Such criteria did not always consider levels of accumulated fatigue. Only 15/279 (5.4%) of the workplaces took the results of checklists into account (Table 1).

Table 1. Implementation of in-person guidance based on varying criteria.

	JSOH (n = 192)		LIHC (n = 87)		Total (n = 279)	
	n	(%)	n	(%)	n	(%)
Overtime working hours > 100 hr/M with worker's claim	178	(92.7)	83	(95.4)	261	(93.5)
Overtime working hours > 80 hr/M with worker's claim	160	(83.3)	82	(94.3)	242	(86.7)
Other company rule*	127	(66.1)	51	(58.6)	178	(63.8)
*e.g.: All the workers who exceeded certain working hours (regardless of their claim)	64	(33.3)	25	(28.7)	89	(31.9)
Taking result of checklist into account	14	(7.3)	1	(1.1)	15	(5.4)
Total number of physicians providingin-person guidance	192	(100.0)	87	(100.0)	279	(100.0)

JSOH: Japan Society for Occupational Health; LIHC: Local Industrial Health Center

Table 2. Physicians using questionnaires or checklists at sites where guidance is provided.

	JSOH (192)		LIHC (87)		Total (279)	
	n	(%)	n	(%)	n	(%)
Using	157	(81.8)	70	(80.5)	227	(81.4)
Not using	26	(13.5)	14	(16.1)	40	(14.3)
No answer	9	(4.7)	3	(3.4)	12	(4.3)
Total number of physicians providingin-person guidance	192	(100.0)	87	(100.0)	279	(100.0)

Table 3. Physicians using questionnaires or checklists at sites where a modified version of guidance is provided.

	JSOH (n = 162)		LIHC (n = 82)		Total (244)	
	n	(%)	n	(%)	n	(%)
Using	97	(59.9)	53	(64.6)	150	(61.5)
Not using	65	(40.1)	29	(35.4)	94	(38.5)
Total number of physicians providing modified version of guidance	162	(100.0)	82	(100.0)	244	(100.0)

At the sites where guidance was provided, 227 physicians used some type of questionnaire or checklist (Table 2). The proportion of physicians using such questionnaires or checklists was 157/192 (81.8%) among JSOH members and 70/87 (80.5%) among LIHC members.

Likewise, for the modified version of guidance, 150 of 244 physicians reporting providing such guidance used checklists to evaluate fatigue (Table 3).

Of the five different types of checklists recommended in the standard manual on guidance, 145/227(63.9%) of physicians used a checklist for accumulated fatigue developed by the Japan Industrial Safety and Health Association. Fewer than 30% of the physicians used other checklists (Table 4). Tendencies related to checklist selection did not differ between JSOH and LIHC physicians.

Table 4.   Types of checklists used by physicians (The standard manual on guidance recommends checklists).

	JSOH (n = 157)		LIHC (n = 70)		Total (227)	
	n	(%)	n	(%)	n	(%)
Checklist A ; for evaluating workload and stress	39	(24.8)	23	(32.9)	62	(27.3)
Checklist B; for evaluating accumulated fatigue	103	(65.6)	42	(60.0)	145	(63.9)
Checklist C; for primary screening of depression	47	(29.9)	18	(25.7)	65	(28.6)
Checklist D; for evaluating working condition of workers whose score was highon checklist B or D	44	(28.0)	17	(24.3)	61	(26.9)
Checklist E; for secondary screening of depression	40	(25.5)	15	(21.4)	55	(24.2)
Other checklists	46	(29.3)	16	(22.9)	62	(27.3)
Total number of physicians using checklists	157	(100.0)	70	(100.0)	227	(100.0)

Table 5.   The proportion of physicians to assess fatigue in guidance sessions and reporting to employers.

	JSOH (n = 192)		LIHC (n = 87)		Total (279)	
	n	(%)	n	(%)	n	(%)
Physicans evaluating workers' fatigue at the guidance	127	(66.1)	60	(69.0)	187	(67.0)
Physicans reporting workers' fatigue condition to their employer after guidance	77	(40.1)	59	(67.8)	136	(48.7)
Total number of physicians providing in-person guidance	192	(100.0)	87	(100.0)	279	(100.0)

Overall, 187 physicians used certain methods to evaluate fatigue at sites where in-person guidance was provided. Of these, 136 physicians reported their results to employers (Table 5). More than half (51.8%) of the members of JSOH evaluated worker fatigue at guidance sessions, although they did not always report the results to employers. The responses did not point to any other methods used to assess fatigue other than questionnaires or checklists.

4   DISCUSSION

Accumulated fatigue is one of the conditions that under law require in-person guidance. However, objectively assessing accumulated fatigue is difficult. According to a bibliographic study reported by Niels *et al.* (2007), fatigue exhibits various characteristics that give rise to varying symptoms, depending on any number of factors causing or associated with the condition. Fatigue is defined variously, and no generic measurement or evaluation instruments have entered common practice. While certain biological mechanisms underlying fatigue have been proposed, no biological markers with a clear relationship to fatigue have been established to date (Payne *et al.*, 2007). For these reasons, assessments of fatigue must be based on cause, quality and quantity of fatigue, subjective complaints, and behavior.

This survey showed that at most workplaces, objective assessments of fatigue were not adopted as an index at the stage of screening workers for guidance, since it is impossible to rule out health impairments based on the results of questionnaires or checklists. The methods used to evaluate fatigue at sites where guidance was provided varied significantly, since fatigue is one of the key factors linking impaired health and long working hours.

Although a textbook on guidance refers to several checklists, more than 60% of the physicians indicated using Checklist B, which was designed and developed to evaluate accumulated fatigue in workers. Simplicity may explain why this checklist was used most frequently; it consists of a mere 20 questions that address both subjective symptoms and subjective sense of work burdens at the primary screening stage.

Also notable is the discrepancy between the rate at which worker fatigue is assessed and the rate at which these results are reported to employers among JSOH. This may be explained by efforts among physicians to protect worker confidentiality from employers. Although Japan has passed legislation that specifies the specific nature of health examinations, it is to be expected that methods for evaluating fatigue would be left to the discretion of professionals.

5   CONCLUSION

Our study showed that no index for fatigue was adopted at the stage at which workers were screened for guidance at most workplaces. However, both JSOH and LIHC members commonly evaluate worker fatigue using checklists at sites where guidance is provided.

REFERENCES

Spurgeon, A., et al., 1997, Health and safety problem associated with long working hours, In Occupational Environmental Medicine, 54: 367–375.
Wada, O., 2002, Occupational and Life -style Risk factors and prevention of "Karoshi", In Occupational Health Review, 14, 183–213.
Checklists for guidance to employees with long working hours (for physicians), 2006, The Occupational Health Promotion Foundation (URL:www.zsisz.or.jp/hoken/doc/m-clist.doc).
Niels, H., et al., 2007, Assessment of fatigue in chronic disease: a bibliographic study of fatigue measurement scales. In Health and Quality of Life Outcomes, 5:12.
Payne J., et al., 2006, Biomarkers, fatigue, sleep, and depressive symptoms in women with breast cancer: a pilot study, In Oncology Nursing Forum, 33(4):775–83.

Ergonomic Trends from the East – Kumashiro (ed)
© 2010 Taylor & Francis Group, London, ISBN 978-0-415-88178-4

Sleep-awake patterns of shift workers on a 4-crew 3-shift system and subjective satisfaction with sleep quality

Masataka Iwane, Shinya Maeda, Koichi Mugitani, Mika Watanabe, Shotaro Enomoto, Chizu Mukoubayashi, Shuya Yukawa & Osamu Mohara
The Wakayama Wellness Foundation, Wakayama City, Japan

1 OBJECTIVE

Poor sleep quality is related to poor physical and psychological health, accidents, difficulty with occupational activities, and conflicts in personal relationships (Doi, *et al.* 2003). Shift work is one of the factors affecting poor sleep quality (Marquié and Foret, 1999). Sleep duration of shift workers is shorter than that of non-shift workers (Drake, *et al.* 2004). On the other hand, sleep behavior and subjective sleep quality of shift workers in comparison to non-shift workers have not been clarified. In this study, we examined the sleep-behavior and sleep-awake cycle of experienced shift-workers with a 4-crew 3-shift rotation, in order to determine the quality and quantity of sleep, particularly pertaining to workers during the night shift.

2 METHODS

Subjects comprised 254 male workers aged from 40 to 59 years at three steel factories located in Kainan City, Wakayama, Japan. The subjects were experienced shift workers with a rotating 4-crew 3-shift system on the following schedule: -MMMMM-HH-AAAAA-H-NNNNN-HH- (M, morning shift (08:00–16:00); A, afternoon shift (16:00–23:15); N, night shift (23:15–08:00); H, holiday). Using a self-report questionnaire, we assayed drinking habits, sleeping and waking hours, as well as subjective complaints regarding sleep quality during the 3 respective shifts. The subjects were asked to rate sleep quality during the three different shifts according to a 3-point scale: good, not so good, bad. Subjects who rated their sleep quality as "good" were defined as good sleepers and those who rated their sleep quality as "not so good" or "bad" were defined as poor sleepers.

Data were expressed as the mean ±SD. One-way ANOVA was used to test differences in the means among groups. The χ^2 test was used to evaluate the differences in the groups. P-value < 0.05 was considered to indicate statistical significance.

3 RESULTS

The mean value and SD of sleep onset, duration, and satisfaction with sleep quality in each shift are shown in Table 1. The sleep duration was longest in the morning shift (7.2 ± 0.9 hours), followed by the afternoon shift (6.7 ± 1.0 hours), and night shift (6.5 ± 1.3 hours). During the morning and afternoon shifts, 65.0 to 69.2% of workers slept soundly; however, only 28.9% of workers reported good sleep during the night shift. The distribution of good sleep groups and poor sleep groups according to age was relatively similar in each shift. Only 21.2% of the workers reported good sleep during all shifts; 29.8% of the workers slept poorly during the night shift, and 18.9% of the workers slept poorly during all shifts (Figure 1).

Table 2 shows sleep behavior and sleep quality during the night shift. Sleep behavior of the night shift group was determined according to the frequency of sleep: once, twice, or three times. The number of one-time sleepers, two-time sleepers, and three-time sleepers was 109 (45.0%), 119 (49.2%), and 14 (5.8%), respectively. There was no significant difference in age among the three

Table 1. Sleep behavior and sleep quality for each shift (vs. morning shift: *; p = 0.0001, **; p < 0.0001. vs. afternoon shift: †; p < 0.05).

Shift (Work hours)	Bedtime	Time at waking	Total sleep period	Satisfaction with sleep n (%), age (mean ± SD)	
				Good	Poor
Morning shift (08:00–16:00)	22:42 ± 54 min	05:48 ± 30 min	7.2 ± 0.9 hrs	165 (65.0%) 49.7 ± 4.0	89 (35.0%) 50.4 ± 4.2
Afternoon shift (16:00–23:15)	01:48 ± 42 min	08:30 ± 78 min	6.7 ± 1.0 hrs*	173 (69.2%) 49.9 ± 4.3	77 (30.8%) 49.9 ± 3.7
Night shift (23:15–08:00)	Dependent upon sleep frequency (see Table 2)		6.5 ± 1.3 hrs**†	72 (28.9%) 49.6 ± 4.2	177 (71.1%) 50.1 ± 4.1

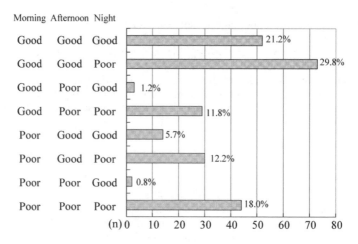

Figure 1. Satisfaction with sleep quality through all shifts.

groups. Total sleep time of the one-time, two-time, and three-time sleepers was 6.4 ± 1.4, 6.6 ± 1.2, and 7.3 ± 0.7 hours, respectively. Meanwhile, the proportion of good sleepers was 38.9, 19.5, and 7.7%, respectively. While the total sleep time was shorter, the proportion of the good sleep group was higher in order of increasing sleep frequency. On an average, the one-time night shift sleepers went to bed later than the other groups.

Table 3 shows the relationships between drinking habits, sleep duration, and sleep satisfaction in each shift. Sleep duration of drinkers was significantly longer than that of non-drinkers during the morning shift (7.3 ± 1.2 hours and 6.9 ± 0.7 hours, respectively, p < 0.05). However, these relationships were not observed during the afternoon or night shifts. Therefore, sleep satisfaction of drinkers tended to be better than that of non-drinkers irrespective of the shift group.

Proportions of drinkers among one-time sleepers, two-time sleepers, and three-time sleepers were 75.2%, 64.3%, and 50.0%, respectively with no significant differences. Therefore, the relationships between the proportion of good sleepers and drinking habits showed no definite correlation (Table 4.).

4 DISCUSSION

Several studies have reported that shift work was related to poor sleep quality (Doi Y, 2005). Although possessing a keen interest in studying the differences among the three different shift conditions, we were unable to assess the actual sleep behaviors of the shift workers within these

138

Table 2. Sleep behavior and sleep quality during the night shift (vs. the first bed time of the other two groups: [††]; $p < 0.001$, vs. the once group: [*]; $p < 0.05$, vs. other two groups: [§]; $p < 0.01$).

Sleep times	n (%) Age	Bedtime	Time at waking	Total sleep period	Satisfaction at the sleep (n (%), age)	
					Good	Poor
Once	109 (45.0%) 50.1 ± 4.2	11:36 ± 78 min[††]	18:00 ± 108 min	6.4 ± 1.4 hrs[††]	42 (38.9%[§]) 50.1 ± 4.0	66 (61.1%[§]) 50.1 ± 4.4
Twice	119 (49.2%) 50.0 ± 4.0	11:00 ± 60 min 18:30 ± 102 min	15:06 ± 96 min 20:54 ± 78 min	6.6 ± 1.2 hrs	23 (19.5%[§]) 50.7 ± 4.2	95 (80.5%[§]) 50.8 ± 4.0
Three times	14 (5.8%) 48.3 ± 2.8	10:36 ± 48 min 15:24 ± 90 min 19:24 ± 72 min	13:24 ± 60 min 17:48 ± 72 min 21:18 ± 66 min	7.3 ± 0.7* hrs*	1 (7.7%[§]) 48	7 (92.3%[§]) 48.0 ± 2.8

Table 3. Drinking habits, duration of sleep and sleep satisfaction (vs. Non-drinkers, *; $p < 0.05$).

	Drinking habits	
	Drinker	Non-drinker
Morning shift	7.3 ± 1.2 hrs*, 70%	6.9 ± 0.7 hrs, 57%
Afternoon shift	6.8 ± 1.4 hrs, 71%	6.6 ± 1.1 hrs, 66%
Night shift	6.6 ± 1.4 hrs, 32%	6.6 ± 1.3 hrs, 21%

Table 4. Drinking habits and sleep satisfaction during the night shift.

Number of sleep during the night shift	Drinking habits Yes	Proportion of good sleep	
		Drinkers	Non-drinkers
Once	75.2%	38.5%	38.5%
Twice	64.3%	27.0%	7.3%
Three times	50.0%	0%	16.7%

studies. In our study, we observed the natural sleep behaviors of experienced shift workers aged 40 years or older. Subjects were not instructed toward a favorable sleep-awake cycle. The shift system of the present study was a 4-crew 3-shift, which consisted of a 5-day slow clockwise rotation. This is the standard shift cycle of the Japanese steel industry. Czeisler *et al.* (1982) found that a clockwise rotation, rather than a counterclockwise rotation, was favored by workers and was consistent with the notion that delaying sleep times was easier than advancing them. Hakola and Harma (2001) reported that a rapidly rotating schedule was more rational since it minimized the time spent in a desynchronized state, while others argued (Sack, *et al.* 2007) that longer runs (more consecutive days) of shift work provided the opportunity to achieve a higher degree of synchronization. On the other hand, Kostreva *et al.* (2002) reported in their review that the best shift schedules adopted a slow, forward-shifting rotation pattern and rotated shifts after 2-week periods.

Drake *et al.* (2004) reported that the sleep period of day shift workers aged from 18 to 65 years was 6.8 ± 1.2 hours, and that of shift workers was 6.7 ± 1.5 hours. Ursin *et al.* (2005) reported that the sleep period of male non-shift workers aged from 40 to 45 years was 6.55 hours, and that of shift workers was 6.41 hours. Our results indicated that the average sleep period during all the shifts was 6.8 hours, and the sleep period during the night shift was 6.5 hours.

Doi (2005) reviewed 11 studies on disturbed sleep in occupational settings of Japanese shift workers. In this article, the author indicated that the prevalence of insomnia among shift workers was from 29 to 38%. In the present study, the prevalence of insomnia was 35.0%, 31.8%, and 71.1% during the morning shift, afternoon shift, and night shift, respectively. Workers who reported good sleep during all shifts were only 21.2%. On the other hand, 29.8% of workers slept poorly during the night shift, and 18.9% of workers slept poorly during all shifts. Thus, results from the present study demonstrate that many workers suffered from poor sleep and the actual prevalence could have been different depending on the manner of the question.

During the night shift, sleep behavior was divided into 3 patterns depending upon sleep times. The one-time sleepers exhibited a significantly shorter sleep time with a significant later bed time. However, prevalence of satisfaction with sleep quality was significantly higher than within the other two groups.

In a retrospective survey of police officers, taking a nap before night shift duty was associated with fewer accidents (Garbarino, et al. 2004). Schweitzer et al. (2006) demonstrated that taking a nap before the night shift, especially when combined with caffeine, improved alertness. In their study, nap start time was approximately 3 to 4 hours before shift start time, and participants were instructed to stay in bed for a minimum of 1 hour and a maximum of 2 hours. In our study, two-time sleepers during the night shift received 84 minutes of nap, approximately 3 hours before their night shift start time without any instruction. This advantage of taking a nap might suggest that two-time sleepers would exert good performance during their work time; however, our study suggests that the sleep satisfaction of this particular group was inferior in comparison to that of one-time sleepers.

Knauth et al. (1980) reported that night shift workers usually went to sleep 1 hour after their shift was terminated, with very little variation between individuals. However, the subjects in the present study exhibited various patterns of sleep behavior during the night shift. This variation may be attributed to the difference in the respective social climates of the night shift workers.

Among three-time sleepers, the sleep periods of the first sleep, second sleep, and third sleep were 168 minutes, 144 minutes, and 114 minutes, respectively. Each sleep period, excluding the first sleep period, was apparently different from the every 90 minutes as the natural sleep-awake cycle. They might have remained in bed all day with very little time spent out of bed. The total sleep time of the three-time sleepers was not different from the other two groups during the morning and afternoon shifts; thus, suggesting that the three-time sleepers were not originally insomniac. Among the 14 night shift three-time sleepers, only 1 worker reported good sleep, implying that this particular group should be instructed toward a more favorable and effective sleeping behavior.

Drinkers tended to report better sleep than non-drinkers; however, there was no significant statistical difference. There was no significant correlation between drinking habits and sleep quality between the two groups. In the present study, drinking habits during the night shift were not surveyed. Further studies are needed to elucidate these points.

5 CONCLUSION

A significant number of shift workers suffered from poor sleep. This study suggests that despite sleep difficulties associated with the night shift, a one-time sleep frequency in accordance with a later bedtime might improve the quality of sleep to some extent.

REFERENCES

Doi, Y. Minowa, M. Tango, T., Impact and correlates of poor sleep quality in Japanese white-collar employees., *Sleep*, 2003:26;467–471.

Marquié, JC and Foret, J., Sleep, age, and shiftwork experience., *Journal of Sleep Research*, 1999;8:297–304.

Drake, C.L., Roehrs, T., Richardson, G. *et al.*, Shift work sleep disorder: prevalence and consequences beyond that of symptomatic day workers., *Sleep*. 2004; 27:1453–62.

Doi, Y., An epidemiologic review on occupational sleep research among Japanese workers., *Industrial health*, 2005;43:3–10.

Czeisler, C.A., Moore-Ede, M.C., Coleman, R.H., Rotating shift work schedules that disrupt sleep are improved by applying circadian principles., *Science*, 1982;217:460–3.

Hakola, T. and Harma, M., Evaluation of a fast forward rotating schedule in the steel industry with a special focus on ageing and sleep., *Journal of Human Ergology*, 2001;30:315–9.

Sack, M.D., Auckley, D., Auger, R.R., Circadian rhythm disorders: Part 1, basic principles, shift work and jet lag disorders. An American academy of sleep medicine review., *Sleep*, 2007;30:1460–1483.

Kostreva, M., McNelis, E., Clemens, E., Using a circadian rhythms model to evaluate shift schedules., *Ergonomics*, 2002;45:739–63.

Ursin, R., Bjorvatn, B., Holsten, F., Sleep duration, subjective sleep need, and sleep habits of 40- to 45-year-olds in the Hordaland Health Study., *Sleep*, 2005;28:1260–1269.

Garbarino, S., Mascialino, B., Penco, M.A. *et al.*, Professional Shift-Work Drivers who Adopt Prophylactic Naps can Reduce the Risk of Car Accidents During Night Work., *Sleep*, 2004;27:1295–1302.

Schweitzer, P.K., Randazzo, A.C., Stone, K. *et al.*, Laboratory and field studies of naps and caffeine as practical countermeasures for sleep-wake problems associated with night work., *Sleep*, 2006;29:39–50.

Knauth, P., Landau, K., Dröge, C. *et al.*, Duration of sleep depending on the type of shift work., *International Archives of Occupational Environmental Health*, 1980;46:167–77.

Chapter 8   Social and organizational ergonomics

Ergonomic Trends from the East – Kumashiro (ed)

The role of culture for successful ergonomics implementation

I.D.P. Sutjana

Departemen of Physiology School of Medicine Udayana University Denpasar Bali

ABSTRACT: The development of knowledge and technology many equipments and method had been produced. The use of equipment or method were mean to help it own for ease to work, reduced the workload, reduced fatigue, increase the productivity and quality of products, increased of benefit, increased of well being, etc. In global period when the accelerated of development information and communication technology, distance to be shorten and the increased of traveling of human. Many equipments or method was carried out or exported from advanced industrial countries to other countries especially to industrial developing countries which are difference in many aspects from the produced countries. Even though the design of equipments and method are base on technology and economics considerations, so many equipments or methods not yet match to the user need or conditions such as the user education, environments, social value, culture value etc. As a result the employees quickly feel fatigues, work related musculoskeletal disorders, work related diseases, accidents, reduced of productivity, production high cost, and at the last reduce of well being. To overcome that problem the design of equipments or method should be meet to the user conditions and value, especially culture consideration. Because culture as a behavior in the human life, they have the value and as a motivated for the human daily activity.

Keywords: ergonomics, equipment and method design, culture.

1   INTRODUCTION

Higher productivity and quality of products, light workload, reduced fatigues, safety, health and well being were hopeless of the workers. To reach that purposes many actions had been done such as using of new equipment or new method, use of personal protective devices, training, etc. In the advance industrial countries easy to do that, but for industrial developing countries there are so difficult, because lower of educations, poverty, lower standard of life, lower of working conditions, etc.

In industrial developing countries the equipment or method are imported from the industrial advance countries, while the equipments or method design dominated by economics and technical considerations, with less consideration to the countries destination condition such as culture. As a result new equipment or method that are implementation to industrial developing countries didn't satisfied the user, because expected to enhance the productivity, safety and health and well-being but otherwise produced the negative impact such as working accident, work related diseases, reduced of efficiency, reduced of productivity, so reduced of wellbeing (Kaplan, 2004).

To overcome of the problems since design of equipment or method which are implementation to other countries especially to the industrial developing countries all aspect included culture value of the destination people should be considered. Holistic approach should be better (Manuaba, 2004)

In this paper should be information the role of culture during the implementation of ergonomic at the rural area in particular Bali rural.

2   DIVERSITY IN THE WORLD

So many countries, so many customs, that is call by the proverb. It is true because so many people in the world, with difference in many aspects such as ethnic, nations, educations, language, pleasure, taboo, body measurement, occupation, habitual, value, custom etc (Chapanis, 2004, Kaplan, 2004). But all of them had the same need such as safety, health, comfort, higher productivity and higher quality of products, benefit and well being.

To overcome their need, they use equipment or method in daily of life. Because very difference of each nation or ethnic in the world, so difference too the design of equipment or method. In case the equipment or method should be use by other people with difference ethnic or costume, their conditions should be considered in the design of equipment or method. Ideally each person design their equipment or method fit to their condition included the community culture, or the design of equipment or method should be fit for mostly of the community.

3 GLOBALISATION AND CULTURE

The human use equipment or method which means to help it own for ease to work, to be stronger, to be longer of reach, to be higher, more precise in work, enhance the productivity, enhance the quality of products, reduced of reject, etc.

During the global period there are increased of free trade, past to nation or countries bridges, shorten of distance, quickly of traveler, exchange of people, etc. The increased of global market not automatically followed by global design, so the equipment or method that are fit for local user not fit for the other user. So many variation of the nation should be considered in design. The nation variation not only physical aspect but included the social aspect or destination of culture (Moray, 2004; Rose, 2004). In global period the think about the destination of culture as the important think for user friendly design to produced user friendly equipment or methods.

4 THE IMPLEMENTATION OF ERGONOMICS IN THE RURAL AREA

The Balinese people especially in the rural area mostly Hindu, so all of people daily activities was dominated by Hindu's culture. For successful in the implementation of ergonomic principle in the rural area of Bali, culture consideration is a must. But if the people mostly Moslem the value of Moslem must be consideration in the implementation of ergonomics.

Some ergonomics objectives which are implemented in the rural area where culture consideration are as follows:

4.1 Design of time schedule

Hotel or other factory which are operation for 24 hours, the employee must be worked for 24 hours too. Because the human can't work fluently for 24 hours, that is divided into 3 shift work each 8 hours. For multi religious employee such as in hotel operation there is Hindu, Moslem, Christian or Buda, which difference holy day. So the regulation of time schedule must be fitted to that holy day. If the Hindu offering on their holy day, such as on full moon day, "Galungan" day, "Nyepi" or silent day, etc. there are off from job, other religious employees take over the job, and vice versa. On the "nyepi" day the people forbidden traveling or go to outside of their houses or work, from early morning to mid night. During the holy day the Balinese Hindu doing many ceremony, so that the employer must be understand for this condition and holy day must be considered in designing of working schedule. But for any reason the employee must be duty during their holy day, they have two days for changes. For Moslem employee near Idul Fitri days must be remember there are need off from job for frying together their family. It need around one week. So in Bali the implementation of time schedule didn't rigid but matching to the culture or local condition. With this methods the hotel or factory can operate and the employees can offering without disturb. With this method all of employee feel happy or this is user friendly method.

4.2 Resting time

For many factories especially for Indonesian employee time for lunch design is from 12.00–13.00, because for Indonesian food only one step. On Friday rest time should be longer from 11.00–13.00, because that time for frying in Mosque for Moslem. But for multinational employee when the employee were combined from Indonesian and other country especially from western country lunch time should be longer from 13.00–15.00, because their habit take lunch with step by step, and talk while lunch, so need longer time.

4.3   Frying place

For Moslem employee fraying place must be built in every working place, airport, office, hotel etc., because their always frying five time a day at the same time although on traveling. At frying time the Moslem people go to the frying place. But for Hindu employee at east working place must be built too fraying place for frying to the God who are look after the working place.

4.4   House developed design

For comfortable of the inhabitant all aspect should be considered included local culture. For Hindu people east and north are the holy place, so at that place should be for fraying place, but don't placed the dirty think such as water closed, sewage disposal. Design of room must be think that the bad room should be placed that the head on the east or north direction. But for Moslem people west direction is holy place, so that the bad room must be design that the head should be on the west direction (Sutjana, 2008).

4.5   Using masker

For polluted factory such as can factory at east Java, the employee should be use maskers for prevention of lung diseases. To selection of the maskers don't think about effective for prevention of dust only, but should be accepted by the employee culture. If the masker didn't fit to the employee culture they should be rejected to use it although it is more effective to prevention of dust. The example plastic masker, if the employee use they look like the pig mouth. For Moslem employee this masker didn't fit their culture because pig was taboo/forbidden for Moslem. So that the employee rejected use that maskers.

4.6   Culture approach in taxi operation

For safety during traveler the Balinese believe for frying for god any where and any time. So for Balinese taxi driver use this believed for safety during driver. Every day their taxi give take "sajen" before drive, and every 210 days on "tumpek landep day" their taxi give "sajen". (Sutjana, 2005). From management experiences the operation cost of taxi in Bali was lower and life span of taxi longer than other areas. So use culture approach for taxi operation can enhance the benefit of taxi business.

4.7   The design of "bade" for carrying the dead body to cemetery

"Bade" was use for carrying the dead body of the Balinese Hindu from their house to the cemetery. The design of "bade" according to body part measurement of the dead body such as the body height, body bread or wide. But some part of the "bade" design use the body part of general community such as shoulder bread, handle height, etc. But recently the "bade" height design according to electric wire height, (les than 5 meters). For special case such as the "bade" of the king Ubud, their "bade" height around 25 meters with weight around 11 tons. To carry out that "bade" need more than 100 persons, and very difficult to do that because the wide of road not more than 8 meters (Sutjana, 2008). To solve that problem the "bade" were carried by the community members changes for a given distant. So the work load was not too heavy, but the culture still accepted.

4.8   Culture approach in handling problem

During bomb blast in Bali 6 years ago, more Balinese people or foreigner was died and injuries. Many sympathy reactions come from other countries. The Balinese reaction from the bomb blast and its impact not by demonstration or other destroyer action but use culture approach with doing frying at the ground zero of bomb blast, may God blest for the kill or injuries people and the bomber should be quickly for arrested.

4.9 *Taboo*

In Bali for land preparation of rice field in many area still using plough. Using plough the farmers believe not good plough on Thursday so that they forbidden or taboo work on Thursday. So if doing research at the rice field using plough, must be remember to avoid design schedule on Thursday (Sutjana, 2008).

5 CONCLUSSION

For daily activity use of equipment or method make the human ease to work, lighter of work load, increased of productivity and enhance the quality of products. But use of equipment or method if didn't fit to the human condition produce accident, work related diseases, increased of rejected products, higher productivity cost. So that the design of equipment or method should be fit to the user condition, not only physical but include the culture aspect. To enhance user friendly equipment or method culture aspect should be considered, or holistic approach is a must in every design equipment or method.

REFERENCES

Chapanis, A. 2004. National and cultural variables in ergonomics. Cultural ergonomics. Edited by Michael Kaplan. Elsevier. Amsterdam 1–30.

Djelantik, A.A.M. 2005. Postcoma Reflections. A Solo Art Exhibition of Water-colour Paintings. Danes Art Veranda.

Kaplan, M. 2004. Introduction: Adding a cultural dimension to human factors. Cultural ergonomics. Edited by Michael Kaplan. Elsevier. Amsterdam. xi

Manuaba, A. 2004. Holistic Ergonomics approach is a Must in Automation to Attain Humane, Competitive, Sustain Work Processes and Products. Denpasar, PhD ergonomics Program, School of Medicince Udayana University.

Moray, N. 2004. Culture, Context, and Performance. Cultural ergonomics. Edited by Michael Kaplan. Elsevier. Amsterdam. 31–60.

Rose, K. 2004. The Development of culture-orientated human machine systems:specification, analysis, and intergration of relevant intercultural variables. Cultural ergonomics. Edited by Michael Kaplan. Elsevier. Amsterdam. 61–104.

Sutjana, DP. 2005. Cultural Approach to Reduced Traffic Accident. 21st Annual Meeting and Conference of the Asia-Pacific Occupational Safety and Health Organization. Denpasar, 5–8 September 2005.

Sutjana, D.P. 2008. Culture ergonomics and it implementation. Culture Ergonomics Seminar, 6 August 2008. Medan North Sumatra.

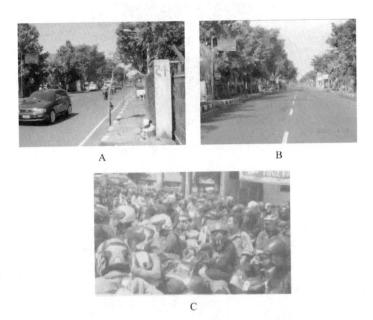

A   B

C

Figure 1. The situation of many Strait in Bali during "Nyepi" day. A. During daily of life more car and any body look past a ways. B. During "nyepi" day, no car, nobody walk. C. The Moslem employee off from job and going to their house ("mudik").

A   B

Figure 2. The employee needs time for praying at the work place or during traveling. A. The praying place at airport for Moslem. B. The driver and passenger must be stop a moment at the temple along the traveling.

A   B

Figure 3. The design of building. A. The height of building not more than 15 m or coconut height. B. The design of bad room should be north or east for Hindu, and west for Moslem.

149

Figure 4.   The employee whose use plastic masker look like pig mouth.

A   B

Figure 5.   The car usually gives "sajen". A. Every 210 days during "tumpek landep" day the owner give "sajen". B. The taxi every day give "sajen" during traveling.

A   B

Figure 6.   The "bade" design for carrying the dad body to the cemetery. A. The height of "bade" lower than electric wire. B. For special person the " bade" height around 25 m with weight 11 ton.

Figure 7.   The family of died people fraying at the ground zero for frying.

Figure 8.   The farmer floughing at the rice field.

Ergonomic Trends from the East – Kumashiro (ed)
© 2010 Taylor & Francis Group, London, ISBN 978-0-415-88178-4

Application of total ergonomic approach model increasing participation and empowerment of ironwork workers at Tabanan Bali

I Putu Gede Adiatmika, Adnyana Manuaba, Nyoman Adiputra & Dewa Putu Sutjana
Ergonomics Postgraduate Programme, Udayana University Denpasar, Bali, Indonesia

1 BACKGROUNDS

Several small-scale industries (SME's) have been able to survive due to global change among others those producing iron or wood. All the products are exported and created new job opportunities for the people (Anonim, 2004). Tabanan is one of the regencies of Bali that has several industries of iron handicraft (Andewi, 1999; Murtiana, 2004). Increase of orders poses a challenge to the workers in regard with fulfilling the number, quality standard, and time schedule. Deadlines for production sometimes cannot be fulfilled by the workers, hence force them to work overtime. To reduce overtime, an ergonomics intervention should be done in order to increase productivity.

The nature of ironwork is monotonous and repetitive, and found some problems such as musculoskeletal complaints, fatigue, low of productivity index and low of income. At the end working overtime was not increased productivity, but the worker get more stress as a whole (Okada, *et al*, 2005). This result indicates that improvement of the working condition is highly recommended in the sense that it should not apply overtime work. But, every improvement should be done wisely and holistically. This approach should be done for sustainability and could minimize the problem in the future.

It's needed special effort based on working condition of SME's especially in Bali. This problem basically came from many aspects such as infestations, management, environment and human resources. Most of them were lack of many resources particularly in human resources. The workers were come from Balinese and without any special skill such as house wife, teenagers, special workers, etc. Most of them were graduated from elementary school, junior and senior high school. They were work daily due to order from their supervisors or the employers without any comment to their conditions. Any improvements that were introduced to them were implemented without any comment or questions. But, sometimes the new improvement was not fully understood due to education of workers conditions. They need such a simple method in ergonomics intervention that is easy to be adopted and implemented.

One of the efforts to increase productivity can be made by means of ergonomic approach. The main objective of ergonomic intervention is to place the employee as the subject of production and to make them work in safe, healthy, comfortable, effective and efficient working condition (Abeysekera, 2002; Manuaba, 2001; 2002; 2003; McLeod, 2003; Phillips, 2000). The effort to increase productivity through ergonomic approach has been done by several researchers in various industries in Java and Bali. The core objective of all the efforts was to decrease stress of employees hence reducing workload, musculoskeletal complaints and fatigue and increase productivity (Adiputra, 2003; Grandjean & Koemer, 2000; Sutajaya, 2000).

However, all the efforts made so far were based on one perspective only that can be called as Standard Model (SM). This model was not encouraged and participation of all stakeholders included the workers. Meanwhile a comprehensive ergonomic way out seems to be critically needed that should be dealt with total ergonomic approach. This kind of approach is needed to minimize arrogance and bad impacts in the future (Manuaba, 2004; 2005; 2006). Total Ergonomics Approach Model (TEAM) is one of method that encourages for their participation of the workers and empowerment them.

Total ergonomic approach focuses on the application of systemic, holistic, interdisciplinary and participatory concept (SHIP approach) with concomitant analyze on the working condition. The designing of the intervention should be based on appropriate technology. Total ergonomic approach encourages and asked for participation of all stakeholders to implement new and appropriate technology. Implementation of total ergonomic approach is approved by many researchers because the method used can create a working condition that is more humane, competitive and sustainable in order to increase productivity and income of employees and employers (Shahnavaz, 2000; Manuaba, 2003; Manuaba, 2004). This model that is called Total Ergonomic Approach Model (TEAM) was designed through comprehensive and holistic process through introduction speech and small group discussion guided by the facilitator (Manuaba, 2003; 2006). Based on the above result, intervention by total ergonomic approach model should similarly be implemented at ironwork industry. That implementation is important to analyze the problem of working condition and to find the solutions.

The objective of this study was to compare the implementation of Standard Model (SM) and Total Ergonomic Approach Model (TEAM) in analyzing the existing problem of working condition at ironwork at Kediri Tabanan Bali.

2 METHOD

The research was done at an ironwork painting industry at Kediri Tabanan during the period of May 2006 – May 2007. Study design was non experimental which was done by cross sectional of data collection. The study population was the employee of an ironwork painting at Tabanan regency. Twenty employees were chosen randomly as sample or respondent.

The working condition was assessed using both the Standard Model (SM) and Total Ergonomic Approach Model (TEAM). Firstly, the working condition assessment was done using SM by questionnaire, observation and interview. After washing out period for one week the working condition assessment was performed again using TEAM which consists of introductory speech and small group discussion. In small group discussion all respondents were involved. Some related stakeholders were encouraged within this discussion based on checklist. The process and result of both models were compared and analyzed descriptively.

3 RESULT AND DISCUSSION

3.1 Sample characteristics

All of the subjects are healthy as indicated by the result of physical examination, blood pressure and resting pulse rate measurements. Nutritional states are normal according to body mass index. Educational background were varies from elementary school to senior high school. They work as ironworkers as a major work, part times or transfer from other job. Age of subject varies from 17 to 50 years old considered as golden productive age.

3.2 Problem investigations process

Problem identification process was done separately between SM and TEAM. The process of SM was started with anthropometric assessment, list of questions using questionnaire, observation and interview. The observers and interviewers conducted the investigations process and asked the respondent.

In the TEAM process some stakeholders such as expert, students, workers, customers, and employers are involved. Assessment was started by introductory speech by an expert about simple ergonomic that can be implemented in SME's. Speech process was continued by Small Group Discussion (SGD) that was guided by the facilitators. During SGD, the process of identification was done step by step, started by problem identification, problem priority, positive sentence, SWOT analysis, work plan and finally make an action plan. Within process, all stakeholders were encouraged to find out any problem in SME's through brainstorming process (Figure 1).

| a. Standard Model | b. Total Ergonomic Approach Model |

Figure 1. Process of problem analyzing and its solutions using Standard Model (SM) and Total Ergonomic Approach Model (TEAM).

Table 1. Comparison of Problem analyze between Standard Model (SM) and Total Ergonomic Approach Model (TEAM).

No	Aspect	SM	TEAM
1.	Problem identify	10	49
2.	Priority of problem		
	– Urgent	1	18
	– Essential	0	12
	– Important	0	11
3.	Positive sentence	0	18
4.	SWOT analyze		
	– Strength	0	5
	– Weakness	0	4
	– Opportunity	0	6
	– Threat	0	4
5.	Work plan	0	9
6.	Action Plan	1	5

Base on the process of problems investigation, SM was able to collect the problem from the respondents only and done by the investigators through questionnaire, observation and interview; meanwhile TEAM could identify, analyze and collect the problem from the workers and other stakeholders and done together by all stakeholders through small group discussion. The TEAM could encourages the workers through participation to analyze the problem holistically based on checklist and finally can do this process by themselves in the future. That is not the case in using the SM.

3.3 *Comparison between Standard Model and Total Ergonomic Approach Model*

The objective of ergonomic intervention on SME's using SM and TEAM are to analyze the problem of working condition and to find out the solutions. Intervention of SM in analyzing the problem was done by one way process between the investigators and the workers. The workers just answer and give any data that was asked by the investigators. Meanwhile TEAM intervention was done holistically and encourages all stakeholders including the workers. In the process of investigation the workers were posed through brainstorming to tell all of their problems of working condition one by one.

Problem investigation using both SM and TEAM looked different of result. The SM just collected lesser problem and action plan, meanwhile the TEAM could collect more problems and more holistic (Table 1). Comparison of both models showed that TEAM seems to be more simply and easy to implement. The model can be used for problem solving process for the problems identified. TEAM could find 49 problems compared with the SM just 10 problems only. It was mean that TEAM could investigate through many perspective related to attendance's horizon.

The TEAM intervention involved implementation of systemic, holistic, interdisciplinary and participatory (SHIP) approached and used of appropriate technology (Manuaba, 2004; 2005;

2006). The problems that were collected such as: too low working station, cumulative fatigue, musculoskeletal complaints and lack of budget. TEAM can collect the data such as: identified the problem, positive sentence, SWOT analysis, improvement and implementation of appropriate technology compare to problem of SM just money, too low work station. It is means that TEAM could analyze comprehensively and empowering the workers how to think and act holistically through participation.

4 CONCLUSION

It is concluded that TEAM could increase participation and empowerment the workers in analyzing the problem and find out the suitable improvement. Application of TEAM could be done by the workers themselves and could maintain sustainability in improvement process. It is suggested to apply TEAM in many other SME's as one simple model in ergonomics intervention.

REFERENCES

Abeysekera, J., 2002. Ergonomics and Industrially Developing Countries. *Journal Ergonomi Indonesia.* June;3(1):3–13.

Adiputra, N. 2003. Kapasitas Kerja Fisik Orang Bali. *Majalah Kedokteran Udayana (Udayana Medical Journal).* Oct;34(120,4):108–110.

Andewi, P.J., 1999. Perbaikan sikap kerja dengan memakai kursi dan meja kerja yang sesuai dengan data antropometri pekerja dapat meningkatkan produktivitas kerja dan mengurangi gangguan sistem musculoskeletal pekerja perusahaan MI Kediri Tabanan. Tesis. Denpasar: Program Pascasarjana Universitas Udayana.

Anonim, 2004. *Monografi Departemen Perindustrian dan Perdagangan.* Available at www.dprin.go.Id/Ind/Statistik. Accessed November 20, 2004.

Grandjean, E., Kroemer, 2000. *Fitting the Task to the Human.* A textbook of Occupational Ergonomics. 5th edition. Piladelphia: Taylor & Francis.

Manuaba, A. 2000. Participatory Ergonomics Improvement at the Workplace. *Jurnal Ergonomi Indonesia.* June;1(1):6–10.

Manuaba, A. 2001. Integrated ergonomics approach toward designing night and shift work in developing countries based on experiences in Bali, Indonesia. *Journal of Human Ergology.* Dec; 30(1–2):179–83.

Manuaba, A. 2002. Integrated Ergonomics SHIP Approach is A Must in Designing Night and Shift Work in Developing Countries, with special reference to Bali, Indonesia. Presented at Night and Shift Work Symposium, Japan.

Manuaba, A. 2003. Holistic design is a must to attain sustainable product, The National Seminar on Product Design and Development Industrial Engineering UK Maranatha, Bandung, 4–5 July.

Manuaba, A., 2004. Designing Humane, Competitive and Sustainable Work System and Product. Keynotes Address. National Seminar on Ergonomics, UPN Yogyakarta, Indonesia, 27 March.

Manuaba, A., 2005. Pendekatan Holistik dalam Aplikasi Ergonomi. *Sosial & Humaniora.* Oct;01(01):1–13.

Manuaba, A. 2006. Macro ergonomics approach on work organizations with special reference to the utilization of Total ergonomic SHIP approach to obtain humane, competitive and sustainable work system and products. Proceeding Seminar Nasional Ergonomi. Surabaya, 21–22 November 2006.

McLeod, I.S., 2003. Real-world effectiveness of Ergonomics methods. Abstrak. *Applied Ergonomics.* Sep;34(5):465–77.

Murtiana, I N, 2004. Wadah-wadah Lilin Kreasinya Jadi Rebutan. Tokoh. 29 Ags – 4 Sept 2004, p. 5.

Okada, N., N. Ishii, M. Nakata, S. Nakayama, 2005. Occupational stress among Japanese Emergency Medical Technicians: Hyogo Prefecture. *Prehospital Disaster Medicine.*20(2):115–121.

Phillips, C.A., 2000. *Human Factors Engineering.* Amerika : John Wiley & Sons, Inc.

Shahnavaz, H., 2000. Role of ergonomics in the transfer of technology to industrially developing countries. *Ergonomics.* July;43(7):903–907.

Sutajaya, 2000. Increasing Productivity of Wood Carving in Peliatan Ubud Gianyar. *Jurnal Ergonomi Indonesia.* June;1(1):15–18.

Ergonomic Trends from the East – Kumashiro (ed)
© 2010 Taylor & Francis Group, London, ISBN 978-0-415-88178-4

Retaining older workers: What strategies do organisations need and when should they be implemented?

Jodi Oakman
Centre for Ergonomics and Human Factors, La Trobe University, Bundoora, Vic, Australia

Yvonne Wells
Lincoln Centre for Research on Ageing, La Trobe University, Bundoora, Vic, Australia

ABSTRACT: An ageing population is creating the need for organisations to change their current attitudes to older workers, and the development of successful organisational strategies to retain older workers will be essential. To support this objective, this study examines the predictors of retirement timing for older workers. Regression analyses of time to retirement and intention to retire soon indicated that different features of the workplace impact on retirement intentions, depending on the time frame. This provides an insight into the changing requirements of employees as they age. Workplace policies need to be flexible so that these requirements can be met, thus enhancing retention rates of older workers.

1 INTRODUCTION

The population is ageing, entailing new challenges and demands that will require a change in mindset at both societal and organisational levels, so that older workers are seen as a valuable resource. A labour shortfall is an inevitable result of this demographic change. In order to operate at full capacity, organisations will need firstly to attract and then to retain older workers (Access Economics Pty Ltd, 2001; Auer & Fortuny, 2000).

Workers' decisions about when to retire are complex and formulated over many years (Karp, 1989). In order to encourage workers to remain at work for longer, an understanding of work and its meaning to mature age employees is necessary. Research evidence suggests two broad groups of factors influencing intention to retire: 'push' and 'pull' factors (Kohli & Rein, 1991). Push factors move individuals towards early retirement; these factors are typically negative in their effects, and include poor health, organisational factors, and work fatigue (Shultz et al.; 1998). In contrast, pull factors are usually positive, tending to increase a person's interest in early retirement; such factors include outside work interests, partner's retirement status, and caring responsibilities. Push and pull factors interact in complex ways to influence a person's intention to retire (Shultz et al., 1998).

An individual's relationship with work changes as they age, but there is little detailed exploration of this in the literature (Griffiths, 2007; Rhodes, 1983). Griffiths (2007) argued that the current literature on work has been based on 'age-free' models and therefore does not account for any changes in requirements from work as we age. She stated that current models of the way in which we attribute meaning and value to work fail to address many such issues, and that there is a need for new models that accurately encapsulate the factors predicting retirement and retirement intentions. Such models are necessary to support policy developments to improve the retention of mature-age workers in the workplace. There is little consensus concerning the strongest predictors of older workers' retirement intentions. In order to develop successful retention strategies for older workers, better understanding of these predictors is needed – particularly of those with the potential to be influenced by organisational policy.

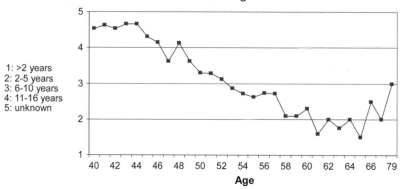

Figure 1.   Intended timing of retirement and age.

2   METHODS

2.1   *Participants*

La Trobe University was commissioned by a large government organisation to assist with the development of strategies aimed at retaining workers who might otherwise be considering retirement. A questionnaire was developed and distributed via the organisation's intranet and responses were received from 332 employees (47.4% response rate). The mean age of the survey respondents was 42.1 years, in comparison to the organisational average of 43.5 years.

2.2   *Survey*

The survey comprised several well established measures: the Work Ability Index (Ilmarinen, 2001), the Copenhagen Psychosocial Questionnaire (COPSOQ) (Kristensen et al., 2005) and the General Health Questionnaire (GHQ12) (Goldberg & Williams, 1988). The WAI was shortened at the request of the employer organisation, and hence is referred to here as the SWAI (Shortened Work Ability Index). Retirement intentions were measured through a single item: "what is your anticipated timing of retirement?".

2.3   *Analysis strategy*

Regression analysis was conducted using SPSS. Independent variables entered into the equation were age, gender, length of service, marital status, dependents, job satisfaction, job demands, social cohesion, job control, SWAI and GHQ. Multiple regression analysis was undertaken for the continuous dependent variable intended timing of retirement. For the dichotomous outcome variable intention to retire soon (within the next five years) logistic regression was undertaken to develop a predictive model.

3   RESULTS

Figure 1 shows the relationship between age and intended timing of retirement (mean), indicating that in general, older employees intend to retire sooner than younger ones.

 The results from the linear multiple regression are presented in Table 1. The overall model significantly predicted intended timing of retirement: $F_{(11, 294)} = 46.91$, $p < .001$. R^2 for the model was .64 and adjusted R^2 was .62. Older workers are likely to intend to retire sooner. Other socio demographic factors associated with a longer anticipated time to retirement were being male, having a shorter length of service, being married or partnered, and having dependents. Of the psychosocial factors, only high job demands remained a significant predictor of a delayed intention

Table 1. Multiple regression showing significant predictors of Intended timing of retirement.

	B	S.E.	Beta	t	Sig.	sri^2
Age	−.12	.01	−.64	−15.21	.00	−.53
Gender	.20	.10	.08	2.02	.04	.07
Length of service	−.04	.01	−.21	−5.56	.00	−.20
Partner	−.35	.12	−.12	−3.03	.00	−.11
Dependents	.20	.10	.08	2.10	.04	.07
Job Demands	.11	.05	.09	2.10	.04	.07
SWAI	.04	.01	.15	3.46	.00	.12
GHQ	.02	.01	.09	2.24	.03	.08
(Constant)	7.77	.70		11.13	.00	

Table 2. Logistic Regression showing significant predictors of intention to retire soon.

Variables	B	S.E.	Wald	Sig.	Odds Ratio	95% C.I.	
Age	.39	.06	45.68	.00	**1.48**	1.32	−1.66
Length of Service	.11	.03	16.30	.00	**1.12**	1.06	−1.18
SWAI	−.14	.05	6.17	.01	**.87**	.78	−0.97
Constant	−18.84	3.95	22.75	.00	.00		

to retire. Workers with high work ability and low mental health (high GHQ scores) were also more likely to intend to retire later.

The results from the multiple logistic regression are presented in Table 2. Of all the predictors entered into the regression equation (as listed in the analysis strategy) only age, length of service and SWAI remained as significant independent predictors of intention to retire soon. For every increase in age of one year, older workers were 48% more likely to retire within the next five years. An increase in length of service of one year meant that workers were 12% more like to retire soon, and for every decrease in work ability rating of one point, workers were 13% more likely to intend to retire within the next five years.

4 DISCUSSION

With a shrinking labour market the retention of employees is gaining greater importance for organisations. There is increasing interest in the psychosocial work environment as organizations strive to maximize the capacity of their workers. This study identified that the predictors which impact an individual's retirement timing changed as they aged. Firstly, older employees were more likely to retire than younger ones, but if an employee remained at work until they were approximately 63, they were less likely to intend to retire within the next five years. It also appeared that at around 55 years of age a change occurred in the intended timing of retirement, indicating that retention strategies should target the attitudes of younger people. These findings are supported by others (Encel, 2008; OECD, 2007) who have reported strong expectations of early retirement from both employees and organisations. To successfully retain older workers will entail changing these attitudes. A more progressive mindset is needed so that individuals work for as long as they are able rather than to a specified age.

Socio demographic factors such as age, gender, having a partner, and having children are much more important when considering time to retirement in the longer term than when it is imminent within the next five years. High job demands – which may indicate more challenging work – are associated with later intended retirement. However, as individuals move closer to their intended timing of retirement, the impact of job demands is reduced. Individuals who are within five years of their intended retirement age are also influenced by their self rated work ability. This last finding is important to organisations, because it implies that strategies to ensure that work is well within older workers' capacities could significantly enhance retention of this group. The design of a high quality work environment does not necessarily favour *only* older workers; good work practices provide a better work environment for all employees (Munk, 2003).

5 CONCLUSIONS

Organisations can influence older workers' retirement intentions. Ensuring that workers can maintain a balance between their personal circumstances and work should enhance the retention of older employees. The most significant impact on the retention of older workers is the currently held perception that once a certain age is reached it is time to retire. Very substantial shifts in attitudes are needed from employees and organisations to successfully enhance retention rates of older workers.

REFERENCES

Access Economics Pty Ltd. 2001. *Population ageing and the economy.* Canberra: Commonwealth Department of Health and Aged Care.

Auer, P., & Fortuny, M. 2000. *Ageing of the labour force in OECD countries: Economic and social consequences*: International Labour Office. Geneva.

Encel, S. 2008. Looking forward to working longer in Australia. In P. Taylor (Ed.), *Ageing labour forces: Promises and prospects.* Bodmin, Cornwall: Edward Elgar Publishing, Inc.

Goldberg, D., & Williams, P. A. 1988. *User's guide to the general health questionnaire.* Windsor, England: National Foundation for Educational Research-NELSON.

Griffiths, A. 2007. Healthy work for older workers: Work design and management factors. In W. Lorretto, S. Vickerstaff & P. White (Eds.), *The future for older workers: New perspectives* (pp. 121–137). Bristol: Policy Press.

Ilmarinen, J. E. 2001. Ageing workers. *Occupational & Environmental Medicine, 58*(8), 546–552.

Karp, D. 1989. The social construction of retirement among professionals 50–60 years old. *The Gerontologist, 29*(6), 750–760.

Kohli, M., & Rein, M. 1991. The changing balance of work and retirement. In M. Kohli, M. Rein, A. Guillemard & H. van Gunsteren (Eds.), *Time for Retirement: Comparative Studies of Early Exit From the Labor Force* (pp. 1–35). Cambridge: Cambridge University Press.

Kristensen, T. S., Hannerz, H., Hogh, A., & Borg, V. 2005. The Copenhagen Psychosocial Questionnaire-a tool for the assessment and improvement of the psychosocial work environment. *Scandinavian Journal of Work, Environment & Health, 31*(6), 438–449.

Munk, K. 2003. The older worker: Everyone's future. *Journal of Occupational Health and Safety-Aust NZ 19*(5), 437–446.

OECD. 2007. *Ageing and employment polices: Australia.* Paris: OECD Publishing.

Rhodes, S. 1983. Age-related differences in work attitudes and behavior: A review and conceptual analysis. *Psychological Bulletin, 93*(2), 328–367.

Shultz, K. S., Morton, K. R., & Weckerle, J. 1998. The influence of push and pull factors on voluntary and involuntary early retirees' retirement decision and adjustment. *Journal of Vocational Behavior, 53*, 45–57.

Effect of goal orientation on the relationship between perceived competence and achievement motivation among high competitive level Japanese university track and field athletes

Yujiro Kawata
Juntendo University Graduate School of Health and Sports Science, Chiba, Japan

Yasuyuki Yamada
Juntendo University of Health and Sports Science, Chiba, Japan

Miyuki Sugiura
Juntendo University Graduate School of Health and Sports Science, Chiba, Japan

Sumio Tanaka, Motoki Mizuno & Masataka Hirosawa
Juntendo University Graduate School of Health and Sports Science, Chiba, Japan
Juntendo University of Health and Sports Science, Chiba, Japan

1 INTRODUCTION

It is necessary for university track and field athletes to continuously practice hard in order to improve their skills and to achieve their goal, while facing the possibility of unexpected injuries, stagnation of their performance and frequent failure. So in order to maintain a high level of motivation to practice, it is essential for coaches to understand athletes' motivation to practice on an individual level, based on an understanding of differences among individuals, such as in regard to their personalities and dispositions. Thus, psychological factors which bring about differences in cognition, affect and behavior in the motivational process should be clarified in order to understand the necessary perspectives for intervention in the university track and field coaching scene. However, until now the psychological factors which bring about differences in cognition, affect and behavior in the motivational process have not been well examined.

In previous studies, in attempting to understand what leads to young people to practice hard in sports domain, scientists often tried to use the framework of achievement goal theory (for recent reviews, see Duda and Hall, 2001, Roberts, 2001). According to this theory, achievement goal orientation, which mainly consisted of task orientation and ego orientation, is a key determinant of motivation-related cognition, affect and behavior (Ames, 1992, Dweck, 1986, 1999, Maehr and Nicholls, 1980, Nicholls, 1984, 1989). Task orientation is seen as propensity to define success and construe one's competence in a way that targets self-focused improvement and mastery, while, ego orientation is seen as tendency to judge one's ability by comparison with others' performance and to regard success as the demonstration of superior ability.

Regarding the psychological factors that bring about differences to the motivational process, such as cognition, affect and behavior, it was also established that goal orientation had effects on the relationship between achievement motivation and achievement behavior among high competitive level university track and field athletes in Japan (Kawata et al., 2008). However, it is still unclear whether goal orientation has effects on the relationship between perceived competence and achievement motivation or not. The purpose of this study was to examine the question of whether goal orientation (task and ego orientation) has effects on the relationship between perceived competence and achievement motivation among high competitive level university track and field athletes in Japan (see Figure 1). If goal orientation (task and ego orientation) can be shown to have a role of making a difference in the motivational process, we will be able to gain important knowledge regarding the coaching of high competitive level university track and field athletes in Japan.

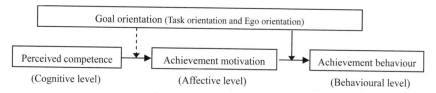

Figure 1. Hypothetical Motivational Model (The effect showed with the broken line was examined).

2 METHOD

2.1 *Participants*

A questionnaire investigation was carried out with 361 high competitive level university track and field athletes from 7 universities which were annually ranked above 8th in the national university track and field athletic competition in Japan. There were 168 (M = 19.4, SD = 1.02) valid respondents, including 141 male (M = 19.4, SD = 0.99) and 27 female (M = 19.6, SD = 0.99).

2.2 *Questionnaire*

The questionnaire was composed of three parts: *Dispositional goal orientation*: Task orientation and ego orientation were assessed using a Task and Ego Orientation in Sport Questionnaire (TEOSQ; Duda, 1989, 1992a). Respondents completed 13 items (7 task items, 6 ego items) with reference to the stem phrase, "I feel most successful when...". An example of a task orientation item is "I do my very best", while an example of ego orientation item is "I beat others". Response options were given on a 5-point Likert scale. The score of task orientation was the sum of the scores of task orientation items. The score of ego orientation was calculated using the same method. *Perceived competence*: Perceived competence was assessed using a perceived competence scale which is a subscale of Intrinsic Motivation Inventory (McAuley, 1991). Perceived competence included three items. Response options were given on a 7-point Likert scale. The score of perceived competence was the sum of the scores of perceived competence items. *Achievement motivation*: Achievement motivation was assessed using an achievement motivation scale (Higuchi, 1996). This scale consisted of five subscales; motive for skills improvement, motive for victory, motive for self improvement, motive for initiative, motive for personal relationship. Response options were given on a 4-point Likert scale. The scores of each subscale were the sum of the scores of each subscale of achievement motivation.

2.3 *Data analysis*

For the analysis, the subjects were classified into following 4 groups: high orientation (high task, high ego; N = 61, 36.3%), task orientation (high task, low ego N = 31, 18.5%), ego orientation (low task, high ego N = 29, 17.2%) and low orientation (low task, low ego N = 47, 28.0%) on the basis of z-scores of task and ego orientation. Two-way factorial ANOVA was chosen to examine the effects of goal orientation on perceived competence and on achievement motivation. Moreover, correlation analysis was carried out to examine the relationship between perceived competence and achievement motivation among the above-mentioned 4 groups.

3 RESULTS & DISCUSSION

In order to examine the effects of goal orientation (task and ego orientation) on perceived competence and on achievement motivation, two-way factorial ANOVA was carried out. As the results, there were significant main effects of task and ego orientation on perceived competence (see table 1). This result indicated that goal orientation (task and ego orientation) had effect on perceived competence. Additionally, there were significant main effects of task and ego orientation on achievement motivation (see table 1). This result indicated that goal orientation (task and ego orientation) had

Table 1. Effect of goal orientation on perceived competence and achievement motivation (ANOVA).

| | | Task orientation | Ego orientation | | | | | | F-value | | |
| | | | low | | high | | total | | | | |
			M	SD	M	SD	M	SD	Task	Ego	NT
Perceived competence		low	9.70	3.35	11.29	3.87	10.29	3.61	4.39*	7.71**	0.01
		high	8.45	4.29	10.10	3.21	9.54	3.67			
		total	9.21	3.77	10.47	3.45	9.88	3.65			
Achievement motivation	personal relationship	low	21.92	4.02	23.85	3.17	22.61	3.83	13.65***	5.06*	1.30
		high	24.68	3.02	25.31	3.43	25.10	3.29			
		total	23.00	3.88	24.86	3.40	23.98	3.74			
	self improvement	low	23.30	4.36	24.96	3.31	23.92	4.06	6.08*	1.81	1.89
		high	25.65	3.44	25.63	3.51	25.63	3.47			
		total	24.25	4.16	25.43	3.45	24.88	3.83			
	skill improvement	low	17.92	4.11	18.81	3.50	18.24	3.90	9.83**	0.03	2.85
		high	20.76	2.72	19.67	3.49	20.02	3.28			
		total	18.99	3.88	19.40	3.49	19.21	3.68			
	initiative	low	15.40	3.84	16.38	2.86	15.74	3.53	0.26	5.95*	0.48
		high	15.29	3.49	17.07	3.33	16.47	3.47			
		total	15.35	3.68	16.86	3.19	16.14	3.51			
	victory	low	10.04	2.16	11.39	1.62	10.54	2.07	0.05	6.57*	2.04
		high	10.60	2.31	10.98	2.07	10.86	2.15			
		total	10.26	2.22	11.11	1.94	10.71	2.11			
	total score	low	88.52	12.96	95.72	11.13	91.06	12.75	9.32**	5.17*	2.46
		high	97.18	9.38	98.50	10.22	98.08	9.92			
		total	91.80	12.41	97.68	10.50	94.94	11.77			

*p < .05, **p < .01, ***p < .001

an effect on achievement motivation. Moreover, there were significant main effects of task and ego orientation on some subordinate factors of achievement motivation. Specifically, task orientation influenced on some factors regarding individual development based on internal criteria, while ego orientation influenced some factors regarding sense of superiority based on external criteria. This result indicated that task orientation and ego orientation play an independent role in achievement motivation. Some studies suggested that task orientation is preferable to ego orientation because task orientation is positively correlated to adaptive cognitive, affective and behavioural outcomes, while ego orientation has predicted maladaptive psychological outcomes (Duda, 2001, Nicholls, 1984, Duda and Nicholls, 1992). However, in this study, it was shown that ego orientation may have an important role in achievement motivation among high competitive level university track and field athletes.

In order to examine the relationship between perceived competence and achievement motivation among the above-mentioned 4 groups, correlation analysis (Pearson) was carried out. As the results, in each group, different patterns of significant correlations between perceived competence and some factors of achievement motivation were found (see Table 2). This consequence indicated that goal orientation (task and ego orientation) had effects on the relationship between perceived competence and achievement motivation. Specifically, perceived competence was positively and negatively related to some factors of achievement motivation depending on the combination of task and ego orientation. According to this result, we can understand that perceived competence may become an enhancing factor or a restraining factor, according to the degree of task and ego orientation.

From an applied perspective, the results of the present research confirm that goal orientation can be one of the psychological factors which bring about differences in motivation-related cognition

Table 2. Correlation analysis between perceived competence and achievement motivation in each group.

		Achievement Motivation					
		personal relationship	self improvement	skill improvement	initiative	total victory	score
High orientation		−0.04	0.26*	0.20*	0.38**	0.17	0.31*
Task orientation	perceived	−0.20*	0.07	0.20*	0.02	−0.01	0.04
Ego orientation	competence	0.07	−0.04	0.25*	0.19	−0.35*	0.11
Low orientation		−0.13	−0.17	0.01	0.27*	0.06	−0.01

*p < .05, **p < .01

and affect. Therefore it is suggested that coaches should consider goal orientation as one of the individual differences which has an effect on the motivational process when they approach athletes' motivation in the university track and field coaching scene. What coaches and managers can attempt to do is to understand individuals' achievement goal orientation and encourage athletes to practice hard according to their individual degree of achievement goal orientation.

A limitation of this study is that it considered only achievement goal orientation as a factor that brings about difference in the cognitive and affective motivational process. Environmental factor such as motivational climate, which is coaches' or team-mate's goal orientation, should be considered in order to more fully understand athletes' motivational process.

4 CONCLUSION

In this study, we examined whether goal orientation (task and ego orientation) has effects on the relationship between perceived competence and achievement motivation among Japanese high competitive level university track and field athletes. Our findings showed that goal orientation (task and ego orientation) had effects on perceived competence and on achievement motivation. In addition, goal orientation (task and ego orientation) had effects on the relationship between perceived competence and achievement motivation.

Therefore, we consider it valid to conclude that goal orientation (task and ego orientation) is one of the psychological factors that have effects on the relationship between perceived competence and achievement motivation among high competitive level university track and field athletes in Japan.

REFERENCES

Ames, C., 1992, Classrooms: Goals, structures and student motivation. Journal of Educational Psychology, 84, pp. 261–271.

Duda, J. L., 1989, The relationship between task and ego orientation and perceived purpose of sport among male and female high school athletes. Journal of Sports and Exercise Psychology, 11, pp. 318–335.

Duda, J. L., 1992, Motivation in sport settings: A goal perspectives approach. In Motivation in sports and exercise, edited by Roberts, G. C., (Champaign IL: Human Kinetics), pp. 57–59.

Duda, J. L and Nicholls, J. G., 1992, Dimension of achievement motivation in schoolwork and sport. Journal of Educational Psychology, 84, pp. 290–299.

Duda, J. L. and Hall, H., 2001, Achievement theory in sport: recent extension and future directions. In Handbook of sport psychology Vol. 2., edited by R. N. Singer, H. A. Hausenblas, and C. M. Janelle. (New York: Wiley), pp. 417–443.

Dweck, C. S., 1986, Motivational processes affecting learning. American Psychologist, 41, pp. 1040–1048.

Dweck, J. L., 1999, Self-theories: their role in motivation, personality and development, (Philadelphia: Psychology Press).

Higuchi, Y., 1996, The relationship between organizational elements and achievement motivation in sport team. The Japanese Journal of Experimental Social Psychology, 36, 1, pp. 42–55 (In Japanese).

Kawata, Y., Yamada, Y., Sugiura, M., Tanaka, S., Mizuno, M., Hirosawa, M., 2008, Relationship between achievement motivation and achievement behaviour among Japanese university athletes: an approach using goal orientation theory. News Letter of Human Ergology, 88, pp. 18–19 (In Japanese) .

Maehr, M. L., and Nicholls, J. G., 1980, Culture and achievement motivation: A second look, In Studies in Cross-cultural Psychology, edited by Warren, N., (New York: Academic Press), pp. 221–267.

McAuley, E., Wraith, S. and Duncan, T. E., 1991, Self-efficacy, perceptions of success, and intrinsic motivation for exercise. Journal of Applied Social Psychology, 21, pp. 139–155.

Nicholls, J. G., 1984, Achievement motivation: conceptions of ability, subjective experience, task choice and performance. Psychological Review, 91, pp. 328–346.

Nicholls, J. G., 1989, The competitive ethos and democratic education. (Cambridge, MA: Harvard University Press).

Roberts, G. C., 2001, Understanding the dynamics of motivation in physical activity: The influence of achievement goals on motivational processes, In Advances in Motivation in Sport and Exercise, edited by Roberts, G. C., (Champaign, IL: Human Kinetics), pp. 1–50.

Ergonomic Trends from the East – Kumashiro (ed)
© *2010 Taylor & Francis Group, London, ISBN 978-0-415-88178-4*

Relationship between depersonalization and self-perceived medical errors

Miyuki Sugiura
Juntendo University Graduate School of Health and Sports Science, Chiba, Japan

Masataka Hirosawa
Juntendo University Graduate School of Health and Sports Science, Chiba, Japan
Juntendo University School of Health and Sports Science, Chiba, Japan
Juntendo University Hospital, Tokyo, Japan

Aya Okada
Juntendo University School of Health and Sports Science, Chiba, Japan
Juntendo University Hospital, Tokyo, Japan

Hideko Aida
Juntendo University Hospital, Tokyo, Japan
Juntendo University Graduate School of Medicine, Tokyo, Japan

Yasuyuki Yamada
Juntendo University School of Health and Sports Science, Chiba, Japan

Motoki Mizuno
Juntendo University Graduate School of Health and Sports Science, Chiba, Japan
Juntendo University School of Health and Sports Science, Chiba, Japan

1 INTRODUCTION

Medical errors made by medical stuffs bring not only negative influence on their professional and personal status, confidence, and function, but also bring patients serious risks and make healthcare costs increase (Kohn et al, 2000; Zhan et al, 2003; Rothschild et al, 2005; West et al, 2006; Kobayashi et al, 2006). In recent years, the mental and physical condition of nurses which include such problems as depression, obsession (Yoshida et al, 2004), burnout (Kitaoka-Higashiguchi, 2005), and fatigue (Carlton and Blegen, 2006) are regarded as factors related to medical errors. Considering the fact that the nursing profession involves various job-related stresses such as time pressure, understaffing, and interpersonal relationships, much wider range of mental problems should be noted in nurses to prevent medical errors.

Depersonalization is characterised by persistent or recurrent episodes of 'detachment or estrangement from one's self.' Although the individual may feel like an automaton or there may be sensation of being an outside observer of one's own mental process, patients have enough ability to recognize the real world during the depersonalization experience (American Psychiatry Association, 2000).

It is well known that depersonalization syndrome accompanies most mental diseases, emotional stress, somatic diseases and exhaustion (Nuller, 1982). Considering the fact that the nursing profession is both cognitively taxing and stressful, it may be an occupation which contains the risk of experiencing depersonalization. Although the existing of the "enough ability to recognize the real world," some neuropsychological studies (Gauralinik et al, 2000; Gauralinik et al, 2007) indicated that, in depersonalization disorder, slower processing speed, poorer perceptual organization, vulnerability to distraction, and problems with immediate recall can be found. These deficits in cognitive functioning seem to prevent nurses from accomplishing their day-to-day business and may make it easy to lead to medical errors.

In hospital health control room, we experienced that some nurses who had depersonalization disorder experienced incidents and accidents, and we suspected depersonalization was one of the essential factors which had an influence on their errors. However, until now, there has been no empirical study which has verified the relationship between depersonalization and medical errors. The purpose of this study is to investigate the relationship between depersonalization and medical errors among nurses working in general hospitals.

2 METHODS

Participants were 1010 female nurses from two general hospitals, one of which was situated in the central business district of Tokyo and the other was in the suburbs of Tokyo. The study was approved by the ethical committee of the Juntendo University hospital system. The investigation was performed in 2007 and nurses participated in this investigation after informed consent was obtained.

We used the following self-report instruments in this study.

The Cambridge Depersonalization Scale (CDS); The CDS comprehensive instrument contains 29 items addressing the complaints associated with depersonalization disorder based on a comprehensive study of the phenomenology of this condition. Each item was rated on two Likert scales for frequency and duration of experience, and the sum of these two scores generated an index of item intensity (range, 0–10). The global score of the scale was the arithmetical sum of all items (range, 0–290). It had a high internal consistency (Cronbach α and split half reliability of 0.89 and 0.92), and a cut-off point of 70 was shown to yield a sensitivity of 75.5% and a specificity of 87.2% by distinguishing depersonalization disorder from panic disorder, generalized anxiety disorder and temporal lobe epilepsy (Sierra and Berrios, 2000). In addition, the CDS has four sub-scales (anomalous body experience, emotional numbing, anomalous subjective recall, alienation from surroundings) (Sierra et al, 2005). The original English scale was translated into Japanese and the Japanese version has also good reliability and validity (Sugiura et al, 2008).

Self-reported medical error; In order to assess the experience of medical errors, we asked participants about whether or not they had experienced incidents or accidents in the past 6 months. Here we adopted the length of 6 months in consideration of the length of depersonalization required in the CDS. A medical incident in the present study was defined as "an error (mistake) that does not reach the patient, or if it reaches the patient, does not result in injury or harm" according to Wu and colleagues (1991)' definition of a "near miss". A medical accident in the present study was defined as "an error (mistake) that brings a patient a complication, injury, or harm resulting from medical management (not from the patient's underlying condition or disease)" according to Jagsi and colleagues (2005)' definition of "adverse event".

Statistical analysis in this study was carried out with the SPSS version 15.0. Firstly, we examined the association between intensity of depersonalization and the experience of incidents and accidents by using the Chi-square test based on the recommended cut-off point in the previous study (Sierra and Berrios, 2000). Secondly, we examined the relationship between symptoms of depersonalization and medical incidents and accidents by using the Kruskal Wallis test and when a significant difference was obtained, the multiple comparison test (Scheffe's test) was applied. In order to estimate this association, we divided participants into 3 groups (non-error, incident and accident groups) according to whether or not they had experiences of incidents or accidents in the past 6 months. The non-error group included nurses who had experienced neither incidents nor accidents, the incident group included those who had experienced only incidents, and the accident group included all those who had experienced accidents regardless of the experience of incident. All tests were two-tailed.

3 RESULTS

In all participants, the prevalence of those who got more than 70 points in the CDS was 2.1%. It was showed that 42.9% of the more than 70 group and 52.0 % of the less than 70 group had experiences of incidents. There was no significant difference between more and less than 70 groups in the experience of incidents ($\chi^2 = 0.68$, df $= 1$, p $= 0.41$). It was also showed that 28.6% of the more than 70 group and 24.3% of the less than 70 group had experiences of accidents. There was

Table 1. Comparison of scores of total CDS and its subscales among 3 groups.

Scale	1.Non-error N = 392		2.Incident N = 372		3.Accident N = 246		χ^2(df = 2)	p	Scheffe
	M	SD	M	SD	M	SD			
Anomalous body experience	2.00	6.25	2.19	5.79	2.53	6.23	5.82	0.05	1 < 3
Emotional numbing	1.68	4.48	1.79	4.38	2.50	5.69	4.60	0.10	
Anomalous subjective recall	2.06	4.18	2.32	4.40	2.50	5.15	2.68	0.26	
Alienation from surroundings	1.97	4.17	2.17	4.10	2.87	5.32	6.66	0.04	1 < 3
Total CDS	8.45	19.96	9.34	19.03	11.20	22.37	7.22	0.03	1 < 3

no significant difference between more and less than 70 groups in the experience of accidents ($\chi^2 = 0.21$, df = 1, p = 0.65).

According to the distribution of experiences of incidents and accidents in nurse who had got more than 70 points in the CDS, we divided three CDS score groups; that is more than 90, between 70 to 89, and less than 70. This result showed that 25.0% of nurses whose CDS score was more than 90, 66.7% of nurses whose score was between 70 and 89, and 52.0% of nurses whose score was less than 70, had experienced incidents in the last 6 months, and 16.7% of nurses whose score was more than 90, 44.4% of nurses whose score was between 70 and 89, and 24.3% of nurses whose score was less than 70, had experienced accidents in the last 6 months.

Table 1 showed that, on the Kruskal Wallis test, there were significant differences among the three groups of non-error, incident and accident groups in scores of total CDS and the "alienation from surroundings" experience. Post hoc tests (Scheffes' test) showed that the accident group scored significantly higher than the non-error group in the score of total CDS and the "alienation from surroundings" experience (p = 0.03, p = 0.04, respectively). Additionally, the accident group had the tendency of higher in the score of the "anomalous body experience" than the non-error group (p = 0.05).

4 DISCUSSION

4.1 Relationship between depersonalization and medical errors

In the present study, the close relationship between depersonalization disorder and medical incidents and accidents seemed not to be confirmed, using the CDS cut off point. It was certain that the rate of accidents in nurses who were over the CDS cut-off point and who were possibly having depersonalization disorder was higher, however, the rate of incidents of these was lower than the other nurses. Here, we should consider not only whether or not they have depersonalization disorder, but also the severity of this disorder and the feature of each of the symptoms of depersonalization.

Regarding the severity of the depersonalization disorder, the rate of experience of incidents and accidents in nurses whose CDS score was more than 90 was the lowest when compared with those whose CDS score was between 70 and 89 and less than 70. Interestingly, in the rate of experience of accidents, those having scores of more than 90 were approximately half of those having less than 70. On the other hand, the rate of experience of incidents and accidents in nurses whose CDS score was between 70 and 89 was the highest, and they experienced accidents approximately twice as frequently as those having scores of less than 70. These results suggest that, among the possible depersonalization disorder group, those whose CDS score is between 70 and 89 solely contribute to the occurrence of self-reported medical errors.

The reason why the nurses who have more severe depersonalization experience fewer medical errors is unknown. However, we can speculate that those nurses with clinical significant depersonalization seemed to have higher levels of anxiety and distraction in cognitive functioning. Under such conditions, they may have expressed much more distress and made inadequate behaviors, and co-workers may also have been able to note their psychological problems and given them some special protect in their job, which prevent them making much job related errors. On the other hand, depersonalization disorder which is expressed by the CDS score between 70 and 89 is considered as slight end of clinical depersonalization. So, it may be much difficult for suffering nurses to express their condition. This may contribute to the problem that support systems cannot adequately function and it may be a cause that nurses whose CDS score is 70 to 89 experience medical errors more frequently.

When we consider the scores of CDS in all nurses, those in the accident groups showed higher than nurses in the non-error group (Table 1). The result indicates the possibility: the summated scores of CDS may explain a certain extent of the medical errors, even in case of it is smaller than the clinical levels of depersonalization, or even in case of the items of CDS represents symptoms fleeting depersonalization-like experience. Therefore, if we find some signs depersonalization-like experience in nurses, we may need to check the total CDS score and grasp their function to work.

On the other hand, occurrence of medical error will give a great shock on nurses concerned and it consider as one of the major stresses for nurses. Regarding the repeatedly described characteristics of depersonalization that was often accompanied by psychological stress (Nuller, 1982), some symptoms of depersonalization may be resulted from stress after an error has occurred. Therefore, when error occurs, we need to pay a particular attention to the appearance of depersonalization symptoms in nurses and to prevent them from making further errors.

4.2 Relationship between symptoms of depersonalization and medical errors

Regarding the symptoms of depersonalization, scores of the "anomalous body experience" tend to be higher in the accident group than in the non-error group and the "alienation from surroundings" experience was significantly higher in the accident group than in the non-error group.

The "anomalous body experience" was composed of items regarding the feeling of out of body experiences, loss of agency feeling, and automatic and mechanical body feeling (Sierra and Berrios, 2000).Under the condition, they are feeling as if their own body is moving automatically and they are not able to control their own behaviour. This suggests that medical errors experienced under the "anomalous body experience," closely connected with subjective feelings of automaton and uncontrollable movement.

Moreover, the "alienation from surroundings" experience was composed of items regarding feelings of not being real, being cut off from the world, and having a veil between oneself and the outside world (Sierra and Berrios, 2000). In cases of the "alienation from surrounding" experience, even if colleagues were able to notice some troubles in that nurse's mind and gave her some useful advice, their words of advice would not reach her mind. Therefore, in case of nurses who experience alienation, it may be more difficult for colleagues to have some way to prevent accident. This indicates that nurses with alienation have "little way to prevent accidents" in their work place and their colleagues need to make additional efforts to prevent accidents than just giving verbal notice.

In conclusion, the factors that have close relation with medical errors were scores between 70 and 89 in the CDS, and symptoms of the "alienation of surrounding" and the "anomalous body experience," and, under these conditions, preventing accidents was very difficult not only by themselves but also by the help of colleagues. Therefore, we need to detect depersonalization in nurses as early as possible, and to prevent nurses from suffering from depersonalization disorder or having depersonalization-like experiences, as well. It is said that depersonalization often accompany emotional stress, somatic disease and exhaustion (Nuller, 1982). Considering nurses' day-to-day practice such as time pressures, understaffing, and excessive work loads, problems in hospital systems may strongly lead nurses to have depersonalization disorder. Therefore, We should pay more attention to improve working conditions from the view points of depersonalization disorder and its related medical errors.

REFERENCES

American Psychiatric Association., 2000, Diagnostic and statistical manual of mental disorders, 4th ed. text revision (DSM-IV-TR). (Washington DC: American Psychiatric Association).
Carlton, G. and Blegen, MA., 2006, Medication-related errors: a literature review of incidence and antecedents. *Annu Rev Nurs Res*; 24: 19–38.

Guralnik, O., Schmeidler, J. and Simeon, D., 2000, Feeling unreal: cognitive processes in depersonalization. *Am J Psychiatry*; 157:103–109.

Guralnik, O., Giesbrecht, T., Knutelska, M., Sirroff, B. and Simeon, D., 2007, Cognitive functioning in depersonalization disorder. *J Nerv Ment Dis*; 195:983–988.

Jagsi, R., Kitch, BT., Weinstein, DF., et al., 2005, Residents report on adverse events and their causes. *Arch Intern Med*; 165: 2607–13.

Kitaoka-Higashiguchi, K., *2005*, Causal relationship of burnout to medical accident among psychiatric nurses. *Journal of Japan academy of nursing science*; 25: 31–40. (In Japanese).

Kobayashi, M., Ikeda, S. and Muto, M., 2006, The estimation of medical cost incurred by incidents and adverse events. *Iryo to Shakai*; 16: 85–96. (In Japanese).

Kohn, LT., Corrigan, JM, and Donaldson, MS., 2000, *To Err Is Human: Building a Safer Health System.* Committee on Quality of Healthcare in America. (Washington, DC: National Academy Press).

Nuller, YL., 1982, Depersonalization -symptoms, meaning, therapy. *Acta Psychiatr Scand*; 66: 451–8.

Rothschild, JM., Landrigan, CP., Cronin JW, et al. 2005, The Critical Care Safety Study: The incidence and nature of adverse events and serious medical errors in intensive care. *Crit Care Med*; 33:1694–1700.

Sierra, M. and Berrios, GE., *2000*, The Cambridge Depersonalization Scale: a new instrument for the measurement of depersonalization. *Psychiatry Res*; 93:153–64.

Sierra, M., Baker, D., Medford, N. and David, AS., 2005, Unpacking the depersonalization syndrome: an exploratory factor analysis on the Cambridge Depersonalization Scale. *Psychol Med*; 35:1523–32.

Sugiura, M., Hirowawa, M., Nishi, Y., et al., 2008 Development of a Japanese Version of the Cambridge Depersonalization Scale and Application to Japanese University Students, *Annual reports 2008 of Juntendo institute mental health*; 93–101.

West, CP., Huschka, MM., Novotny, PJ., et al., 2006, Association of perceived medical errors with resident distress and empathy: a prospective longitudinal study. *JAMA*; 296:1071–1078.

Wu, AW., Folkman, S., McPhee, SJ. and Lo, B., 1991, Do house officers learn from their mistakes? *JAMA*; 265: 2089–2094.

Yoshida, Y., Otsubo, T., Takenaka, K., et al., 2004, The factors related to accident proneness among hospital nurses. *Clinical Psychiatry*; 46: 723–30. (In Japanese).

Zhan, C. and Miller, MR., 2003, Excess length of stay, charges, and mortality attributable to medical injuries during hospitalization. *JAMA*, 290:1868–1874.

Ergonomic Trends from the East – Kumashiro (ed)
© 2010 Taylor & Francis Group, London, ISBN 978-0-415-88178-4

Difference of the relationship between coping and burnout with respect to Typus Melancholicus and other personalities: A study of Japanese nurses

Yasuyuki Yamada
Juntendo University School of Health and Sports Science, Chiba, Japan

Masataka Hirosawa
Juntendo University School of Health and Sports Science, Chiba, Japan
Juntendo University Graduate School of Health and Sports Science, Chiba, Japan

Miyuki Sugiura & Yujiro Kawata
Juntendo University Graduate School of Health and Sports Science, Chiba, Japan

Motoki Mizuno
Juntendo University School of Health and Sports Science, Chiba, Japan
Juntendo University Graduate School of Health and Sports Science, Chiba, Japan

1 INTRODUCTION

According to the definition of burnout syndrome (burnout), burnout can be understood according to the three symptoms of "Emotional Exhaustion (EE)," "depersonalization (DP)," and "Reduction of Personal Accomplishment (RPA)" as the results of chronic stress (Maslach & Jackson, 1981; Starrin *et al.*, 1990). Furthermore, we can regard it as an universal phenomenon which is observed in various countries and occupations. On the other hand, the features of the dynamic process of burnout differ according to the individual. Hence, in order to gain more practical knowledge for individual intervention in cases of burnout, clarifying the individual factors which characterize its dynamics is necessary.

Under these circumstances, we have focused on nursing burnout especially and explored a concept which determines the features of burnout dynamics from the individual perspective. In our previous studies, we used Tellenbach's (1961) melancholic type of personality (in German, Typus Melancholicus; TM) as one of the individual factors which characterize Japanese nurses' burnout and examined the effects of TM on the burnout dynamics (Yamada *et al.*, 2007a, 2007b, 2007c). In these studies, it was clarified that TM had the two traits of "sthenic in work (SW)" and "asthenic in human relationships (AHR)," and both personality traits were shown to have effects on the process of perception of stressors, adoption of coping strategies, and degree of burnout, individually. In addition to these findings, these studies also imply a possibility that four personality types of TM (having SW/AHR), SW (lack of AHR), AHR (lack of SW), and Anti-TM (lack of SW/AHR) characterize the features of burnout dynamics among Japanese nurses. However, this has not been examined in detail.

Therefore, the purpose of this study was to clarify the characteristics of burnout dynamics among Japanese nurses in the four personality groups, individually. In particular, this study focuses on the process of coping with burnout symptom as one of the elements involved in burnout dynamics and examines the differences in this process. For this purpose, this study established two hypotheses; (1) four personality traits characterize the adoption style of coping, and (2) four personality traits characterize the kinds of coping strategies relating to burnout. If our hypotheses are supported, we may be able to understand the characteristics of burnout dynamics from the personality perspective. Furthermore, we believe that our attempt is practical not only for organizational intervention but also for individual intervention toward burnout syndrome among Japanese nurses.

Table 1. Factor structure of KS.

TM factors	Items	Factor Loadings	
		F1	F2
SW	3 I have a strong sense of responsibility.	0.67	0.03
($\boxtimes = .73$)	2 When I start something, I always finish it thoroughly.	0.65	−0.07
	14 I like to arrange my belongings.	0.55	−0.06
	15 I am neat.	0.54	−0.07
	4 I give importance to my social duty.	0.49	0.12
	1 I like to work.	0.44	−0.08
	9 I give importance to common sense.	0.41	0.37
	13 I am rather cheerful.	0.27	0.01
	12 I sometimes get excited easily.	0.16	−0.04
AHR	7 I am rather timid.	−0.21	0.66
($\boxtimes = .71$)	6 I would rather avoid confrontation with somebody.	−0.06	0.60
	8 I am nervous about what other people think of me.	−0.05	0.54
	11 I do not like to be conspicuous.	−0.16	0.52
	10 I would not do something extreme.	0.16	0.49
	5 I cannot say no, when someone asks me to do something.	0.20	0.46
	Eigenvalue (First Solution)	3.05	2.56
	Percent of Variance (%)	15.8	28.4

2 METHODS

The participants of this study were Japanese female nurses from two university hospitals. This research was carried out with female Japanese nurses in 2007. Valid data were 1007. Their mean age was 27.5 (SD = 6.6). This study was approved by the ethical committee of the university hospital system.

Moreover, the questionnaire of this study was composed of following three scales. First scale was Kasahara's Typus Melancholicus Scale (KS, Kasahara, 1984). This scale has been adopted as one factor structure based on Tellenbach's (1961) original concept of TM. However, our empirical study confirmed that there were two factors, SW and AHR, among the KS items, according to statistical evidence. Hence, this study applied the two factor structure to KS. Other scales were Maslach Burnout Inventory (MBI, Maslach & Jackson, 1981) and the Brief Scales for Coping Profile (BSCP, Kageyama et al., 2004).

3 RESULTS

3.1 Factor analysis of KS

First, we examined factor the structure of KS and elucidated four typical personality groups based on the scores of KS. A factor analysis was conducted using promax rotation (principal factor method) for all items (Table 1). Basically, the results of this study also supported Yamada et al.'s (2007a) two factors structure with enough eigen values (>1.0) and factor loadings (>0.40). However, in addition to Q12 and Q13 that were omitted from the two factor structure in Yamada et al.'s (2007a) study, Q9 was omitted in this study because this item showed closed factor loadings toward both factors. Hence, although Q9 was omitted, this study also treated KS as having a two factor structure, consisting of "sthenic in work (SW)" and "asthenic in human relation (AHR)" traits. Furthermore, we categorized three level personality groups of SW and AHR on the basis of 33% of the whole sample and elucidated typical samples of four personality groups. As the results, 161 of TM (having SW/AHR), 102 of SW (lack of AHR), 121 of AHR (lack of SW), and 81 of Anti-TM (lack of SW/AHR) samples were elucidated from 1007 samples.

Table 2. Comparison of the scores of MBI and BSCP among four groups using a one–way ANOVA.

BSCP Factors	I .TM		II .SW		III .AHR		IV. Anti-TM		ANOVA	Maltiple Comparison
	M	SD	M	SD	M	SD	M	SD	F-value	Hochberg's GT2
Changing Mood	6.7	2.6	7.2	2.6	6.7	2.6	6.9	2.7	1.0	
Active Solution for Problems	9.1	1.8	9.3	1.8	7.8	1.5	7.8	1.7	23.2*	III, IV<I, II***
Avoidance and Suppression	4.7	1.8	4.1	1.3	5.5	1.9	5.0	1.9	11.8*	I<III**, II<I*, III***, IV**
Changing a Point of View	6.5	2.2	7.4	2.3	6.1	2.0	6.6	2.0	6.9*	I<II*, III<II***
Seeking Help for Solution	6.3	1.5	6.4	1.3	6.0	1.5	6.0	1.5	1.9	
Emotional Expression	5.3	1.4	5.3	1.4	5.3	1.6	5.5	1.6	0.4	

Emotional Expression: Emotional Expression Involving Others,* $p < .05$, ** $p < .01$, *** $p < .001$

3.2 Comparison of the BSCP scores among four groups using a one-way ANOVA

In order to examine the hypothesis 1, this study compared the scores of BSCP among four personality groups using the one-way ANOVA (Table 2). For the results, we observed significantly meaningful differences in the scores of "active solution for problems," "avoidance and suppression," and "changing a point of view" ($p < 0.05$). In detail, TM and SW groups showed higher scores in the "active solusion for problems" than AHR and Anti-TM groups ($p < 0.001$). In the comparison of the "avoidance and suppression," the AHR group showed higher score than the TM one ($p < 0.01$), and the SW group showed the lowest score among the four groups (SW $<$ TM/AHR/Anti-TM, $p < 0.05$). Finally, the SW group showed higher scores in the "changing a point of view" than the TM and AHR groups ($p < 0.05$). Hence, these results supported the hypothesis 1.

3.3 Comparison of the correlations between BSCP and MBI scores among four groups

In order to examine the hypothesis 2, this study compared the relationships between BSCP and MBI scores using Pearson's correlation coefficient. As the results, meaningful differential features were observed between four personality groups (Table 3). Concretely, in the TM group, "active solution" showed negative correlation with RPA ($r = -0.27$, $p < 0.001$). Furthermore, "seeking social support" showed positive correlation with EE ($r = 0.23$, $p < 0.01$) and negative one with RPA ($r = -0.29$, $p < 0.001$). In the SW group, "active solution for problems" and "changing a point of view" showed negative correlations with RPA ($r < -0.20$, $p < 0.05$). Moreover, "avoidance and suppression" and "emotional expression involving others (emotional expression)" showed positive correlation with DP and the total score of MBI ($r > 0.20$, $p < 0.05$). In the AHR group, "avoidance and suppression" showed positive correlation with DP ($r = 0.20$, $p < 0.05$). In the anti-TM group, "avoidance and suppression" showed positive correlation with total score of MBI ($r = 0.23$, $p < 0.05$). Furthermore, "emotional expression" showed positive correlations with EE, DP, and total score of MBI ($r > 0.20$, $p < 0.05$) and negative one with RPA ($r = -0.31$, $p < 0.01$). Hence, the hypothesis 2 was supported.

4 DISCUSSIONS

4.1 Theoretical findings of this study

The results showed that our hypotheses were supported. Namely, it was clarified that four personality traits of TM, SW (lack of AHR), AHR (lack of SW), and Anti-TM characterized the adoption of coping style and the effects of coping with burnout. In previous studies, burnout studies have chiefly paid attention to its symptomatic aspects, and have built a definition based on the symptoms which can be observed in every worker. However, this study clarified the existence of different types of dynamics in the burnout process. This finding was important because it clarified some issues which exist in recent burnout studies. One of the issues is that when we evaluate burnout with the above

Table 3. Correlation coefficients between BSCP and MBI scores among four groups.

Personality Groups	MBI Factors	BSCP Factors					
		Changing Mood	Active Solution for Problems	Avoidance and Suppression	Changing a Point of View	Seeking Help for Solution	Emotional Expression
TM	EX	0.07	0.16*	0.14	−0.04	0.23**	0.12
	DP	0.03	−0.08	0.14	0.07	0.06	0.19*
	RPA	−0.18*	−0.27***	−0.12	−0.13	−0.29***	−0.16*
	MBI	−0.09	−0.13	0.08	−0.11	−0.02	0.06
SW	EX	−0.06	0.19	0.19	0.15	0.04	0.21*
(lack of AHR)	DP	−0.02	0.01	0.21*	0.07	0.02	0.25**
	RPA	0.02	−0.22*	−0.02	−0.21*	−0.07	0.00
	MBI	−0.05	−0.02	0.24*	−0.01	−0.01	0.30**
AHR	EX	0.02	0.17	0.15	−0.01	−0.09	0.12
(lack of SW)	DP	0.04	0.07	0.20*	0.10	−0.06	0.13
	RPA	−0.04	−0.19	−0.04	−0.06	−0.07	−0.04
	MBI	0.01	0.06	0.19*	−0.01	−0.14	0.14
Anti-TM	EX	0.21	0.04	0.14	−0.01	0.13	0.34**
	DP	0.06	−0.01	0.21	0.05	−0.06	0.27*
	RPA	−0.19	−0.10	−0.03	−0.11	−0.08	−0.31**
	MBI	0.09	−0.05	0.23*	−0.07	0.03	0.22*

EX: Emotional Exhaustion, DP: Depersonalization, RPA: Reduction of Personal Accomplishment, MBI: Total Score of MBI, Emotional Expression: Emotional Expression Involving Others, $*p < .05$, $**p < .01$, $***p < .001$

mentioned scale, we cannot identify the types of burnout processes, which will make it difficult for us to obtain some useful findings for an effective individual approach. Furthermore, when we consider burnout as a phenomenon which can be seen in nurses having a particular personality and which has its own characteristic process, we will fail to grasp or make less of the similarly measured phenomenon in nurses who have other personalities. Hence, we suggest that employing a categorization of the features of burnout dynamics based on the personality traits surrounding TM will provide one of the meaningful approaches to support individually burned out nurses and to provide an elaborate understanding of the burnout phenomenon. Next, we will discuss the features of the four personality groups' burnout dynamics in detail.

4.2 Features of burnout dynamics among four groups

According to the results of the TM group, they tended to adopt problem solving behaviors and hated avoidance coping under stressful conditions. Their coping behavior for problem solving was revealed to increase in their feeling of personal accomplishment and decrease in emotional exhaustion. However, taking their personality trait into consideration, it can be assumed as a feature of burnout dynamics that their tendency to seek the feeling of accomplishment brings emotional exhaustion because their SW and AHR traits force them to make excessive efforts for problem solving. The SW group also showed unique results. This group tended to use the problem solving behavior while changing their point of view and this contributed to their maintenance of a feeling of accomplishment. Moreover, they hated the adoption of coping behaviors which do not connect with problem solving directly, such as avoidance and emotional expression. Hence, it can be assumed that they don't give up on solving their problems easily. However, when they are forced to adopt avoidance and/or emotional expression coping behaviors because their effort is not rewarded for a long time, they suffer from more serious burnout symptoms.

On the other hand, the AHR group didn't prefer to adopt problem solving coping behaviors and tended to adopt avoidance coping behavior frequently. These results indicated that the AHR group complains of burnout easily, without making enough efforts to solve their problem. Especially, in this group, avoidance coping behavior connected with depersonalization directly. According to conventional studies, the depersonalization disturbs the ability to provide polite nursing service with kindness toward patients. Hence, it is said that the quality of their nursing job was easily influenced by stressful conditions. Similar to the AHR group, the Anti-TM group frequently didn't adopt problem solving behavior. Furthermore, they also tended to adopt avoidance coping and it was connected to burnout. In addition to these results, in this group, adopting a coping of emotional expression involving others affected their risk of burnout. Therefore, these results indicated that, in the AHR and Anti-TM groups, promoting their own effort to solve their problems was one of the effective approaches to prevent burnout.

5 CONCLUSIONS

This study aimed to clarify the characteristics of involved in burnout dynamics among four personality groups associated with TM. As the results, the features of adoption of coping and its effects on burnout were different among four personality groups based on TM factors. Hence, we can say that focusing on these personality features is meaningful for providing a detailed understanding burnout dynamics.

REFERENCES

Maslach, C. and Jackson, S. E. (1981) The measurement of experienced burnout. *Journal of Occupational Behavior*, 2, pp. 99–113.

Starrin, B., Larsson, G., and Styrbon, S. (1990). A review of critique of psychological approaches to the burn-out phenomenon. *Scand J Caring Sci*, Vol.4, No.2, pp. 83–91.

Tellenbach, H., 1961. *Melancholie*. Springer, Berlin. (In Germany)

Yamada, Y., Hirosawa, M., Miyuki, S., Nishi, Y., Tanaka, S., and Mizuno, M. (2007a). Relation between Typus Melancholicus and burnout syndrome among Japanese nurses. *The Proceedings of the 3rd International Symposium on Work Ability*, CD-ROM.

Yamada, Y., Sugiura. M, Nishi, Y., Hirosawa, M., Tanaka, S., and Mizuno, M. (2007b). Effect of Typus Melancholicus on cognition of job stressor among Japanese nurses. *The 18th Japan- China-Korea Joint Conference on Occupational Health, Program and abstracts*, pp. 131–132.

Yamada, Y., Sugiura. M, Mizuno, M., Tanaka, S., and Hirosawa, M. (2007c). Relation between Typus Melancholicus and stress coping among Japanese nurses. *News letter of Human Ergology*, No.86, p. 31. (in Japanese)

Kasahara, Y. (1984). Depression in general practice. *Japan J. Psychosom Med*, Vol.24, pp. 6–14. (in Japanese).

Kageyama, T., Kobayashi, T., Kawashima, M., and Kanamaru, Y. (2004). Development of the Brief Scales for Coping Profile (BSCP) for workers: Basic information about its reliability and validity. *Journal of Occupational Health*, Vol.46, No.4, pp. 103–114. (in Japanese)

Safety and security concerns among Singapore elderly towards home monitoring technologies in smart homes

Alvin Wong, Jenny Ang & Zelia Tay
Singapore Institute of Manufacturing Technology, Agency for Science, Technology and Research, Singapore

Jamie Ng, Koh Wei Kiat & Odelia Tan
Institute for Infocomm Research, Agency for Science, Technology and Research, Singapore

Martin G. Helander
School of Mechanical and Aerospace Engineering, Nanyang Technological University, Singapore

ABSTRACT: Smart home technologies are developed to support older adults in maintaining independence. These technologies have the potential to help older adults perform activities required to live independently or to help caregivers in providing them care at a distance. We hope that through the use of smart home technologies to provide home-based healthcare and enhanced safety and security, the quality of the older adults' life could be improved. To understand their concerns towards home monitoring technologies, contextual inquiry (Beyer and Holtzblatt, 1998) and in-depth interview techniques was utilized in our research. The first interview was conducted to understand the older adults' daily activities and concerns at home. Follow-up studies were conducted to understand their thoughts on safety and security, where a video-based fall detection prototype was demonstrated in a smart home environment. In this paper, we present our findings on elderly attitudes towards the safety and security in their homes, as well as privacy concerns and adoption attitudes towards monitoring technologies.

Keywords: Smart home, safety, security, elderly, perceptions, attitudes, home monitoring

1 INTRODUCTION

By 2030, one in five residents in Singapore will be 65 and above, up from the current one in twelve. (United Nations Population Division, 2006). As the population of older adults increase in Singapore, it is important to pay attention to safety and independence needs of the elderly. As people age, they may suffer from impaired cognitive and physical functions, a decline in activity and poor health, leading to a dependency on others in daily life (Mann, 2005). The implementation of smart home technologies to support older adults seems a logical and useful plan that could be pursued. There has not been widespread deployment of smart home solutions among older adults. One of the main reasons for the delay in adoption is the lack of in-depth understanding of them and their personal and cognitive requirements. If smart home technology designers apply a predominantly innovation driven approach without the appropriate regard for users needs, these technologies are less likely to be adopted. User-driven design and user evaluation during the development of new smart home technologies is critical for successful adoption of such technologies for older adults.

The Agency of Science and Technology Research (A*STAR) in Singapore, is developing a smart home to study and design technologies that address user needs. The Silver Industry Conference and Exhibition (SICEX) was held in January 2008, where a temporary smart home was setup to elicit feedback on its potential. A survey was first conducted focusing on Singaporeans' understanding of smart homes, their attitude towards certain technologies and how they categorize the importance of their needs in their homes. The fall detection monitoring device was rated among the top 3 product in a list of A*STAR technologies (Koh *et al.*, 2008). Safety and Security needs were listed top among other needs such as Relaxation and Health.

Safety and security at home is an integral part of our lives. Many Singaporeans feel the need for living safely and securely in their own home (Koh *et al.*, 2008). For example, before opening the main entrance door, it is common for people to identify visitors at their doorsteps. Other safety issues include prevention of fire, explosion or falls at home. The risk of falls increases with aging. Falls are common among older adults and often results in long-term injury or death. In Singapore, fall is one of the leading causes of death and disability among the older adults. 20% of deaths are caused by falls occur among those over 65 (Singapore Health Promotion Board Website, 2007). These observations have led to the development of fall detection devices to detect or alert the caregiver of such fall events (Kelly *et al.*, 2003). There are different types of fall detection technologies, ranging from video based, acoustic or vibration analysis, intelligent telecare systems, and worn devices (White, 2005 and Perolle *et al.*, 2006). Wearable fall detectors like accelerometer or help buttons are currently available. However, older people often forget to wear them or may be unconscious after a fall. Video monitoring technology came about to overcome these problems. It captures images, which may compromise privacy, and it is a major concern for a home environment (Caine *et al.*, 2006).

2 METHOD

The overall aim was to describe aspects of older adults' safety and security. Contextual inquiry was utilized with in-depth interview as an alternative method of identifying user needs for the fall detection prototype system (Beyer and Holtzblatt, 1998).

2.1 *Study Phase I*

Following the survey at SICEX, a series of interviews were conducted at the participants' homes to understand their daily activities and their concerns while living at their own home. 10 participants from the SICEX survey were chosen and their age range from 56 to 81 years old. Each interview session lasted on average one and half hours and the participants were given S$25 shopping vouchers as incentive.

2.2 *Study Phase II*

The objective of the second phase was to delve deeper on elderly thoughts on safety and security, and identify the factors that were important for installing a video-based fall detection device, such as the system demonstrated in A*STAR's smart home (STARhome). Six of the participants from Phase I (four males and two females), were invited to STARhome. They were short listed based on their interest for potential solutions that may meet their needs and their experiences in using computers. The incentives for all participants were the same as Phase I for their time and effort.

3 RESULTS AND DISCUSSION

As seen in Table 1, five out of the 10 respondents (50%) commented on their fear of falling. To prevent trips and falls some of the elderly would keep their floor free from clutter. Some highlighted that wet and slippery floors in the bathroom are the leading reason for falls. Another concern highlighted was forgetting to turn off the gas stove. One respondent conveyed his fears and distress when there were several instances when he forgot to turn off the stove and the pot caught fire. This can be addressed by an induction cooker which is a flameless cooking appliance. Induction cookers are safer to use than conventional stoves because there are no open flames and the "element" itself reaches only the temperature of the cooking vessel; only the pan becomes hot. In this case, the stovetop stays cool: that means no burned fingers or hands, for children or especially for any older adults in the household. And for kitchens that need to take into account special needs, such as wheelchair access. Hence, induction is good for both safety and convenience.

Of the 10 respondents, three raised concerns on the security of their neighbourhood. Their comments came about despite their feedback that there were security cameras in their flat's void

Table 1. Respondents' feedback on safety and security concerns in their home.

Comments made by respondents	Percentage of people (N = 10)
Fear of falling down	50%
Forgetting to switch off the gas stove	50%
Feels that neighborhood is safe and secure	40%
Worried about security in neighborhood	30%
Having a maid is not reliable	20%
Wanted a surveillance camera to monitor parents	20%
Would keep home free from clutter	20%
Would like to have a alarm trigger to notify medical practitioners	20%
Does not trust internet banking	20%
Checks if the main door is locked before sleeping	20%
Would switch off electrical appliance before leaving home	20%
Would like to have security guard in neighborhood for older adults	20%
Would like one child to stay with or close to them	10%
Home should be equipped with security system	10%
Checks for the stove to be turned off every night before going to bed	10%

Table 2. Understanding of the monitoring device versus their needs at Phase II study.

Comments made by respondents	People (N = 6)	Ratio
Felt that the system could also be used for security monitoring and as form of surveillance	5	83%
Felt that the system would be beneficial to older adults who are handicapped, senile or suffer from dementia	4	67%
Prefers having maids to assist their needs rather than utilizing technology	2 (Females)	33%
System would be good for families with young children or babies. They will be able to look out of their kids' well being remotely	1	17%

decks and elevators. Three respondents had no qualms about security in the neighbourhood, while four respondents did not mention about it. Security concerns about respondents' neighbourhood is dependent on whether the elderly is staying in a good neighborhood or if there were any recent crime near their living environment. As such, some respondents mentioned that technology has the potential to provide some form of reassurance and help alleviate their fear of insecurity in their own homes. Feeling safe is also attributed to the location of the respondent's residential area and thus could affect their mindset and feeling of security.

From Table 2, most respondents mentioned that the video-based system function as a security monitoring device rather than a fall detection device. In addition, it could also help monitor other home activities. Installing a video camera could ensure that both safety and security surveillance needs are covered at home and having a camera at the main door has the potential to help an older adults or a person with physical disabilities to remotely know who is at the door.

Previous studies have examine potential privacy concerns that older adults may have about using visual sensing devices in their home as well as factors that may mitigate those concerns. In addition, older adults tended to be more concerned with privacy of space than information privacy (Caine et al., 2006). Our results (Table 3) show that, the older adults were concerned about if contents would be broadcast, and who would have access to these images. The person or older adult being monitored is perturbed by the fact that their privacy is being compromise.

From our interview, person being monitored feels comfortable to share images with their love ones and medical caregiver as opposed to their thoughts on privacy. These images could include

Table 3. Concerns with monitoring device at Phase II study.

Comments made by respondents	People (N = 6)	Ratio
Had concerns over intrusion of privacy and was uncomfortable to have it installed in their own house. All were not comfortable to install a camera in their toilets.	6	100%
Would not mind if the streaming was sent to someone close to them.	5	83%
Thought that if a fall image was captured, the last image could be sent to the caregiver	2	33%
Would consider installing the device in his parent's home for monitoring mother's health well being.	1	17%

footages of the instances prior to the fall or emergency, which would allow the relevant authorities to access it before making any recommendation. Caregivers value the usage of a tele-monitoring device to reassure them that their loved ones are safe and secure. It would allow them to view the activities taking place at their home. However, the caregiver would be expected to monitor the surrounding consistently, and this may not be an effective way of assessment. Perhaps a less intrusive form communication could be adopted to keep the caregivers informed of the well-being of their cared one more effectively.

Apart from privacy concerns, many of our respondents mentioned that reliability of the system is important and designers must ensure that the alarms put in place are trustworthy. The system picking up falls or emergency situations must be reliable for usage. Respondents had concerns relating to maintainability, serviceability, cost, service time and ease of getting help for support of system.

4 CONCLUSION AND FUTURE WORK

The results of this study indicate that homes today do not offer sufficient safety and security to older adults who live by themselves and suffer from impaired cognitive and/or physical ability. Older adults' perception of technology indicated that the new technology geared toward enhancing the quality of life of elderly in their homes would be welcomed by many. However receptivity was directly influenced by need and social support factors, as well as by one's level of concern for problems that could be alleviated through the use of technology. Systems designers and researchers should identify the parties and relevant people to contact at times of fall or emergencies and investigate on the effectiveness and reliability of call centers response to emergencies.

For future research, the outcome of fall detection measures could use signal detection theory to mitigate the limitations of false alarms and misses. Also, a study of Slovic's risk perception would be interesting. A better understanding of how people respond to risk would allow designers to come up with effective safety technologies for homes. There are comparisons which show that risk perception is greatly affected by cultural variables (Slovic, 2000). It would be interesting to compare Singapore with other countries and understand the similarities and differences in attitudes on risk perception, which would be a topic for another study.

REFERENCES

Beyer, H. and Holtzblatt, K. (1998). *Contextual Design: Defining Customer-Centered Systems.* (San Francisco: Morgan Kaufmann.)
Caine, K.E., Fisk, A.D and Rogers, W.A. (2006). Benefits and privacy concerns of a home equipped with a visual sensing system: A perspective from older adults. *Proceedings of the Human Factors and Ergonomics Society 50th Annual Meeting.*
Health Promotion Board, Singapore, Elderly Heath – Accidental Injuries (2007). http://www.hpb.gov.sg

Kelly, E., Brownsell, S. and Hawley, M.S. (2003). Falls and Telecare Evaluation. *Assistive Technology – Shaping the Future*, IOS Press 803–807

Koh, W.K., Ng, S.L.J., Wong, H.Y.A., Yap, M.Y.D., Tay, Y.C. and Helander, M.G. (2008) Users' Perception toward Smart Home Healthcare Technologies – A Survey in Singapore. *Proceedings of IEEE Healthcom Conference.*

Perolle, G., Fraisse, P., Mavros, M. and Etxeberria, I. (2006). Automatic fall detection and activity monitoring for elderly, *Proceedings of MEDETEL*

Slovic, P. (2000). *The Perception of Risk.* (London: Earthscan Publications, Ltd.)

United Nations Population Division. *World Population Prospect*: The 2006 Revision. http://www.un.org

White, M. (2005). Injury Prevention and Health Promotion. *In Smart Technology for Aging, Disability, and Independence,* edited by Mann, W. C., (John Wiley & Sons, Inc., Hoboken, New Jersey.)

Chapter 9   Ergonomics in occupational health 2

Ergonomic Trends from the East – Kumashiro (ed)
© 2010 Taylor & Francis Group, London, ISBN 978-0-415-88178-4

Improvement of the screen plate lifting method in the fabric printing section

C. Theppitak, S. Palee, W. A-Ga, Y. Ariyadech, N. Tripetch & B. Srikuttanam
*School of Occupational Health and Safety, Suranaree University of Technology,
Nakhon Ratchasima, Thailand*

ABSTRACT: Workers in the fabric printing section have to lift many screen plates per day for fabric printing. The shape and size of these plates have cause difficulty and present a hazard for lifting. The aim of this study was to improve the screen plate lifting method in this section in order to reduce feelings of fatigue and increase feelings of satisfaction for subjects. A total of 31 fabric printing workers aged between 20 and 49 years participated in this study. The method that the worker's used to lift the screen plates was observed. A questionnaire was used to measure subjective fatigue associated with lifting the screen plates. The data from observation and the questionnaire was used to design a screen plate lifting instrument. The instrument was designed using the ergonomic and engineering principles in order to decrease improper lifting posture, increase convenience and effectiveness for lifting screen plates. After using the new-designed instrument, a questionnaire measuring feelings of satisfaction in addition to subjective fatigue was administered. The questionnaire also asked workers on the average period of time to lift screen plates before and after using the instrument. The results show that the subjective fatigue was reduced when using the screen plate lifting instrument as opposed to manual lifting ($p < 0.001$) especially in the middle back, lower back, wrist, hand and fingers. Using the screen plate lifting instrument reduced subjective fatigue and the average period of time for lifting each screen plate.

Keywords: Screen plate; Lifting; Fabric; Subjective fatigue; Printing

1 INTRODUCTION

Manual handling especially lifting represents a major occupational safety and health risk in industry. Musculoskeletal and low back disorders are often attributed to overexertion of the body during manual handling tasks (Plamondon et al., 2006, Lin et al., 2006). Heavier objects require more energy to handle and can cause whole-body and local muscle fatigue. As an employee becomes fatigued, they will be more likely to make errors, use improper lifting techniques, and cause an accident that can result in back injury (Kevin, 2000). Each year, injuries caused by manual handling contribute significantly to industry in terms of medical costs and lost productivity. Work-related musculoskeletal disorders (MSDs) are the most common types of occupational ill health in Thailand. The number of employees who have been injured by manual handling tasks has been increasing every year. Fabric printing in Thailand is an example of a job that requires a significant amount of manual handling of screen plates every day. The weight of a screen plate is in the range of 8.95 to 17.10 kg. The number of plates that a worker lifts per day can be 57 plates per printing machine. Moreover, the shape and size of these plates is large and heavy and can cause difficulties in lifting. If these tasks are performed repeatedly or over long periods of time, they can lead to fatigue, injuries and MSDs. Hsiang (1992) found that back injuries due to manual lifting are a serious problem in terms of human suffering and cost. The aim of this study was to improve the screen plate lifting method in the fabric printing section in order to reduce feelings of fatigue and increase feelings of satisfaction for subjects.

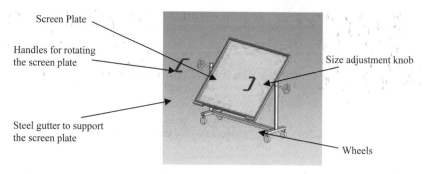

Figure 1.   The screen plate lifting instrument.

2   METHODS

2.1   *Subjects*

A total of 31 workers (18 male and 13 female) from the fabric printing section of a silk and fabric company in Thailand participated in this study. Their ages ranged from 20 to 49 years. Data from the annual employment health check and health condition questionnaire completed by the subjects indicated that they had no medical history of serious back injuries or any recent discomfort.

2.2   *Methods*

The method that the subjects used to lift the screen plates was observed. The subjective fatigue questionnaire (modified from the Nordic questionnaire) was used to compare the improvement in screen plate lifting methods before and after intervention. Subjective fatigue was measured by an 8 point Likert scale, where 0 meant no pain or fatigue and 7 meant pain or extreme fatigue where medical attention is required. The data from observation and questionnaire before improvement was used to design a screen plate lifting instrument. For the design procedure, a meeting was held with the factory manager, a leader of the fabric printing section and engineering to define the effective use of an instrument in a real situation. After using the newly designed instrument, the feelings of satisfaction questionnaire was used. This questionnaire was created based on the principles of safe manual handing (Chawanithikun and Noptepkangwan, 1997). This was measured using a 5 point Likert scale, where 1 corresponded to least feelings of satisfaction, and 5 the most satisfaction. There were a total of 13 topics regarding the feeling of satisfaction as follows: 1. Reduction in trunk flexion; 2. Reduction in lifting the hands up to shoulder height; 3. Reduction in reaching the hands forward; 4. Reduction in trunk twisting; 5. Reduction in rotating wrists during screen plate lifting; 6. Reduction in exerting force in lifting; 7. Reduction of the distance travelled; 8.Reduction in the risk of accident during lifting; 9. Reduction in number of members used for lifting; 10. Convenience of use; 11. Decrease of period of time of the lift; 12. Increased effectiveness of the lifting and; 13. Overall image of the feeling of satisfaction. The average period of time to lift the screen plates was compared the before and after-using the instrument. Statistical analysis was completed using a paired t-Test to compare the subjective fatigue during use of the screen plate lifting instrument to that during manual lifting.

3   RESULTS

3.1   *Screen plate lifting instrument design*

The screen plate lifting instrument was 150 cm in height and an adjustable width of 91.5 to 210 cm depending on the size of the screen plates. There was a steel gutter to support the screen plate, and holders for rotating the screen plate forward and backward. The instrument had four wheels for transport.

Table 1. The screen plate lifting method before and after improvement.

Before Improvement	After Improvement

Step 1. Moving the screen plate from the storage area

Step 2. Laying the screen plate down in front of the printing machine

Step 3. Lifting the screen plate up to the level of the printing press surface

Step 4. Pushing the screen plate down onto the printing press

Step 5. Moving the screen plate out of the printing machine

3.2 The screen plate lifting method before and after improvement

Before the improvement, subjects had to lift screen plates from a storage area, transport it to the printing press manually, lay it down on the floor in front of the printing machine, lift it up to the level of the printing press surface and push it down onto the printing press. After the improvement, subjects were able to pull a screen plate down onto the steel gutter of the lifting instrument, push it from a storage area to the front of the printing press, and push the screen plate forward onto the surface of the printing press. The results from the working posture assessment indicated that workers had problems with lifting heavy screen plates. They experienced spasms at the wrists and hands when lifting screen plates because of the unsuitable holders of the screen plate. Before the improvement, workers had to over-extend their arms due to the large size of screen plates, and flex forward in order to lay it down on the floor in front of the printing machine. The introduction of the screen plate lifting instrument removed the need to lift and carry the heavy screen plates and reduced extreme exertion by substituting pushing the screen plate instead of lifting. Moreover, over-extension of the arms was eliminated due to suitable holders for rotation of the screen plate.

Score of the feeling of fatigue

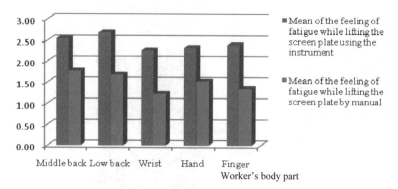

Figure 2.   Comparison of the subjective fatigue of the worker's body parts during lifting using the screen plate the lifting instrument and during manual lifting.

3.3   *Comparison of the subjective fatigue of the worker's body parts during lifting using the screen plate the lifting instrument and during manual lifting*

The results from statistic analysis using a paired sample t-test showed that the subjective fatigue of workers' body parts while lifting the screen plate using the instrument were significantly less than that of the manual lifting ($p < 0.001$) especially in the middle back, lower back, wrists, hands and fingers.

3.4   *Feelings of satisfaction when using the screen plate lifting instrument*

The results from the feelings of satisfaction questionnaire showed that workers were mostly satisfied at a moderate level in ten topics; reduction in reaching hands forward, reduction in trunk twisting, reduction in lifting hands up to shoulder height, reduction in rotating wrists during screen plate lifting, reduction in exerting force in lifting, reduction risk of accident during lifting, reduction in trunk flexion, reduction in the number of members used for lifting, reduction of the lifting distance and decrease in the period of lifting time. The topics that workers were less satisfied in were convenience of instrument use and effectiveness of the instrument for lifting. The mean of feelings of satisfaction in overall image when using the screen plate lifting instrument was 2.97.

3.5   *Lifting time when manual lifting and using the screen plate lifting instrument*

There were four steps in lifting the screen plate as follows; 1. Moving the screen plate from the storage area; 2. Lifting the screen plate up and pushing it down onto the printing press; 3. Moving the screen plate out of the printing machine; 4. Moving the screen plate back to the storage area. Using the instrument could reduce the time spent in lifting the screen plate up and pushing it down onto the printing press, moving the screen plate out of the printing machine and moving the screen plate back to the storage area. This instrument could not decrease the time spent in moving the screen plate from the storage area to in front of the printing press. However, the result from calculation the average time for lifting 57 screen plates per day/worker showed that the use of the screen plate lifting instrument reduced the transport of the screen plate to 2 minutes and 51 seconds.

4   DISCUSSION

Manual handling is one of the most common hazards that contribute back pain and injuries for workers in many workplaces. Risk factors of lifting task that we should consider are weight of the load, location of the load in horizontal and vertical distance, shape and size of load, frequency of lifting, period of time for lifting, working area, unbalance of lifting, stability of lifting and

Table 2. The feelings of satisfaction when using the screen plate lifting instrument.

| Topics | The percentage of satisfaction feeling | | | | | |
	Least 1	Less 2	Moderate 3	More 4	Most 5	Mean
1. Reduction in trunk flexion	3.2	3.2	80.6	9.7	3.2	3.1
2. Reduction in lifting hands up to shoulder height	0.0	12.9	61.3	22.6	3.2	3.2
3. Reduction in reaching hands forward	0.0	6.5	61.3	29.0	3.0	3.3
4. Reduction in trunk twisting	0.0	6.5	67.7	22.6	3.2	3.2
5. Reduction in rotating wrists during screen plate lifting	3.2	19.4	45.2	25.8	6.5	3.1
6. Reduction in exerting force in lifting,	3.2	12.9	58.1	19.4	6.5	3.1
7. Reduction of distance for movement	6.5	16.1	58.1	12.9	6.5	3.0
8. Reduction risk of accident during lifting	6.5	19.4	38.7	29.0	6.5	3.1
9. Reduction in the number of members used for lifting	6.5	16.1	51.6	22.0	3.2	3.0
10. Convenience of use	6.5	19.4	51.6	19.4	3.2	2.9
11. Decrease in the period of lifting time	6.5	16.1	51.6	25.8	0.0	3.0
12. Effectiveness of the instrument	6.5	19.4	61.3	12.9	0.0	2.8
13. Overall image of the feeling of satisfaction	3.2	19.4	54.8	22.6	0.0	3.0

Table 3. Lifting time when using the screen plate lifting instrument in each working step.

| Work Steps | Time | |
	Manual Lifting (Sec.)	Using Instrument (Sec.)
1. Moving the screen plate from the storage area	15	18
2. Lifting the screen plate up and pushing it down onto the printing press	7	5
3. Moving the screen plate out of the printing machine	9	7
4. Moving the screen plate back to the storage area	15	13
Total time of use	46	43
Average time for lifting 57 screen plates per day/worker	2622	2451

working environment (Plamondon et al., 2006). Several of these aspects were covered in this study as follows: weight, shape and size and the frequency of lifting screen plates. The use of the screen plate lifting instrument solved these problems.The results from the working posture assessment indicated that workers had problems with lifting heavy screen plates. They experienced spasms at the wrists and hands when lifting screen plates because of the unsuitable holders of the screen plate. Before the improvement, workers had to over-extend their arms due to the large size of screen plates, and flex forward in order to lay it down on the floor in front of the printing machine. The use of the screen plate lifting instrument could removed the need to lift and carry the heavy screen plates and reduced extreme exertion by substituting pushing the screen plate instead of lifting. Moreover, over-extension of the arms was eliminated due to suitable holders for rotation of the screen plate. Moreover, the results showed that subjective fatigue in workers body parts while lifting the screen plate using the instrument were significantly less than that of the manual lifting especially in the middle back, lower back, wrists, hands and fingers. The instrument achieved this by reducing the amount of trunk flexion, trunk twisting, lifting hands up to shoulder height, reaching the hands forward, exerting force in lifting and rotating wrists during screen plate lifting. When using the screen plate lifting instrument, the mean feelings of satisfaction of workers was of a moderate level. This was because this instrument was a prototype and a limited amount of time was allowed for familiarization. In addition, the period of time for manual lifting and using this instrument was not different. In Step 1 when moving the screen plate from the storage area, lifting the screen plate using the instrument took more time than that of manual lifting by 3 seconds. From observation and interview, workers had to lock and move the screen plate carefully. Furthermore, two of four wheels of this instrument could not propel easily. Therefore, it took more time for moving. However, total time for use of the screen plate lifting instrument per worker per day reduced the transport of the screen plate to 2 minutes and 51 seconds. This prototype instrument needs future improvement in the aspects of the convenience in moving and to further decrease the time in using the instrument.

5 CONCLUSION

In this study, we designed the screen plate lifting instrument in order to reduce subjective fatigue and increase feelings of satisfaction for subjects After using the newly-designed instrument, the subjective fatigue in the workers body parts while lifting a screen plate using the instrument were significantly less than that of the manual lifting especially in the middle back, lower back, wrists, hands and fingers. The feelings of satisfaction when using the screen plate lifting instrument was average, and the use of the screen plate lifting instrument reduced the transport of the screen plate to 2 minutes and 50 seconds. The unfamiliarity of using the instrument may have contributed to the minimal decrease in the lifting time.

REFERENCES

Chawanitikun, C. and Noptepkungwan, N., 1997, Safely Manual Handling, Department of Labour protection and welfare, Ministry of Labour. Thailand.

Hsiang, S., 1992, Simulation of manual materials handling, Doctorate Dissertation, United States: Texas Tech University, TX.

Kevin, S., 2000, Lessons for Lifting and Moving Materials, WISHA Services Division, Department of Labor and Industries, United States: Olympia, WA.

Lin, C.J., Wang, S.J., and Chen, H.J., 2006, A field evaluation method for assessing whole body biomechanical joint stress in manual lifting tasks, *Industrial Health*, 44, 604–602.

Plamondon, A., Delisle, A., Trimble, K., Desjardins, P. and Rickwood, T., 2006, Manual materials handling in mining: The effect of rod heights and foot positions when lifting "in-the-hole" drill rods, *Applied Ergonomics*, 37(6), 709–718.

Ergonomic Trends from the East – Kumashiro (ed)
© *2010 Taylor & Francis Group, London, ISBN 978-0-415-88178-4*

Comparison of work characteristics for evaluating musculoskeletal hazards of atypical works

Jong-Hun Yun
Korea Atomic Energy Research Institute, Korea

Hyeon-Kyo Lim
Chungbuk National University, Korea

1 INTRODUCTION

In order to prevent musculoskeletal disorders (MSDs), lots of researches are underway. The majority of them concentrate on highly repetitive works such as assembling parts in automobile manufacturing companies. On the other hand, works of mechanics or nurses have such characteristics that are infrequent but diverse so that cycle time or proportion of their work elements would be difficult to be estimated.

In spite of that, many researchers assumed that those works were similar to repetitive works and adopted the same assessment techniques that used for repetitive works. Of course, however, there would exist differences between them, and it would be so natural that those differences would make biases in assessment results.

Therefore, this research aimed to confirm work characteristics and similarity between jobs to draw methodology for assessing ergonomic hazards of musculoskeletal disorders, and to examine whether the same assessment techniques that used for repetitive works can be applied to atypical works.

2 METHODS

To collect work characteristics related with musculoskeletal disorders, a hospital, an automobile-part manufacturing company, and maintenance shops for automobile were selected.

In case of atypical works, it would be difficult to specify a worker's task because all the workers have to carry out diverse tasks regardless of his/her official title. Therefore the works of interest were selected with the base of self-assessment on risk factors (higher than 10 points: subjectively "heavy") and REBA assessment (higher than 4 points; "requires work improvement soon"). As a consequence, 56 works were selected from hospital works, and 28 works were selected from maintenance shops. On the contrary, in case of automobile-part manufacturing company, 15 works were selected with the same criteria were selected that assigned to individual worker.

Meanwhile, a questionnaire survey was conducted which consisted of questions about personal characteristics on one hand, and those about MSD symptoms on the other hand. And the results were compared with ergonomic assessment results with the 3rd Quantification Technique that were developed by C. Hayashi.

3 RESULT

3.1 Complaining MSD symptom

In case of hospital, complain of back-aches were as high as 60.4%, and followed by shoulder and neck with 58.0% and 46.8%, respectively. It was presumably attributed to standing posture throughout work hours and treating patients.

Table 1.　Result of investigation of risk factors.

	Work intensity	Work frequency	Total score
Hospital	4.7 ± 0.5	2.2 ± 0.7	12.7 ± 1.7
Maintenance shops	4.2 ± 0.5	2.6 ± 0.5	13.8 ± 3.4
Automobile-part manufacturing company	2.5 ± 0.7	4.0 ± 0.8	9.9 ± 2.4
t-value	11.80	5.08	4.32
significance level	0.00	0.00	0.00

Table 2.　Result of REBA evaluation.

		Action Level			
		1	2	3	4
hospital	frequency	9	40	51	16
	%	2.9	26.1	50.0	*20.9*
maintenance shops	frequency	1	27	48	45
	%	0.8	22.3	39.7	37.2
automobile-part manufacturing company	frequency	0	12	3	0
	%	0.0	80.0	20.0	0.0

REBA Action level	REBA score	Risk level	Action (including further assessment)
0	1	Negligible	None necessary
1	2~3	Low	May be necessary
2	4~7	Medium	Necessary
3	8~10	High	Necessary soon
4	11~15	Very High	Necessary now

However, in case of maintenance shops, shoulder-aches were ranked at top with 68.6%, and followed by neck and foot/ankle with 57.1% and 50.7%. This fact was not independent of that most mechanics had to spend much time with standing and overhead postures.

In case of automobile-part manufacturing company, as maintenance shops showed, shoulder-aches were ranked at top with 70.3%, and followed by neck and back with 61.8% and 58.8%, respectively. It was presumably attributed to repetitive work while standing and watching downward.

3.2　Result of ergonomic assessment

1) Investigation of risk factors

In hospital and maintenance shops as atypical work sites, work intensity was high while work frequency was low. It was caused by the fact that atypical work is non-repetitive and work method is can be changed by work objects and object condition. On the contrary, in automobile-part manufacturing company, work intensity was low while work frequency was high, as expected.

2) Result of REBA assessment

Table 2 shows the result of REBA assessment. Action level 3 and 4 occupied more than 70% in atypical work sites. But in automobile-part manufacturing company, Action level 2 was ranked at top with 80.0%, and followed Action level 3 with 20%. It meant that many works in atypical work sites have workload mainly due to work postures as compared with works in typical work sites.

3) Result of investigation of job stress factors

In the high risky group of MSDs, stress factors, compared with the normal group, were 'work environment', 'trouble in personal relation', and 'managerial system'. And workers in atypical work sites were relatively under the stress by job demands such as 'time load', 'short rest', 'increased work', and 'multitasking' than workers in typical work sites.

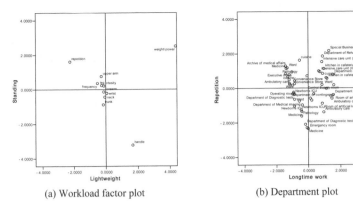

(a) Workload factor plot   (b) Department plot

Figure 1. Distribution of workload characteristic factors in hospital.

3.3 *Assessment through the 3rd Quantification*

1) Hospital
   (1) Analysis on work characteristics
       Fig.1 shows the analysis result of the 3rd Quantification Technique on similarity in distribution of work characteristic factors and workload factors among 56 works in hospital. As shown on the surface consists of axis 1 and axis 2, three points of 'repetition', 'weight', and 'handle' were placed far from the origin, and most other points were distributed within the triangle composed of those three points. It meant that the general workload factors of workers in hospital were mainly due to 'postural load', 'work intensity', and 'work frequency', and the three kinds of workload factors of 'repetition', 'weight', and 'handle' could be observed differently according to departments and works.
   (2) Complaining MSD symptom
       Comparing with body parts of MSD symptom and department of hospital, axis 1 divided the department into two groups. These departments plotted above the axis 1 have to handle heavy objects, whereas these departments plotted below the axis 2 have to handle light objects.
       Axis 2 divided department into two groups – standing works and sedentary works. Workers of departments plotted on right surface of graph had a rest place. On the other side, other workers of department plotted on left surface of graph was standing and working longtime and moving in a wider area.
       Work characteristics of the 1st quadrant was mainly for patients, and required careful treatment, therefore there were many complaints of back-aches and wrist/finger-aches. Heavy things in the 2nd quadrant were material, and workers did mainly standing works, therefore the workers complained of knee-aches and food/ankle-aches. In the 3rd quadrant, there are complaints of neck-ache, it caused by standing and repetitive works rather than weight materials. And finally, the works in the 4th quadrant were sedentary work with shoulder-aches.
2) Maintenance shops
   (1) Analysis of work characteristic
       As the analysis result of the 3rd Quantification Technique on similarity in distribution of work characteristic factors and workload factors among 28 works in maintenance shops, on the surface consists of axis 1 and axis 2, three points of 'repetition', 'weight', and 'leg' were placed far from the origin, and most other points were distributed within the triangle composed of those three points. It meant that the general workload factors in maintenance shops were 'postural load', 'work intensity', and 'work frequency', and the three kinds of workload factors, 'repetition', 'weight', and 'leg' could be observed differently according to works.

195

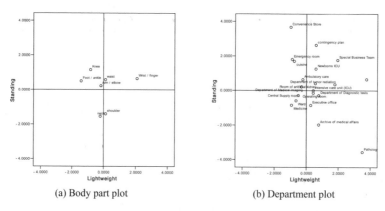

| (a) Body part plot | (b) Department plot |

Figure 2.   Distribution of MSD symptom factors in hospital.

(2) Complaining MSD symptom
  Shoulder and neck, foot/ankle gathered against axis 1 prove that these parts were not independent of standing and overhead postures. Back-aches were caused by bending the body at
  hood work, and gathered with neck and shoulder on axis 2. Therefore work characteristics
  in maintenance shops can be divided into overhead works and hood works.
3) Automobile-part manufacturing company
  (1) Analysis of work characteristic
    As the analysis result of the 3rd Quantification Technique on similarity in distribution
    of work characteristic factors and workload factors among 15 works in automobile-part
    manufacturing, the workload factors are 'work intensity, 'work frequency', and 'repetition',
    and 'a workload of weight materials' depends on the 'presence of handles' and 'the wrist
    posture'.
  (2) Complaining MSD symptom
    Unlike atypical work site, body part points of MSD symptom in automobile-part
    manufacturing company distributed widely over the graph.
    The point of neck was placed near the origin on graph, it meant that neck-ache was the most
    frequent symptom and can be concurrently observed with other body part aches. Other body
    part aches except neck could be observed differently according to works.

4   DISCUSSION

In case of atypical work sites such as a hospital and maintenance shops, general workload factors were mainly due to 'postural load', 'work intensity', and 'work frequency' as shown fig.1,
and divided into 'repetition' and 'weight'. On the contrary, in an automobile-part manufacturing
company, 'repetition' was also general workload factor, and workload of 'weight' was observed
in works of 'grinding', 'cutting', and 'plating' related to presence of handles. It meant that workload factors of 'postural load', 'work intensity', and 'work frequency' are common in every work
site, and workload factors mainly divide into 'repetition' and 'weight'. The 'repetition' was one of
general workload factors in automobile-part manufacturing company, but in atypical work sites,
'repetition' was specific factor with 'weight'.

In hospital and maintenance shops as atypical work sites, the results, high work intensity and low
work frequency, was caused by the fact that atypical work is non-repetitive and a working method is
can be changed by work objects and the object condition. And as for the result of REBA assessment,
action level 3 and 4 occupied more than 70% in ergonomic hazard assessment of atypical work
sites. It meant that many works in atypical work sites have workload by work posture rather than
repetition.

In the high risky group of MSDs, stress factors, compared with the normal group, were 'work
environment', 'trouble in personal relation', and 'managerial system'. It proved that MSDs would

Table 3. Relation of workload factors among each work site (%).

		Hospital	Maintenance shops	Automobile-part manufacturing company
Physical factors	Intensity	○	○	
	Frequency			○
	Trunk posture	○	○	
	Neck posture	○	○	
	Leg posture			
	Weight/Power	○	○	
	Upper arm posture			
	Forearm posture			○
	Wrist posture	○	○	
	Handle	○	○	
	Repetition			○
Psychological factors	Work Environment		○	○
	Job demands	○		○
	Self-regulation	○		○
	Trouble in personal relation			
	Instability		○	○
	Managerial system	○		○
	Inappropriate compensation	○		○
	Organization culture			

be affected by stress. And workers in atypical work sites were relatively under the stress by job demands such as 'time load', 'short rest', 'increased work', and 'multitasking' than workers in typical work sites.

Table 3 shows relation of workload factors among work sites. Relative high workload factors were marked by '○'. The results showed that workload in atypical work sites were high and similar in physical workload factors. Especially noticeable was high work frequency in automobile-part manufacturing company but not in atypical work sites, so it would not be reasonable to adopt assessment technique considering 'repetition' factors of importance like ANSI Z-365 at non-repetitive work in atypical work sites. On the contrary, psychological factors were different from work sites.

The work frequency in atypical work sites requires careful consideration when assessing the musculoskeletal hazards. In order to assess the posture load, lots of researches are underway. Some researchers asserted that the postures were selected by sampling technique, and other some researchers asserted that the highest load postures were selected. But in atypical work sites, the work frequency was not periodic and work postures were changed usually. If sampling techniques were used, the results could be underestimated, and most time and cost would be used to catch many work postures.

On the other hand, workers in a hospital, back-ache was the most frequent symptom as the complaints of MSD symptom, followed by shoulder, foot/ankle while workers in an automobile-part manufacturing company and maintenance shops complain of should-ache. It meant that the complaining of MSD depends on each work sites characteristics not between typical and atypical works.

5 CONCLUSION

In this research, work characteristics and workload factors were analyzed with the help of the 3rd Quantification Technique. The results showed that work characteristics and factors of MSDs in atypical work sites were quite different from those in typical work sites.

Based on analysis results, it's difficult to adopt ergonomic assessment techniques that used for typical works at atypical works because mainly the work frequency was not considered. And when assessing posture load of atypical work, it's appropriate to assess posture of the highest load for lowest risk.

Throughout this research, to evaluate musculoskeletal hazards need to consider not only physical factors such as work posture and repetition, but also work characteristics and psychological factors, and have to know the differences between work sites as well as between typical and atypical work sites.

REFERENCES

Yong-Goo Lee, Yong-Hee Lee, "A Comparison Study for the Quantification Analysis", *Journal of Statistics in Korea*, No.1, pp.35–57, 1994.

Hyung-Jin Rho, "A Study on the Analysis of Causal Relation about Categorical Data", *Journal of the Korean Institute of Office Automation*, Vol.5, No.2, pp.143–151, 2000.

Bon-Ean Koo, Keun-Sang Park, Chang-Han Kim, "Analysis of Musculoskeletal Disorders for Labor of In-standardization Work; on the Subject of the Centering-Work in a Shipbuilding Industry", *Journal of the Ergonomics Society of Korea*, Vol.26, No.2, pp.113–122, 2007.

I.Kloimuller, R.Karazman, H.Geissler, I.Karazman-Morawetz, H.Haupt, "The relation of age, work ability index and stress-inducing factors among bus drivers", *International Journal of Industrial Ergonomics*, Vol.25, pp.497–502, 2000.

Hignett & McAtamney, L., "Technical Note: Rapid Entire Body Assessment(REBA)", *Applied Ergonomics*, Vol.31, pp.201–205, 2000.

Alexander Cowell McFarlane, "Stress-related musculoskeletal pain", *Best Practice & Research*, Vol.21, No.3, pp.549–565, 2007.

Myung-Hoe Huh, Quantification Methods I, II, III, IV, Freedom Academy, 1992 (in Korean).

A comparison of the workloads of fish sorting tasks on three types of trawlers by the OWAS method

Hideyuki Takahashi
National Research Institute of Fisheries Engineering, Fisheries Research Agency, Kamisu, Ibaraki, Japan

Shuji Hisamune
Takasaki City University of Economics, Takasaki, Gunma, Japan

1 INTRODUCTION

The number of fishery workers in Japan has been decreasing for some time, and aging of the workers is becoming a major problem (Fisheries Agency, 2007). Possible causes for the decline in the number of workers are the work environment, poor management, and a decrease in fisheries resources. Moreover, fisheries work is dirty, dangerous, and demanding, and the accident rate is among the highest of all industries (Hisamune, 1999). Therefore, we must establish a safe and comfortable fisheries work environment so that aging fishermen can continue to work and younger workers can be more easily recruited. However, very few quantitative investigations of actual fisheries work conditions have been conducted, and thus the nationwide work conditions must first be evaluated to draw up the improvement policies of the place-of-work.

In this paper, we report an example of the work conditions on trawlers. Trawl fisheries are very important in Japan because the small and offshore trawl-fishery (single boat trawl) catches represent the third and fourth highest catch tonnage, respectively (Ministry of Agriculture, Forestry and Fisheries, 2008). Moreover, the work accident rate in the trawl fishery is the highest among all types of fishery (Hisamune *et al.*, 2003).

Sorting tasks comprise a necessary trawling fish activity because many species are collected from the seabed, and fish of economical value must be separated from other catch. As the sorting tasks require the most time in the tasks onboard (Takahashi, 2006), identifying proper sorting procedures is important to improve the work environment. The objective of this study was to identify differences in the sorting procedures of three types of trawl fisheries based on a quantitative workload analysis.

2 MATERIALS AND METHODS

We conducted video imaging studies of the sorting tasks on a 160 Gross Tonnage (GT) offshore trawler (Hokkaido), an 80 GT offshore trawler (Aomori), and a 9.7 GT small trawler (Chiba) (Table 1). The 160 GT trawler had a conveyer belt in the cabin, and the crew stood next to the conveyor as they sorted the fish catch (Figure 1(a)). On the 80 GT trawler, workers stooped or

Table 1. Outline of each objective ship and investigation time.

Ship	Port of registry	Number of crew	Investigation time
160 GT offshore trawler	Wakkanai, Hokkaido	18	June, 1998
80 GT offshore trawler	Hachinohe, Aomori	8	Nov, 1998
9.7 GT small trawler	Choshi, Chiba	3	May, 2005

Figure 1. Example images of the sorting tasks on the 160 GT offshore trawler (a), 80 GT offshore trawler (b), and 9.7 GT small trawler (c). Positions of the crew were shown by white arrows.

squatted and sorted fish on the deck of the ship (Figure 1(b)), while on the 9.7 GT trawler, the fish catch was placed on the deck and then moved to a simple sorting table where the crew sat on simple chairs and sorted the fish (Figure 1(c)). Although we had already performed a preliminary analysis of the work on these trawlers (Hisamune *et al.*, 2003; Takahashi, 2009), a reanalysis was conducted using a single method to facilitate strict comparisons.

The Ovako Working-posture Analyzing System (OWAS) method, one of the simplest and most widely used methods (Karhu *et al.*, 1977), was applied. In the OWAS method, the information of the postures of upper body, upper limbs and lower limbs, and the weight of objects handled by a worker are required to evaluate a workload. Worker posture was chosen from a list of categorized postures (Figure 2), and the weights of the objects were classified into those that weighed less than 10 kg, 10–20 kg, or more than 20 kg. Once the postures and the weights were identified, we estimated the workload using action categories (ACs) (Table 2) using JOWAS software (Seo, 2002). Still pictures were extracted from the video records with a 1-s snap reading method. A work posture for each body part and the weight of the object being handled were chosen, and the AC was estimated for

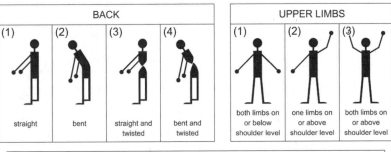

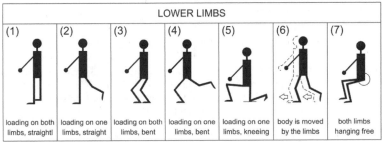

Figure 2. Categorized work postures used in the OWAS method (based on Karhu *et al.*, 1977).

Table 2. Action categories used to judge the physical load of the work posture by the OWAS method (based on Kant *et al.*, 1990).

Action Category	Harmfulness of work posture
AC1	Normal posture: NO ACTIONS REQUIRED;
AC2	The load of the posture is slightly harmful: actions to change the posture should be taken IN THE NEAR FUTURE;
AC3	The load of the posture is distinctly harmful: actions to change the posture should be taken AS SOON AS POSSIBLE;
AC4	The load of the posture is extremely harmful: actions to change the posture should be taken IMMEDIATELY.

Table 3. Profile of the results of the work posture analysis.

Ship	Number of observed crew	Number of data (N)
160 GT offshore trawler	6	2269
80 GT offshore trawler	5	2126
9.7 GT small trawler	2	5515

the worker in the still image. The same procedures were repeated for all crew in all still pictures, and the frequencies of the ACs were estimated.

3 RESULTS

Determining the workers involved with sorting on the 160 GT and 80 GT trawlers was difficult because many intermittently joined the tasks. The crew whose working postures could be identified is shown in Table 3. In the case of the 9.7 GT trawler, the analysis was focused on two crews they

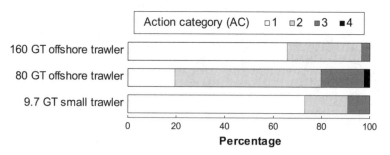

Figure 3.   AC frequencies in sorting task on each trawler.

were clearly engaged in a sorting task. Figure 3 shows the average AC frequencies for all of the crews on each trawler. The crew on the 80 GT trawler had the highest frequency of work postures with the strong degrees of improvement-demand (AC 3 or higher, 20%). The frequency was the lowest for the crew of the 160 GT trawler (4%), and the frequency of the crew on the 9.7 GT trawler was midway between those of the two offshore trawlers (9%).

4   DISCUSSION

According to the results, the use of suitable machines or equipment controlled the high workloads. The crew on the 160 GT trawler could stand when using the conveyer belt to sort fish. The crew on the 9.7 GT trawler could sort in a sitting posture, although they had to install the sorting table and chairs in advance. In contrast to these methods, the crew on the 80 GT trawler had to stoop or squat with deep forward bending of the upper body, which increased their workload.

To improve working conditions on the large trawlers, equipment for sorting fish should be considered in the design phase of ship construction. Although incorporating large machines or equipment in small trawlers is difficult, simple devices such as a sorting table and chairs might be useful. It is possible that these types of simple devices are introduced not only in new boats but also existing boats. However the equipments should be ergonomically designed since the unsuitable use of them may increase workloads or possibilities of work accidents on the contrary. For example, Anderson and Örtengren (1974) reported that the physical load on the lower back is higher in a sitting posture without proper lumber support than in a standing posture. Small trawlers usually have limited deck space, so installing the equipments so as to prevent obstruction of the onboard work is also necessary.

The decrease in fisheries resources has had a deleterious effect on many kinds of fishing in Japan, and the trawling industry has suffered greatly. Because trawl fisheries are critical in providing various fish species to markets, in addition to supporting the traditional food culture of Japan, continued study is needed to sustain this fishery sector.

ACKNOWLEGMENTS

We are thankful to all the fishermen and the staffs of the fishermen's cooperative associations for their kind cooperation with our researches. Our gratitude is also expressed to Mrs. Mayumi Izu for her devoted help of the analysis.

REFERENCES

Anderson, G.B.J. and Örtengren, R. (1974), Myoelectric back muscle activity during sitting, In *Scandinavian Journal of Rehabilitation Medicine* (London, Taylor and Francis), Supplement 3, pp. 91–108.
Fisheries agency (2007), Fisheries of Japan – 2006/2007, In Homepage of Ministry of Agriculture, Forestry and Fisheries (http://www.maff.go.jp/).

Hisamune, S. (1999), The ergonomics study of fishing boat's workers – II –the behavior of the trawler workers-, In *The Journal of Japan Institute of Navigation* (Tokyo, Japan Institute of Navigation), 101, pp. 253–258. (in Japanese).

Hisamune, S., Kimura, N. and Amagai, K. (2003), Ergonomic study on the behavior of trawler workers analysis of work movement line and working posture. In *Fisheries Engineering* (Kamisu, Japanese Society of Fisheries Engineering), 40(2), pp. 151–158.

Kant, I., Notermans, J.H.V. and Borm, P.J.A. (1990), Observations of working postures in garages using the Ovako Working Posture Analyzing System (OWAS) and consequent workload reduction recommendations, In *Ergonomics* (London, Taylor and Francis), 33(2), pp. 209–220.

Karhu, O., Kansi, P. and Kuorinka, I. (1977), Correcting working postures in industry: A practical method or analysis, In *Applied Ergonomics* (Amsterdam, Elsevier), 8(4), pp.199–201.

Ministry of Agriculture, Forestry and Fisheries (2008), Statistics of fishing and aquaculture productions in Heisei 19 year (corrected edition), In Homepage of Ministry of Agriculture, Forestry and Fisheries (http://www.maff.go.jp/). (in Japanese).

Seo, A. (2002), The Ovako working-posture analyzing system software JOWAS ver. 0.9, In Homepage of ergonomics and occupational health (http://homepage2.nifty.com/aseo/index.html) (in Japanese).

Takahashi, H. (2006), Quantitative analysis of fishery works using a stationary video observation, In *Proceeding of the Annual Meeting of the Japanese Society of Fisheries Engineering* (Kamisu, Japanese Society of Fisheries Engineering), pp. 271–274. (in Japanese).

Takahashi, H. (2009), Work analysis on a small trawl fishery at Choshi city, Chiba prefecture, In *Fisheries Engineering* (Kamisu, Japanese Society of Fisheries Engineering), 46(1), pp.1–8. (in Japanese).

Chapter 10   Cognitive ergonomics 2

Ergonomic Trends from the East – Kumashiro (ed)
© *2010 Taylor & Francis Group, London, ISBN 978-0-415-88178-4*

Evaluation of three 3D virtual reality display devices

Hung-Jen Chen, Ping-Yun Cheng, Yung-Tsan Jou & Chiuhsiang Joe Lin
Department of Industrial and Systems Engineering, Chung Yuan Christian University, Chung Li, Taiwan

1 INTRODUCTION

Virtual Reality (VR) is a modern technology that offers huge potential impact for improving people's lives at home, work and school. Since introduced in the 1960's, interest in Virtual Reality has escalated as people find uses for VR in architecture, interior design, medicine, education, training and many other industries. Initially, more attention was paid to the development of the technology and its potential application uses, so until recently very little research has been reported on usability and ergonomic issues associated with VR.

Virtual Environments (VE) can be described as computer-generated immersive surroundings where participants feel they are part of the simulated world and can interact intuitively with on-screen objects. The hardware of a typical immersive VR system includes a computer, a display device and a hand-held input device. As computer technology progresses, there are plenty of display devices on market that could be adapted to present 3D virtual environments. However, the performance enhancement, comfort and fatigue of using these display devices to present a 3D VE are still indeterminate.

The 3D stereo television (3D TV) technology introduced by Philips in 2006 is an industry breakthrough. This autostereoscopic 3D display provides a stereo image that does not require users to wear special 3D glasses. However, in an interactive Virtual Environment, the impact of the 3D TV on human motor performance is still unclear. For this reason, we apply Fitts' Law to this evaluation.

Fitts' Law (Fitts, 1954) is well known as an effective quantitative method for modeling user performance and evaluating manual input devices. Fitts' Law states that the time to move and point to a target of width W at a distance A is a logarithmic function of the spatial relative error (A/W), as shown in Equation (1),

$$MT = a + b \log_2(2A/W) \tag{1}$$

where *a* and *b* are empirical constants determined through linear regression. The log term is called the index of difficulty (ID) and describes the difficulty of the motor task (bits).

Some variations of the law have been proposed, such as the Shannon formulation (MacKenzie, 1989). This equation (Equation (2)) differs only in the formulations for ID.

$$MT = a + b \log_2(A/W + 1.0) \tag{2}$$

The benefit of the Shannon formulation is that it provides the best statistical fit, reflects the information theorem underlying Fitts' law, and always gives a positive ID. The throughput (TP) proposed by ISO (ISO, 2000) is the most familiar performance measurement of information capacity of human motor system. This approach combines the effects of the intercept and slope parameters of the regression model into one dependent measure that can easily be compared between studies (Soukoreff and MacKenzie, 2004). As shown in Equation (3), MT is the mean movement time for all trials within the same condition, and $ID = \log_2(D/W + 1)$.

$$Throughput = ID/MT \tag{3}$$

Fitts' Law is an effective method for evaluating input devices. However, it has rarely been applied in comparing visual display devices. Fitts' pointing paradigm is a visually-controlled task, hence either input devices or display devices could influence performance of human motor system. In order to test whether Fitts' Law is applicable to compare display devices, Chun et al. (2004) used Fitts' Law to evaluate four 3D stereo displays, all with the same haptic input device. The results of this study demonstrated that Fitts' Law was applicable to the evaluation of visio-haptic workstations.

With the exception of performance measurement, subjective questionnaires were usually used to measure discomfort that resulted from display devices in the virtual reality environment. In 1997, Howarth and Costello proposed a Simulation Sickness Questionnaire (SSQ) to compare the effects of using a Head Mounted Display (HMD) versus a Visual Display Unit (VDU). The results showed that HMD was rated with higher scores than VDU in fatigue, headache, nausea and dizziness.

In summary, the present study attempts to evaluate the 3D TV and two other display devices, the HMD and projection display, via Fitts' task and Simulator Sickness Questionnaire. The results of this study will show how each these three displays influences human motor performance.

2 METHODS

2.1 *Subjects*

Ten right-handed adult subjects aged 20 to 30 years old (six males and four females, mean age = 26.7 years, SD = 2.8 years) participated in this experiment, All subjects were undergraduate Industrial and Systems Engineering students of Chung Yuan Christian University. All subjects had normal or corrected-to-normal vision with no other physical impairments.

2.2 *Apparatus*

Three display devices were used to present a virtual reality interactive environment. The first display device tested was the 3D stereo television (Philips 42" WOWvx). The 3D TV is used to get a distinct image into each eye of the viewer. From that point, the viewer's brain takes over, processing each image in the same way that it processes the images it receives from the three-dimensional world. The resolution of 3D TV was 1280×1024 pixels and refresh rate was 60 Hz. It was located 1.5 meters in front of subjects and the field of view was estimated at 32.8 degree. The second display device used a head mounted display (i-glasses VIDEO 3D Pro), which was worn like ordinary goggles. The resolution of HMD was 800×600 pixels, refresh rate was 60 Hz, and the field of view was 26 degree along the diagonal. The third display device was a Benq PB6100 projector with a light output of 1500 ansilumens. At a three-meter distance, it projected a 1340 mm × 1010 mm image with 1680 mm diagonal on a projective screen. The resolution was 800×600 pixels, refresh rate was 60 Hz, and the field of view is 25.2 degree.

The experiment environment created by EON professional 5.5 software was a 50 cm × 50 cm × 50 cm cube. The cube background was set to white, and two dark gray tiles were drawn in the vertical plane at the right and left screen. The purpose of the cube was to enhance the depth perception in our virtual reality environment. The sphere-shaped cursor was motivated by VR SPACE Wintracker and was attached on subjects' base proximal phalange of medius. There were three red sphere targets located respectively in X, Y and Z directions.

2.3 *Procedure*

The experiment was a within subjects experiment conducted with ten subjects. The factors were three display devices (3D TV, HMD and projection display.), three movement directions (X, Y and Z direction) and three levels of ID (2, 3 and 4 bits). Movement direction Y was into the screen, X was horizontal across the screen and Z was vertical across the screen. IDs were produced by a constant target distance of 19 cm and three target diameters. The C:D ratio was always $1 : 1$. Therefore, a hand movement in a particular distance resulted in an equal cursor movement.

At the start of each condition, subjects were allowed practice trials to familiarize tasks. Following this, subjects carried out all 27 experimental conditions in a random order to minimize the effect of learning and fatigue. Subjects were instructed to point as fast as possible while still maintaining high accuracy. Movement time began when the subject pressed the start point and did not stop until the subject correctly clicked inside the targets. Each condition consisted of ten pointing trails. The average movement time and throughput of these ten pointing trials were recorded for ANOVA analysis.

The majority of the symptoms were assessed on a seven-point rating scale. These were: general discomfort, fatigue, boredom, headache, eyestrain, sweating, claustrophobia, disorientation, nausea and difficulty concentrating. The scale endpoints were described on the questionnaire as "no symptoms" (1) and "severe symptoms" (7); the intermediate scale points (3) and (5) were described as "slight" and "moderate", respectively. The remaining symptoms on the questionnaire were reported simply as being either present or absent. These were: vomiting, dizziness, faintness, exhilaration, stomach awareness, confusion, and a non-specific category "other symptoms". These scores were then collected and transferred to the Minitab software for the statistical analysis.

3 RESULTS

3.1 *Dependent Variables Analysis*

3.1.1 *Movement time*
Mean movement time for the three display devices: 3D TV, HMD and projection display respectively were 3689, 4324 and 3467 ms, which proved that there were significant differences between display devices in movement time ($F_{2,234} = 34.31$, $p < 0.01$). The movement time for projection display was the fastest; HMD movement took the longest time. Further post-hoc Duncan tests showed that the movement time of HMD was significantly longer than 3D TV and projection display. The main effect for movement directions ($F_{2,234} = 35.04$, $p < 0.01$) and IDs ($F_{3,234} = 190.82$, $p < 0.01$) were also significant in movement time. Movement time was 4323 ms for the highest Y direction, followed by Z direction (3709 ms), with X direction measuring lowest at 3448 ms. Results showed that movement time increased with the level of indices of difficulty (2820 ms for 2 bits, 3075 ms for 3 bits and 4912 ms for 4 bits). Post-hoc Duncan showed that the movement time was different between the three movement directions and between the three levels of IDs statistically.

3.1.2 *Throughput*
Throughput computed for the projection display at 0.9144 bps was the highest. For 3D TV and HMD, measurements were 0.8610 and 0.7320 bps respectively. The main effect for display devices was clearly significant ($F_{2,234} = 35.01$, $p < 0.01$). Further Duncan test results showed that there was no significant difference between projection display and 3D TV, but both were obviously higher than HMD.

The ANOVA results also revealed significant differences between movement directions ($F_{2,234} = 46.99$, $p < 0.01$) and IDs ($F_{2,234} = 9.87$, $p < 0.01$). The post-hoc Duncan test showed that Y direction (0.7135 bps) was statistically lower than X (0.9210 bps) and Z (0.8729 bps) directions. Throughput for three levels of IDs were 0.7799, 0.8523 and 0.8752 bps in order, which increased with the level of ID. The post-hoc test showed that the throughput obtained at 2 bits was statistically lower then it obtained at 3 and 4 bits.

3.2 *Simulation sickness questionnaire*

The means and standard deviations of the responses on six of the seventeen SSQ questions are shown in Table 1. The remaining eleven questions, which participants didn't experience, are excluded in this table. The results of the questionnaire analysis show that all of these six responses are statistically significant between display devices. Subjects clearly rated the discomfort, fatigue, boredom, headache and eyestrain with higher scores while using HMD than while using 3D TV and projection display. The mean responses for these six questions were lowest for the projection display, which implies that subjects considered the projection display more comfortable, because it did not induce fatigue, headache and eyestrain. Discomfort, fatigue and eyestrain scores for 3D TV were near the midpoint, suggesting that similar to the projection display, the 3D TV did not induce headache.

Table 1. Simulator sickness questionnaire rating of three display devices on 5-point scale.

Variable	3D TV	HMD	Projection	P-value
General discomfort	$2.90(1.02)^b$	$4.60(0.54)^c$	$1.00(0.00)^a$	0.000
Fatigue	$2.89(0.77)^b$	$4.44(0.50)^c$	$1.08(0.27)^a$	0.000
Boredom	$1.84(0.54)^b$	$2.26(0.53)^c$	$1.02(0.15)^a$	0.000
Headache	$1.06(0.23)^a$	$3.01(0.44)^b$	$1.00(0.00)^a$	0.000
Eyestrain	$2.68(0.90)^b$	$4.47(1.03)^c$	$1.09(0.39)^a$	0.000
Difficulty concentrating	$1.99(0.57)^b$	$2.79(1.27)^c$	$1.09(0.32)^a$	0.000

1. Data are presented as mean (SD).
2. a, b, c groups frm the post-hoc Duncan test.

It could be concluded from the questionnaire data that the SSQ assessment showed significant difference between the three display devices. Based on the results, subjects showed a preference for projection display.

4 DISCUSSION AND CONCLUSION

Based on the results, there were significant differences between display devices on human motor performance. It is surprising that the projection display obtained the best performance and highest preference among three display devices. The HMD received the worst results in both experiment and SSQ assessment. Chun *et al.* (2004) considered that HMD did not provide good stereo sense and only 800×600 resolution with a small field of view. This uncomfortable configuration made subjects feel fatigue and eyestrain. The 3D TV was as good as projection display on human motor performance. This means that the 3D TV provided enough depth cues; however, this technique, which delivers a distinct image into each eye, also resulted in middle level of discomfort, fatigue and eyestrain.

REFERENCES

Chun, K., Verplank, B., Barbagli, F. and Salisbury, K., 2004, Evaluating haptics and 3D stereo displays using Fitts' law. In *Proceeding of the 3rd IEEE Workshop on HAVE*, pp. 53–58.
Fitts, P., 1954, The information capacity of the human motor system in controlling the amplitude of movement. *Journal of Experimental Psychology*, Vol. 47, pp. 381–391.
Howarth, P.A. and Costello, P. J., 1997. The occurrence of virtual simulation sickness symptoms when an HMD was used as a personal viewing system, *Display*, Vol. 18, 2nd, pp. 107–116.
ISO, 2000. *Ergonomic requirements for office work with visual display terminals (VDTs) Part 9: Requirements for non-keyboard input devices.* Geneva, Switzerland: International Organization for Standardization.
MacKenzie, I. S., 1989. A note on the information-theoretic basis for Fitts' law. *Journal of Motor Behavior*, Vol. 21, pp. 323–330.
Soukoreff, R. W. and MacKenzie, I. S., 2004, Towards a standard for pointing device evaluation: Perspectives on 27 years of Fitts' law research in HCI. *International Journal of Human-Computer Studies*, Vol. 61, pp. 751–789.

Ergonomic Trends from the East – Kumashiro (ed)
© 2010 Taylor & Francis Group, London, ISBN 978-0-415-88178-4

Perception and response of driver by driving simulator

Hidetoshi Nakayasu
Faculty of Intelligence and Informatics, Konan University, Kobe, Japan

Nobuhiko Kondo
CELL Institute for Educational Development, Otemae University, Nishinomiya, Japan

Tetsuya Miyoshi
Faculty of Business and Informatics, Toyohashi Sozo University, Toyohashi, Japan

1 INTRODUCTION

It is not unimportant for a driver who drives an automobile safely to perceive hazard which is associated with accidents in a traffic situation (Lalley, 1984 and Brown, 1988). This hazard perception is a needed function to take on hazard information at early stage (Soliday, 1974 and 1975 and Renge, 1998), and the most of necessary information is obtained from visual information processing and attention (Miura, 1992).

A new experimental paradigm is performed in this paper for the measurement and analyses of time histories of eye movements and vehicle response in simulated driving. The aim of this work is to offer the knowledge for hazard information by relevance between several traffic situations and human factors in order to forecast a human behavior such as visual attention during driving work. It is studied that the relation between histories of vision motion by the eye tracking system and mechanical behavior of vehicles at the same time.

2 EXPERIMENTAL SETUP

2.1 Driving simulator

The automobile driving simulator (DA-1102: HONDA Motor Co. Ltd.) was used to simulate driving scenarios. The schematic aspect of DS and driver warred with eye tracking system are shown in Figure 1. This simulator is typically used for driver trainees or novices to learn safe driving techniques. The simulator is implemented on an advanced 6-axis motion base (sway-motion device) that can simulate closely automobile dynamics in real traffic situations. The visual scenes of driving a road from the viewpoint of a driver are presented with audio on a frontal screen and three mirror-CRTs that are synchronized with the behaviours of vehicle as a response of driver. This simulator enables one to drive a vehicle in scenario courses such as urban road, rural road, and highway and so on. Experimental studies were performed at an urban road scenario shown in Figure 2. This course consists of several traffic events; merging, passing and crosswalk with pedestrian, where dangerous situations were installed. The mechanical behaviour of vehicle such as current position, and degrees of accelerator pedal and brake pedal can be recorded by the sampling rate of each 10 msec.

2.2 Eye tracking system

During simulated driving, eye movements in the driving road scene were recorded by a head-mounted eye tracking system (EyeLink II: SR Research Ltd.). Figures 3 and 4 show a head set and its block diagram with control and host PCs of the system. As shown in Figure 3, two eye cameras in head set track the right and left eyes separately. The range of trace of eye cameras is within ±30 degrees in horizontal axis and ±20 degrees in vertical axis with accuracies of 0.01(deg.) in vertical and 0.022(deg.) in horizontal resolutions. The sampling rates are 500 Hz in pupil one recording and 250 Hz in pupil and corneal reflection recording.

Figure 1.   Schematic aspect of DS and eye tracking system.

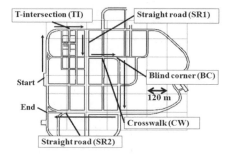

Figure 2.   Scenario of urban road course for driving in DS.

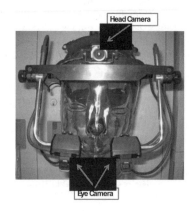

Figure 3.   Head set of eyelink II.

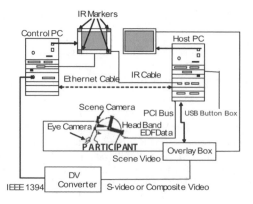

Figure 4.   Block diagram of eye tracking system.

On the other hand, a scene camera on a head band in head set also records visual scene that the participants' view. The scene camera generates 30 frames of a scene image per second where the resolution of the flame is 720 and 480 pixels.

It is seen in Figure 4 that the system consists mainly of several components; EyeLink Host PC, EyeLink Control PC, Scene Camera, Overlay box, and DV converter. The scene image is also streamed to the overlay box, in which participants' gaze position is superimposed onto the visual scene.

2.3 *Experimental procedure*

Participants were seated in the driving simulator and drove the simulator in automatic transmission (AT) mode. After the participants practiced driving in the simulator for two or three runs on the urban road, an experimental session was conducted. In the experiment, participants were instructed to drive safely on the driving courses. On the urban road course the participants drove about 60 km/h. No specific instruction about eye movements was given.

In this study, horizontal and vertical eye movements of the participants were recorded from both eyes with 250 Hz sampling rate in pupil and corneal reflection recording. Before the experimental session, the participants were asked to fixate on nine points presented on the recording system monitor to calibrate the eye tracking system. At the same time, the physical data of vehicle such as current position, angle of wheel, degrees of depth of accelerator pedal and brake pedal can be recorded by 10 msec sampling rate. All data were stored on the computer and analyzed off-line by computer programs that drew scan paths of eye movements and traced the trajectories of the vehicle.

2.4 *Participants*

Participants were four males from 21 to 23 years old (mean age of 21.8 ± 0.8) that had not experienced until they drove DS. All participants had regular class automobile licenses in Japan with normal or corrected-to-normal vision. All participants received an explanation of the experiments and gave informed consent before participation.

3 EXPERIMENTAL RESULTS

3.1 *Eye movements*

Figure 5(a) and (b) show the time histories of eye movement of a participant when he turns left and right respectively at the intersection of urban road in Figure 2. In the figures, there were little eye movements in vertical axis (Y-axis) though the typical eye movements appeared more than 15 degrees in horizontal axis (X-axis). It is recognized in both turns that typical patters of eye movements in horizontal axis appeared between start and end of the turns. It is also found that there were several times of a set of eye movements appeared in horizontal axis from the beginning to the end of turns.

3.2 *Vehicle movements*

Figure 6(a) and (b) show the trajectories of vehicle movement in left and right turns for the same participant in 3.1. It is seen that the overshooting was occurred in the early stages of the left turn though the smooth trajectory was plotted in right turn without overshooting and undershooting. It is important that the trajectories of the turn will be corresponded to the time histories of eye movements by synchronizing data.

3.3 *Gaze points in front view from driver's seat*

Figure 7(a) and (b) show the movements of gaze point of driver in front view from driver's seat during left and right turns. As pointed out in 3.1 eye movements, it is confirmed that there are several iterations of same eye movements found during the operation of vehicle turning. It is suggested that this means the eye movements during turn will be divided into several kinds of unit work of eye

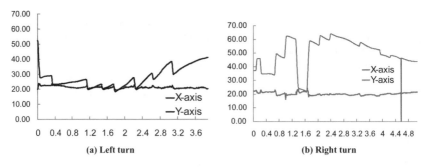

(a) Left turn   (b) Right turn

Figure 5.   Time histories of coordinates of eye movements in the process of turning.

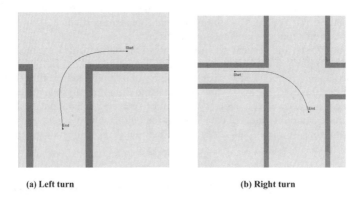

(a) Left turn   (b) Right turn

Figure 6.   Trajectory of vehicle.

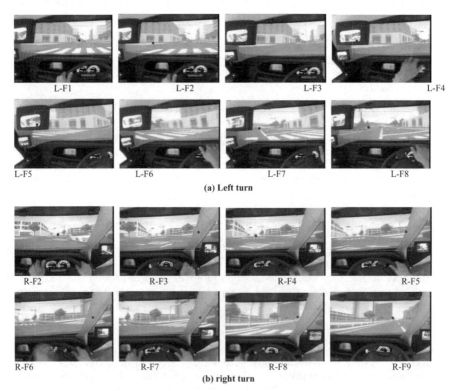

Figure 7.   Fixation points in vision fields.

movement during turning task. It is important to clarify what this typical eye movement requires and why the eye movements are divided into small tasks. It is important to solve these questions that eye movements during turn will be corresponded to the time histories of eye movements under the synchronous version.

ACKNOWLEDGEMENT

The authors express thanks to MEXT since a part of this work was supported in part by MEXT ORC (2004-2008). In addition, it is noted taht this work was also supported by the Hirao Taro Foundation of the Konan University Association for Academic Research, Japan.

REFERENCES

Brown, I.D. and Groger J.A., 1988, Risk perception and decision taking during the transition between novice and experienced driver status, In *Ergonomics,* Vol.31, pp.585–597.

Lalley, E.P., 1982, *Corporate uncertainty and risk management,* (Risk Management Society Publishing).

Miura, T., 1992, Visual Search in Intersections – An Underlying Mechanism –, In *IATSS RESEARCH,* Vol.16, pp.42–49.

Renge, K., 1998, Drivers' hazard and risk perception, confidence in safe driving, and choice of speed, In *IATSS RESEARCH,* VOL.22, pp.103–110.

Soliday, S. M., 1974, Relationship between age and hazard perception in automobile drivers, In *Perceptual & Motor Skills,* VOL.39, pp.335–338.

Soliday, S. M., 1975, Development and preliminary testing of a driving hazard questionnaire, In *Perceptual & Motor Skills,* VOL.41, pp.763–770.

Ergonomic Trends from the East – Kumashiro (ed)
© *2010 Taylor & Francis Group, London, ISBN 978-0-415-88178-4*

Effects of visual information given to driver on driving control performance

Yuko Tsutsumi & Toru Miyazaki
Vehicle Testing & Research Dept. Mazda Motor Corporation, Japan

Yoshifumi Miyahama
Powertrain Technology Development Dept. Mazda Motor Corporation, Japan

Kazuo Nishikawa & Takahide Nouzawa
Vehicle Testing & Research Dept. Mazda Motor Corporation, Japan

Hiroyuki Izumi & Masaharu Kumashiro
Department of Ergonomics UOEH, Japan

1 INTRODUCTION

Driving behaviour consists of a behavioural sequence of cognizing information from outside world such as road conditions, and information on position and conditions of one's vehicle, followed by selecting individual actions and then performing the operation (Doi, 2004). Thus, information from outside world and information on position and conditions of one's vehicle being correctly cognized by the driver is an important key in order to appropriately drive under the ever-changing road conditions. Approximately 90 percent of the information necessary for driving a vehicle is regarded as visual information (Hartman, 1970). Visibility is a very important performance in a vehicle because it may affect driver's operation. Therefore, our mission as an automotive manufacturer is to develop a vehicle with a visibility, which can accurately provide drivers with visual information necessary for driving.

Aiming to offer a safe and an enjoyable car life, this study attempts to clarify how visual information given to a driver may affect one's driving control performance and to define its mechanism, by comparing the behaviours of two vehicles (namely, vehicle A and vehicle B) with different front visibility driven on a winding road.

2 EXPERIMENT

The test subjects were two healthy male who drive routinely, without visual impairment or any other disease that may interfere with one's driving ability. Two vehicles with different front visibility, vehicle A and B were selected for the test. Vehicle A represents a vehicle with typically good visibility. The specifications of each vehicle on front visibility are shown in Table 1.

A winding road was selected as the test track. This is because sports driving, including winding driving, is the condition that most requires accuracy to the driver in deciding the next driving behaviour based on the cognized vehicle position and external information shown in the front visibility. The vehicles' behaviour was observed during driving by recording the lateral distance between the front wheel and the centre line with a CCD camera, to be used as an index for driving control performance (Figure 1). Drivers' eye-gaze trace was also monitored to observe the impact on gaze behaviour with different visual information, using an eye-tracking camera. The subjects were instructed to drive with 50 km/h vehicle speed.

Table 1. Specifications of the front visibility of the test vehicles.

	Vehicle_A	Vehicle_B
Eye-point Ground Clearance (mm)	1117	1040
Visibility Angle, Front Up (degree)	19.7	18.2
Visibility Angle, Front Down (degree)	2.9	5.1
Lateral Visibility Angle: Dr side (degree)	24.2	26.3
Lateral Visibility Angle: Pass side (degree)	56.9	61.7
A-pillar Obstruction Angle: Dr side (degree)	4.6	6.6
A-pillar Obstruction Angle: Pass side (degree)	6.1	7.9

Figure 1. Video recording to measure the lateral distance between the front wheel and centre line – distance (a).

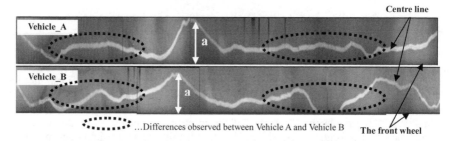

Figure 2. Difference of the changes of distance (a) between the two vehicles.

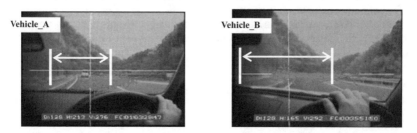

Figure 3. Lateral eye movement during driving: vehicle A vs vehicle B.

3 RESULTS

The observation data on driving behaviour was compared to study the impact on driving control performance. The results of both drivers showed a trend that driving behaviour of vehicle A, with generally better visibility, was much smoother and driving was also stable. Figure 2 shows the result of subject I as a representative.

Data of eye-gaze trace during driving was compared to study the impact on gaze behaviour. The results of both drivers showed a trend eye movement smaller with vehicle A that has generally better visibility. Figure 3 shows the result of subject I as a representative.

Figure 4. View of centre line through windshield: vehicle A vs vehicle B.

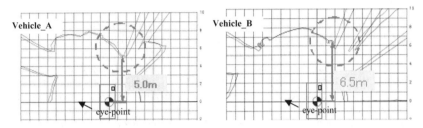

Figure 5. Visibility chart: vehicle A vs vehicle B.

4 DISCUSSION

The important keys for the realization of smooth and stable driving are accurate cognition of information from outside world such as road conditions, information on position and condition of one's vehicle, and unerring decision of driving behavioural target based on the information. Therefore, the relation between the cognition of visual information necessary for driving and the front visibility was studied. According to the analysis of the gaze behaviour data of the drivers captured with an eye-tracking camera, following relations were discussed:

- Comprehension of vehicle position vs bottom contour of windshield
- Comprehension of road condition ahead vs eye-point, A-pillar location and angle
- Comprehension of vehicle speed vs A-pillar location and angle

4.1 Relation between vehicle position cognition and bottom contour of windshield

Focusing on the view through the windshield, difference between vehicle A and B on how the centre line could be seen was identified (Figure 4). When driving vehicle A, the centre line is much visible from windshield at the foot of A-pillar than with vehicle B. Further study was conducted using visibility chart, which is an index that shows the range of road surface visible to driver. As a result, it was clarified that the closest point from the foot of A-pillar where the road surface is visible from the drivers' eye-point was approximately 1.5 m closer in case of vehicle A than that of vehicle B (Figure 5). In addition, vehicle A had a wedge-shaped design at the bottom of the windshield near the foot of A-pillar. This was considered to help focus the drivers' eye to a certain position where the centre line is being sighted in the windshield and to help maintain a certain lateral distance between the vehicle and the centre line.

Although further study is required regarding how the bottom contour of a windshield affects the driver in cognizing one's vehicle position, it is indicated that the design of the windshield has some relation to the accuracy in cognizing one's vehicle position against the road, and has impact on smoother and more stable driving.

4.2 Relation between comprehension of road condition ahead and eye-point

Looking into the point of gaze during driving, difference was identified between vehicle A and B (Figure 6). With vehicle A, the drivers were gazing above the vanishing point of road surface and

219

Figure 6.   Point of gaze during driving: vehicle A vs vehicle B.

were viewing far ahead, while they were gazing around the vanishing point or lower and viewing a closer area with vehicle B. From this, it is considered that when driving vehicle A, it was easier to cognize the condition of the road ahead and to determine the next behaviour target at an earlier stage, which made sufficient time to move on to the next action, resulting in a smooth operation. But with vehicle B, the delay in cognizing of the road condition ahead and making decision of the next behaviour target, seems to lead to unsmooth operation. Also, the visual angle in checking the circumstance ahead and the amount of lateral displacement of eye movement might have increased when trying to sight closer area in case of vehicle B. A possible factor of this result may be the eye-point ground clearance of vehicle A being higher than that of vehicle B by approximately 75 mm, consequently raising the gazing point higher with vehicle A. Meanwhile, the comparison of the drivers' eyeball movement suggests another factor that with vehicle B, the drivers' were trying to gaze as far as possible with their eyeballs turning upward from under their brows. The drivers' eye-point were possibly at a location where they had to look from under, meaning that they might have been in a driving posture where it was difficult to gaze farther even if they want to. It is said that the direction of human eyes is typically biased to lower than horizontal direction and this is the best condition for the human eyes to capture the visual object at ease. Based on this, vehicle B has a visibility that may possibly fatigue the driver's eyes. This indicates that the point of gaze has an impact on driving control and the driver's eye-point, which determines the gazing point, is related to driving posture. Further study is required in order to clarify the relation between gazing point and driving posture and to discuss a visibility that enables early cognition of front road condition and poses less stress to the driver's eyes.

4.3   Relation between comprehension of road condition ahead or vehicle speed and A-pillar location/angle

The views of the landscape ahead at bends were different with vehicle A and B. The road condition at the exit of the bend was visible with vehicle A, while it was not visible with vehicle B as the A-pillar obstructed the visibility. The obstruction angle of A-pillar indicating the degree of interruption was larger with vehicle B compared to that of vehicle A. The possible cause for this may be the A-pillar location in vehicle B close to the driver's eye-point. The interruption of the view of road condition ahead may cloud the decision for the next behaviour target, induce frequent steering correction, impede driving performance and increase driving workload. The obstruction of visibility may also affect the sensibility of vehicle speed. It is said that what is most important in speed perception of human is visual information (Yamanaka, 1990). Visually perceived speed could be described with the degree and vector of the visibility angle rate as a flow of road and surrounding landscape passing by. In other words, visibility obstruction interrupting the flow of landscape may reduce the accuracy of the speed perception and lead to a misjudgement of vehicle speed. This may also cloud the decision for the next behaviour target, induce frequent steering correction, impede driving performance and increase driving workload.

5   CONCLUSION

In this study, the behaviour of two vehicles with different visibility was compared to clarify the impact of visual information given to a driver on driving performance and to define its mechanism.

As a result, differences were identified between the two vehicles. The results showed that the bottom contour of the windshield, eye-point, A-pillar position and angle might have an impact on the comprehension of one's vehicle position and speed or on the accuracy of the cognition of external information, and ultimately on driving control performance. Further work is required in order to examine these findings in detail and to develop a level of vehicle visibility that can provide smooth and stable driving.

REFERENCES

Doi, S., 2004, Recent and Future Trends of Driving Performance Evaluation, In *Journal of Society of Automotive Engineers of Japan* Vol. 58 (12), pp. 4–9.
Hartman, E., 1970, Driver vision requirements, In *Society of Automotive Engineers*, Technical paper Series, 700392, pp. 629–630.
Yamanaka, A. *et al.* 1990, Illustrated Ergonomics, Japan, edited by Noro, K., (Japan: Japanese Standards Association), pp. 314–315.

Chapter 11   Ergonomics in occupational health 3

Ergonomic Trends from the East – Kumashiro (ed)
© 2010 Taylor & Francis Group, London, ISBN 978-0-415-88178-4

Variations in esophageal temperature with distance from nostrils

Seichi Horie, Shoko Kawanami, Makiko Yamashita, Nozomi Idota & Takao Tsutsui
Department of Health Policy and Management, Institute of Industrial Ecological Sciences,
University of Occupational and Environmental Health, Japan, Kitakyushu, Japan

Yasuhiro Sogabe & Koichi Monji
Bio-information Research Center, University of Occupational and Environmental Health,
Japan, Kitakyushu, Japan

1 INTRODUCTION

Ideally, efforts to prevent heat stroke at workplaces should seek to reduce the temperature and humidity of the working environment. At real-world workplaces where improving the thermal environment itself may be impractical, feasible alternatives may include shortening working hours, rotating workers, or improving work styles. Ideally, such alternatives should be assessed to ensure that they do in fact reduce core body temperatures among workers, given individual characteristics.

Of the various measurements of internal body temperature, the most important is rectal temperature, which most closely reflects the temperatures of the visceral organs. However, a drawback of this index is latency following exposure to or removal from heat stress. Esophageal temperature may represent a superior index for detecting the peak value of internal body temperatures, since it quickly reflects changes in body temperature due to circulating blood. In theory, the optimal point for measuring esophageal temperatures is just behind the left ventricle and aorta. An earlier study performed in Canada suggests that the optimal insertion length of an esophageal probe can be predicted from a subject's height by placing the tip at the level of the eighth and ninth thoracic vertebrae (Mekjavić and Rempel, 1990). We performed the study reported here to clarify differences in esophageal temperatures and to identify the optimal measurement distance from the nostrils among East Asian subjects.

2 METHODS

The study examined two healthy volunteers: Subject A, a 42 y.o. male measuring 167 cm in height, and Subject B, a 37 y.o. female measuring 160 cm in height. After entering a climatic chamber (Model TBL-15FW5CPX, Tabaiespec) controlled at 35°C dry bulb and 60% relative humidity, the subjects were asked to insert six copper-constantan thermocouples covered with a polyethylene tube (Figure 1) through the pharynx into the esophagus at distances of 13, 28, 33, 38, 43, and 50 cm from the nostrils. One subject was also asked to insert two thermocouple probes together, one thick and one thin, into the rectum; the thick thermocouple probe, covered in vinyl resin, measured 3.47×1.91 mm diameter; the thin thermocouple probe, covered in polyethylene resin, measured 1.05×1.05 mm diameter (Figure 2). We measured skin temperatures at the frontal plane, chest, the left forearm, the dorsal surface of the hand, and at the thigh, calf, and feet. Electrocardiograms ("Bioview 1000", NEC) were monitored throughout the experiment. The experiment adhered to the protocol adopted in the preceding study (Horie, *et al.*, 2002), involving a repeated exercise and rest cycle with the subject exposed to heat, intended to simulate actual work conditions and heat stress. After a brief period to allow temperatures to stabilize, the subjects were asked to pedal a bicycle ergometer ("Corival 400", LODE) to maintain a work rate of 75 W continuously for 20 minutes.

Figure 1.   Thermocouples used to measure esophageal temperatures at six different locations.

Figure 2.   Cross-sectional photographs of thermocouples used to measure rectal temperatures: thick (vinyl resin, left) and thin (polyethylene resin, right).

The subjects were then asked to repeat the exercise two additional times after a 15-minute rest period. We sampled esophageal temperatures at x cm distance (Tes-x); rectal temperatures (Tre-thick, Tre-thin); skin temperatures at seven different points on the skin (Tsk-x); and Heart Rates (HR), at intervals of 15 ms, monitored outside the chamber, and recorded the data to a computer (Mate NX, NEC) every 5 seconds using an analog-to-digital converter (Remote Scanner DE1200, NEC). Mean skin temperatures were calculated by the formula of Hardy and Dubois.

3   RESULTS

The experiments were performed without incident, with neither subject exhibiting any adverse effects from heat exposure.

All indices increased during exercise periods. Most indices declined during rest periods, except for Tes-13. The peak respective HRs were 154 and 171 for Subject A and Subject B, respectively.

Highest body temperatures were recorded at Tes-50 with Subject A, ranging from 37.4°C to 38.4°C, and at Tes-43 with Subject B, ranging from 37.6°C to 38.7°C (Figures 3, 4). Tes-33 and Tes-38 were mostly within 0.1 to 0.4°C of the highest temperature. However, T-28 was 0.7 to 0.9°C lower and Tes-13 much lower than Tes-28. The Tre of the male subject with the thin tube (Tre-thin) showed continuously higher temperatures than with the thick tube (Tre-thick) (Figure 5).

Elevations in Tre values trailed elevations in Tes and Tsk values. Declines in Tre were less marked than those of Tes or Tsk. Both Tre-thin and Tre-thick changed within a smaller range than Tes, and peak values were 0.3°C lower than for Tes-50. Tsk moved similarly to Tes, but with greater amplitude, although all Tsk values were continuously lower than the highest Tes (Figures 6, 7). The highest Tsk was recorded at the forehead of both subjects, followed by the forehand in Subject A and the chest in Subject B. The lowest Tsk was recorded at the chest with Subject A and at the forearm with Subject B. The distribution of Tsk values differed between the two subjects.

4   DISCUSSIONS

Except for Tes-13, Tes values measured at various distances from the nostrils changed in a similar pattern during and after exercise, indicating that these temperatures truly reflect esophageal

226

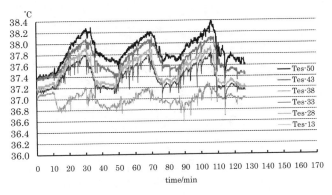

Figure 3.　Esophageal temperature of male subject measured at 13, 28, 33, 38, 43, 50 cm from the nostrils.

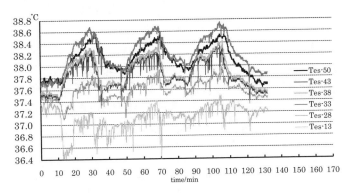

Figure 4.　Esophageal temperatures of female subject measured at 13, 28, 33, 38, 43, and 50 cm from the nostrils.

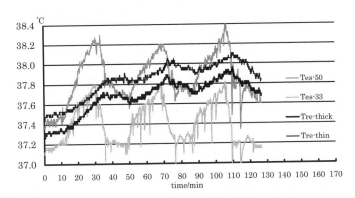

Figure 5.　Rectal temperatures of male subject measured with thick and thin thermocouple probes.

(rather than stomach) temperatures. Tes-13 may reflect temperatures in the pharyngeal region. Suppressed elevation or temporary declines in Tes-13 during exercise are likely attributable to the influx of ambient air at 35.0°C from the open mouth of the subjects, resulting in cooling of the thermocouples themselves.

The measured values for Tes clearly differed with varying distance from the nostrils. Tes at the lower region of the esophagus demonstrated higher temperatures, presumably due to the temperature of blood-rich organs such as the left ventricle and aorta located just behind that point. The differences between the two subjects in the depth at which the highest temperature was recorded are likely

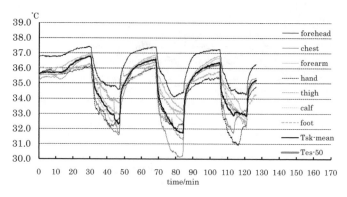

Figure 6.   Skin temperatures of male subject as measured at seven different locations.

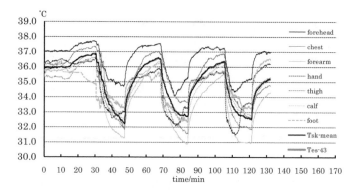

Figure 7.   Skin temperatures of female subject as measured at seven different locations.

due to anatomical differences in the length of the esophagus, which is also related to overall body size. At the same time, this suggests we may be able to predict the optimal insertion length of an esophageal probe based on subject height. A formula proposed in a report from Canada (Mekjavić and Rempel, 1990) gives an optimal depth of 37.9 cm for Subject A and 36.3 cm for Subject B. The lengths observed in our study were longer, perhaps due to anatomical characteristics specific to East Asians.

The thickness of the materials coating a thermocouple appeared to affect measured rectal temperatures by up to 0.2°C, a non-negligible difference. It may be advisable to adjust for the thickness of the material covering the copper-constantan thermocouple when comparing such temperatures.

Tes responded faster and rose much higher than Tre after initiation of exercise, indicating that Tes at the lower region of the esophagus may be a better index than Tre for early detection of heat strain. This observation has also been reported in other studies (Lefrant *et al.*, 2003). Our findings in a preceding study showed that the time to peak temperature and thereafter to declines in temperature following the cessation of exercise was less for Tes than Tre (Lee *et al.*, 2000). This is explained by the lag in temperature elevations in organ tissue compared to circulating blood. Although rectal temperature is currently regarded as the gold standard among indices of core body temperature, it may be possible to predict the development of heat stroke even earlier by assessing esophageal temperatures.

Inter-individual differences in Tsk appear to be due to differing patterns of sweating at different points on the skin.

Finally, due to the limited number of study participants, caution is in order when generalizing these results.

5 CONCLUSIONS

Esophageal temperatures differed significantly—up to 0.9°C—with varying distance from the nostrils. Esophageal temperatures measured at 43 to 50 cm from the nostrils may reflect heat strain more quickly than rectal temperatures. Anatomical characteristics specific to East Asians may require a modified formula for identifying optimal points at which to measure esophageal temperatures.

REFERENCES

Horie, S., Tsutsui, T., Sakata, S., Monji, K., and Sogabe, Y., 2002, Optimum room temperature during rest periods provided between repetitive exercise under heat stress. In *Environmental Ergonomics X*, edited by The Organizing and International Program Committees of the 10th International Conference on Environmental Ergonomics, (Fukuoka, Kyushu Institute of Design), pp. 81–84.
Lee, SM., Williams, WJ., Fortney Schneider, SM., and Schneider SM., 2000, Core temperature measurement during supine exercise: esophageal, rectal, and intestinal temperatures. In *Aviation, Space, and Environmental Medicine*, Vol.71(9), pp. 939–945.
Lefrant, JY., Muller, L., de La Coussaye, JE., Benbabaali, M., Lebris, C., Zeitoun, N., Mari, C., Saïssi, G., Ripart, J., and Eledjam, JJ., 2003, Temperature measurement in intensive care patients: comparison of urinary bladder, oesophageal, rectal, axillary, and inguinal methods versus pulmonary artery core method. In *Intensive Care Medicine*, Vol.29(3), pp. 4141–418.
Mekjavić, IB. and Rempel, ME., 1990, Determination of esophageal probe insertion length based on standing and sitting height. In *Journal of Applied Physiology,* Vol.69(1), pp. 376–379.

Ergonomic Trends from the East – Kumashiro (ed)
© *2010 Taylor & Francis Group, London, ISBN 978-0-415-88178-4*

Chernobyl radioactive contamination survey by imaging plate

Shinzo Kimura
Human Engineering and Risk Management Research Group, National Institute of
Occupational Safety and Health, Kanagawa, Japan

Satoru Endo
Quantum Energy Applications, Graduate School of Engineering, Hiroshima University,
Hiroshima, Japan

Sarata K. Sahoo
Department of Radiation Dosimetry, National Institute of Radiological Sciences,
Chiba, Japan

Yoshikazu Miura
Department of Public Health, Dokkyo Medical University, Tochigi, Japan

Tetsuji Imanaka
Research Reactor Institute, Kyoto University, Osaka, Japan

ABSTRACT: The Chernobyl Nuclear Power Plant (CNPP) accident occurred on April 26th, 1986, at Chernobyl, Ukraine (formerly USSR) leading to changed public perception of nuclear risk. Among others, agricultural and environmental impacts pose serious problems. Radioactive contamination in Chernobyl and surrounding areas of cesium-137 and strontium-90 persists in soil posing a threat to land plants. Here we report use of a photostimulable phosphor screen imaging technique to detect radioactive contamination that may affect the workers in the exclusion zone at Chernobyl. Utilizing samples from the exclusion zone, including plants, and fish, imaging plate technology was applied to show photo-stimulated luminescence from contaminated samples on imaging plate. Our results suggest the importance of imaging plate technology in monitoring the potential risks to human population.

1 INTRODUCTION

The worst ever accident in the history of nuclear power plants occurred on April 26th, 1986, at Chernobyl, Ukraine, formerly a part of the USSR. It is referred to as the "Chernobyl Nuclear Power Plant (CNPP) accident" (An international advisory committee., 1991, Mould, R. F., 2000). The CNPP accident exposed most of the population of the northern hemisphere to various degrees of radiation. Due to this, after 1986, the public perception of a nuclear risk was changed to a great extent. Other than the obvious and much studied health impact, the agricultural and environmental impacts, relatively unstudied, still pose a serious problem. At present, 22 years after the CNPP accident, contamination is still a major problem in Chernobyl and the surrounding areas originally included in the exclusion zone. Cesium-137 (^{137}Cs, gamma- and beta-emitter), which has a half-life of 30.1 years, is the most important radionuclide from Chernobyl's catastrophic explosion, and is present at high concentrations in the 0–5 cm soil layer. Another damaging residual radionuclide, strontium-90 (^{90}Sr, beta-emitter), which has a half-life of 29.1 years, also persists in the soil layer. Both of these radionuclides pose a potential threat to plant life in the region.

This report describes the use of a photostimulable phosphor screen imaging technique to detect radioactive contamination in environmental radiation monitoring. This technology was confirmed with samples obtained from the CNPP accident. The sample used was leaves of wormwood

(*Arte misiavulgaris L.*) and fern (*Dryopteris filix-max* CL. Schoff) plants, the bark of pine (*Pinus* species) tree, and the scale of a fish (Carp). The imaging plate technology is well known for many striking performances in two-dimensional radiation detection. Since imaging plate comprises an integrated detection system, it has been extensively applied to surface contamination distribution studies. In this study, plant samples were collected from high- and low-contaminated areas of Ukraine and Belarus, which were affected due to CNPP accident and exposed for analysis by the imaging technique. Samples from the highly contaminated areas revealed the highest photo-stimulated luminescence on the imaging plate. Based on the results obtained, the importance of imaging plate technology in environmental radiation monitoring has been suggested.

2   SAMPLING AND EXPOSURE TO IP

Plant leaves were prepared for exposure to an IP (BAS-MS2025, Fuji Photo Film Co., Ltd., Tokyo, Japan) by pressing the leaves between pages of a notebook and subsequent drying in clean air. The exposure was carried out for 12, 24, or 72 h in a shield box, as mentioned in the figure legends. The IP plate was analyzed using an IR Bio-Imaging Analyzer (BAS-2000, Fuji Photo Film Co., Ltd.). Photo-Stimulated Luminescence (PSL) intensity was also calculated and presented wherever necessary.

3   GAMMA AND BETA RAY MEASUREMENTS

Gamma-ray measurement was performed before the decomposition of samples, using a Ge-detector (CANBERRA GC1318, 1.9 keV half bandwidth), coupled to a multichannel analyzer (Laboratory Equipment Co., MCA-48F). During measurement, one channel was set to 0.5 keV so that measurement up to 4,000 keV was possible. The spectrum of beta-ray emission was measured for the soil sample. Digestion by an acid bomb was performed in a closed vessel (PTEF vessels) microwave unit (MLS 1200 mega, Italy) using a mixture of HNO_3-HF-$HClO_4$ for soil samples [Sahoo, S.K., Kimura, S., *et al*, 2002,]. The soil is extracted in the presence of strontium carrier and filtered. Calcium and strontium are then concentrated by precipitation as oxalates. With clay soil, kieselgur is added as a filter aid and, because of the presence of appreciable amounts of aluminum in the extract, the preliminary concentration of calcium and strontium is best affected by carbonate precipitation from sodium hydroxide solution. The carbonates are then dissolved in acid and alkaline earths precipitated as oxalates. The oxalates are ignite, the residue dissolved in dilute acid, any residual iron and aluminum removed as hydroxides, and the calcium and strontium precipitated as carbonates and weighed. The strontium is then separated from calcium by successive precipitations as nitrate and purified by two barium chromate scavenges and an iron hydroxide scavenger. Yttrium carrier is added and, after storing for a known length of time, preferably at least 14 days, separated, mounted and counted [IAEA, 1989]. The provided strontium was measured by low background beta-ray spectrometer (ALOKA LBC-471Q).

4   RESULTS AND DISCUSSION

Sample collection points were at the Masany ecology research center (10 km from the CNPP) in Belarus; CNPP circumference areas (1.78 km and 9.63 km from the CNPP) in Ukraine. Fern, wormwood, thistle, raspberry, mulberry, hop and moss plants were collected, and an index plant was selected based on the clarity of leaf structure, size and availability in all sampling areas mentioned above. Consequently, two kinds of index plants, namely wormwood and fern were selected. Wormwood is widely distributed from the subarctic to the subtropical zones, whereas the ferns are predominantly present in the dry sandy soils in and around Chernobyl.

The radioactive nuclide, which existed in these plants, was measured by a Ge-detector and a low background beta ray spectrometer. These radionuclides were identified as [137]Cs (half-life 30.07 year) and [90]Sr (half-life 28.78 year). This was compared with the same contaminated soil samples from the area and found to be 67.9 ± 0.024 Bq and 3.18 ± 0.33Bq for ^{137}Cs and ^{90}Sr, respectively. In this study, we have selected indicator plants such as wormwood and fern to check the IP technique.

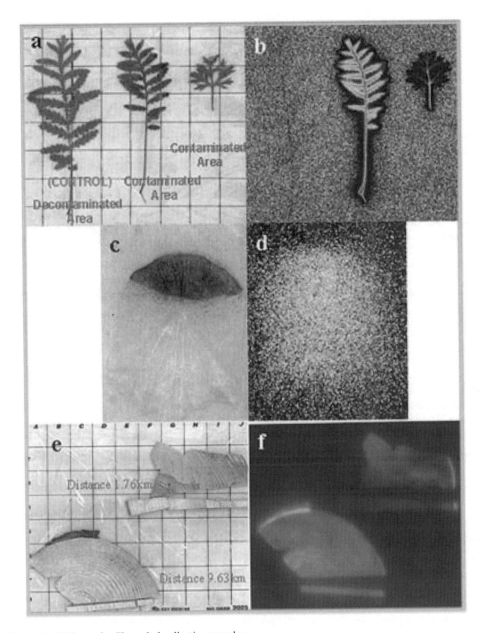

Figure 1.   IP image by Chernobyl collecting samples.

The influence of ^{40}K, which exists as a naturally occurring ring background radioactive source, was distinguished by checking the image in real time (data not shown). In order to investigate the distribution state of ^{137}Cs and ^{90}Sr, ^{137}Cs was separated using various shield materials, for example, since it was found that ^{90}Sr had reached radiation parallel with ^{90}Y, a radiation shield was used for ^{90}Y. Results presented in Figure 1(a, b) show ferns from Masany (both decontaminated and contaminated areas) exposed to the IP. It can be clearly seen that the leaves of the ferns from the contaminated areas show very high radioactivity, particularly in the veins (dark red color), strongly suggesting that these radionuclides had moved into the plants from the soil. In total contrast, control plants from the decontaminated area did not reveal any radioactivity. These initial data indicate the usefulness of IP in visually determining radioactivity.

Cesium and the strontium dissolve in water. The fish was collected in a polluted pond of Chernobyl. The cesium is concentrated to muscular tissue. In addition, the strontium concentrates to the bone and scale. We paid attention to the scale and observed the concentration of strontium. The strontium was found to spread around the scale from the growth center of the scale (Fig. 1. c, d). The pine was collected at about 1.78 km and a spot of 9.63 km from Chernobyl. As for the characteristic of the tree, radioactive nuclide was found to be concentrated by the bark. The spotting pattern in the rings also agreed with the diffusion direction of the radioactive nuclide at the time of the CNPP accident [Imanaka T., 1998]. After 8 years of the accident, the sample which we used was collected. In the annual ring at the time of the accident, strong radioactivity was detected in comparison with the annual ring of other years (Fig. 1. e, f).

5   CONCLUSION

Sample selection of plants is very simple at any accident site. It does not require any special treatments or sample preparation. Results can be obtained within a few hours (0.2~1 h), ascertaining the nature of the contamination due to radiation. For example, Chernobyl exclusion zone samples can be detected within 10 minutes. Measurement of environmental radioactivity using an IP can serve as an in expensive and rapid method/technique to check surface contamination in case of an emergency. Moreover, it can also be used for monitoring the long-term effects caused by a nuclear accident.

ACKNOWLEDGEMENTS

The authors kindly thank Dr. R. Rakwal (National Institute of Advanced Industrial Science and Technology, Tsukuba, Ibaraki, Japan) for helpful discussions. This works was supported by grants from the Ministry of Education, Culture, Sports, Science and Technology, Japan, and Ministry of Foreign Affairs, Japan (KAKENHI No. 20402001 & 22406019).

REFERENCES

An international advisory committee., 1991, The international Chernobyl project technical report. IAEA. Vienna.

Mould, R. F., 2000, Chernobyl Record. The Definitive History of the Chernobyl Catastrophe, Institute of Physics Publishing, Bristol, UK.

Sahoo, S.K., Kimura, S., Watanabe, Y., Shiraishi, K. and Masuda, A., 2002, Detection of ^{236}U and variation of uranium isotope composition in the soil samples affected by the JCO critical accident. *Proc. Japan. Acad.*, **78**:196–200.

IAEA, 1989, Measurement of radionuclides in food and the environment. Technical Reports Series No. 295, IAEA, Vienna.

Imanaka T., 1998, Research Activities about the Radiological Consequences of the Chernobyl NSP accident and Social Activities to Assist the Sufferers by the Accident. KURRI-KR-21, KUR Report of the Research Reactor Institute of Kyoto University.

Chapter 12   Physical ergonomics

Ergonomic Trends from the East – Kumashiro (ed)
© 2010 Taylor & Francis Group, London, ISBN 978-0-415-88178-4

Understanding diaper wearing and fitness of sizing system for infants

Bor-Shong Liu
Department of Industrial Engineering and Management, St. John's University, Tamsui, Taipei, Taiwan

Ching-Wen Lien
Nursing Department, Taipei Veterans General Hospital, Taipei, Taiwan

Chih-Hung Hsu
Department of Industrial Engineering and Management, Hsiuping Institute of Technology, Taichung, Taiwan

1 INTRODUCTION

Diapers are commonly worn from birth until a person is toilet trained. Diapers evolved in the 1960s from a double layer of cotton folded into a triangle and attached with safety pins to products that include a top sheet of plastic and cellulose pulp core. More recently, the core was changed to gel for its absorbency potential. Product innovations include the use of superabsorbent polymers, re-sealable tapes and elastic waist bands. Gel also prevents skin from becoming super hydrated, and as a result diaper rash has been on the decline in recent decades. However, it still occurs, and in most cases, diaper dermatitis is caused by the diaper itself (Gorgos, 2006). Diaper rash or diaper dermatitis is one of the most common skin disorders in infants and toddlers (Liptak, 2001). The etiology is multifocal and a diaper rash may present in various conditions in the pediatric community (Borkowski, 2004). Diaper dermatitis usually occurs as a primary reaction to irritation by urine, feces, moisture, or friction (Van Onselen, 1999). Distribution patterns may vary, but irritative dermatitis typically involves the convex surfaces where the skin is in greatest contact with the diaper. Irritative dermatitis usually spares the inquinal folds, and may be mild red, shiny, and with or without papules (Liptak, 2001; Wysocki and Bryant, 1992).

Thus, care and management of diaper dermatitis can present a challenge for pediatric nurses and care providers. Prevention of diaper rash is achieved through maintenance of skin integrity to prevent damage to the stratum corneum, the skin's barrier. Keeping the baby dry, which entails frequent diaper changes, is the ideal way to both treat and prevent irritant diaper dermatitis (Jorden *et al.*, 1986). Diapers should be changed and the area cleaned and allowed to dry as soon as possible after soiling or wetting. The skin should be exposed to the open air for 5 to 10 minutes following each diaper change. The use of plastic pants should be avoided or at least limited and diapers should not be too tight. Changing of diaper brands may be considered with chronic diaper rash breakouts. Irritants should be avoided or removed by washing with warm water and cotton balls and patting dry.

In addition, the argument has been that diaper type (cloth versus paper) is an important factor in controlling the spread of fecal material and therefore a major contributor to the spread of enteric pathogens in the environment (Kubiak *et al.*, 1993). Previous studies have suggested that paper diapers are more effective than cloth diapers in constraining fecal spread (Kubiak *et al.*, 1993; Van *et al.*, 1991). Apart from diaper types, and materials, the sizing system of diaper is another important factor for wearing comfortable. However, the sizing systems of disposable diaper are divided into stages and sold based on three distinct groupings in Seaddlers, Cruisers and Baby Dry (e.g. Pampers Diapers). The sizing systems of diaper are only classified according to growing stage and infant's weight. If an infant is six months with weight of seven kg, how should pick up the diaper size to coordinate with her or him (Size 1: Small or Size 2: Medium?). Obviously, selection of diaper size for infants is a problem for parents and care providers.

There are several strategies to improve the fitting comfort of products to wear on body: sizing, adjustable products, traditional customisation, and mass customization. In population, accommodation, sizing has been aimed at improving the product to either get better fit, or broaden the range of befitting users (Li, 2006; Nui et al., 2006). Anthropometry is considered the very ergonomic core of any attempt to resolve the dilemma of 'fitting the tasks to the human' (Sanders and McCormick, 1993). As products are designed for specific types of consumers, an important design requirement is selection and efficient utilization of the most appropriate anthropometric database (Wickens et al., 2004). Furthermore, a garment sizing systems are used to fit different groups of the population based on demographic anthropometric data. Persons of the same subgroup have the same body shape characteristics, and share the same garment size (Ashdown, 1998; Chung et al., 2007). With the difference in body dimensions and morphological characteristics, different body shapes can be generalized to a few figure types (Ray et al., 1995). Emanuel et al. (1959) recommended the use of the difference in figure types as the classification of ready-to-wears, and developed a set of procedures to formulate standard sizes for all figure types. According to this system, people of all figure types were first classified into one of four body weight groups (i.e. super heavy, heavy, normal, light. These body weight groups are further subdivided by stature (i.e. tall and short). People were, thus, divided into eight categories based on similar heights and weights.

Standard sizing systems of garment are very crucial issue, play an even important role for garment manufacturing industry. Under traditional production procedures, garment manufacturers have never developed standard sizing systems for the market, finally resulting in heavy stock burden to garment manufacturers. Standard sizing systems can correctly predict numbers of items and ratio of sizes to be produced, resulting in accurate inventory control and production planning (Hsu et al., 2007). Furthermore, the purpose of this study was to provide product designs with the anthropometic dimensions of baby and children and analyze along with demographic data, including gender and age. The second purpose was to understand the diaper wearing problems and evaluate the fitness of size system for infants to compare the dimensions of various diapers with anthropometic database and recommend appropriate solutions for design.

2 METHODS

The present investigation consisted of two stages. In the first stage, a questionnaire was used to survey about determine preferences and problems with respect to worn diaper for infants and children. In the second stage, provide product designs with the anthropometic dimensions of infants and children. Finally, recommend the appropriate solutions for sizing systems of diaper with anthropometric database.

2.1 Questionnaire survey

The study was approved by the Research Ethics Committee of the researcher's institution. A total of 300 parents which were infants and children from newborn to 3 years old have been interviewed. The questionnaire was designed to determine background information, age of infant or children, size of worn diaper currently, and discomfort problems. In addition, participants filled the questions as followed: how are the descriptors and statements related to choice in diapers for infants. Factors were including price, material, breathability, sides design, fitness, wetness indicator and absorbability etc. Ranking of descriptors are (1) very important, (2) important, (3) ordinary, and (4) not important.

2.2 Methodology of measurement

Three hundred subjects who aged from newborn to 3 years old were separated into eight stratifications (i.e. under one month, 1–3 months, 3–5 months, 5–7 months, 7–9 months, 9–12 months, 1–2 years, and 2–3 years old) for further analysis. Registered nurses in Tamsui Health Center have conducted the anthropometric survey. Altogether six anthropometric characteristics were measured (Figure 1).

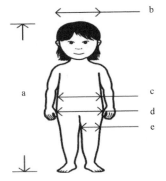

Stature;
Head circumference;
Waist circumference;
Buttocks circumference;
Thigh circumference;
Body weight.

Figure 1. Anthropometric measurements.

2.3 *Data analysis*

All data were coded and summarized using SPSS 13.0 software for Windows. The frequencies procedure provides statistics that are useful for describing opinions of questionnaires. Furthermore, Analysis of variance (ANOVA) was used to investigate the effects of age and gender on anthropometric dimensions of infants. Where statistically significant differences were determined, the Duncan post hoc test was performed.

3 RESULTS

3.1 *Survey for questionnaire*

Results revealed that reports of participants about impression and fitness problems while wearing diaper. About 38% of participants had no impression on the body when wearing diaper. Further, only 35% and 26% of participants denoted well-suited for waist and thigh respectively. In addition, results of questionnaire revealed that about 22.7% and 29.3% of participants considered the sale price as very important and important factor respectively when choosing the diapers (Table 1). In addition, more than half participants considered the material, side design and cutting fitness as the very important factor for choosing diapers. These were 58% and 60.7% of participants denoted the very important factors for breathability and absorbability. Only 39% of participants reported the wetness indicator as the extremely important factor.

3.2 *Effect of age on dimensions*

Results of ANOVA showed that these were not significant differences in all dimensions between genders. By contrast, all dimensions were significant differences between age groups ($p < 0.001$). Further, Duncan's post hoc test revealed that mean stature could be divided into seven subsets ($p < 0.05$). Mean head circumference were the shortest for the age under one month (35.6 cm), followed, in ascending order, by the 1–3 months (38.5 cm), 3–5 months (40.2 cm), 5–7 months (41.5 cm), 7–9 months (43.2 cm) and 9–12 months (44.3 cm), 1–2 years (46.5 cm), 2–3 years old (48.7 cm). The mean waist circumference were also lowest for age under one month (34.6 cm), followed, in ascending order, by the 1–3 months (38.7 cm), 3–5 months (41.8 cm), 5–7 months (43 cm), 7–9 months (44.9 cm), 9–12 months (46.5 cm), 1–2 years (47.2 cm), 2–3 years old (49.5 cm). In addition, mean buttocks circumference could be divided into six subgroups ($p < 0.05$). The mean thigh circumference could stratify six subgroups: subgroup 1 (age under one month, 16.7 cm), subgroup 2 (1–3 months (20.3 cm), 3–5 months (21.7 cm)), subgroup 3 (3–5 months (21.7 cm), 5–7 months (21.9 cm)), subgroup 4 (7–9 months (23.4 cm), 9–12 months (23.9 cm)), subgroup 5 (1–2 years (26 cm)), and subgroup 6 (2–3 years old, 28.3 cm). Finally, the mean body weight could stratify seven subgroups: subgroup 1 (age under one month, 3.3 kg), subgroup 2 (1–3 months, 5.4 kg), subgroup 3 (3–5 months, (6.9 kg), 5–7 months (7.2 kg)), subgroup 4 (7–9 months, 8.1 kg),

Table 1. Factors of contribution for choosing diaper.

	Sale price	Material	Sides design	Breathability	Cutting fitness	Absorbability	Wetness indicator
Very important	22.7%	56%	49.3%	58%	49.7%	60.7%	39%
Important	29.3%	38%	44.3%	38.3%	37.7%	33.3%	28%
Ordinary	39.3%	5.3%	5.3%	3.3%	12.3%	5.3%	22.3%
Not important	8.7%	0.7%	1.1%	0.4%	0.3%	0.7%	10.7%

Table 2. Measurement dimensions by age categories.

Age Groups	Stature (cm)	Head circumference (cm)	Waist circumference (cm)	Buttocks circumference (cm)	Thigh circumference (cm)	Weight (kg)
under one month	51.7 a*	35.6 a	34.6 a	32.7 a	16.7 a	3.3 a
1–3 months	56.8 b	38.5 b	38.7 b	39.5 b	20.3 b	5.4 b
3–5 months	62.4 c	40.2 c	41.8 c	43.5 c	21.7 bc	6.9 c
5–7 months	64.9 c	41.5 d	43.0 cd	43.6 c	21.9 c	7.2 c
7–9 months	69.2 d	43.2 e	44.9 de	47.4 d	23.4 d	8.1 d
9–12 months	72.4 e	44.3 e	46.5 ef	48.3 df	23.9 d	9.0 e
1–2 years	81.0 f	46.5 f	47.2 f	50.5 f	26.0 e	10.9 f
2–3 years	88.6 g	48.7 g	49.5 g	53.1 g	28.3 f	13.0 g

* a, b, c, d, e, f, g indicated the results of Duncan's multiple comparison tests.

Table 3. Sizing system of diapers for infants.

Sizing systems	Categories	Waist Circumference (cm)	Stature (cm)	Percent
1	Thin-Short	<36	<56	4.3%
2	Medium-Short	36.1–42	<56	6.3%
3	Medium-Medium	36.1–42	56.1–73	18.7%
4	Slight plump-Medium	42.1–48	56.1–73	9.0%
5	Slight plump-Tall	42.1–48	>73.1	6.7%
6	Plump-Medium	>48.1	56.1–73	23.0%
7	Plump-Tall	>48.1	>73.1	23.0%

subgroup 5 (9–12 months, 9 kg), subgroup 6 (1–2 years, 10.9 kg), and subgroup 7 (2–3 years old, 13 kg). The means and standard deviations and infant-dimension percentiles by age group are presented in Table 2.

3.3 Sizing system of diapers

Each figure type was classified into several size subgroups based on the control dimensions and size interval. The waist circumference and stature were chosen to represent the control dimensions, because these are the most important anthropometric variables in diaper size. Present study drew a distribution graph of seven categories, with stature as the X-axis and waist circumference as the Y-axis (Figure 2). By studying the distribution in detail, and coordinating our findings with the judgment of the domain experts, the standard sizing systems for seven categories were developed. Furthermore, the categories of sizing systems could be denoted the seven sub-groups by waist circumference and stature (Table 3). The waist types could be divided into four types (i.e. thin, medium, slight plump, and plump). In addition, stature could be divided into three types (i.e. short, medium, and tall). A total of 91% of samples have been fit on proposed sizing system.

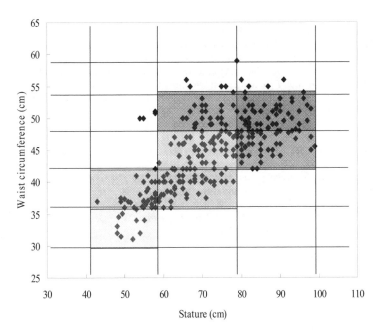

Figure 2. The distribution graph of stature versus waist circumference for seven categories.

4 CONCLUSION

Results of analysis of variance showed that all dimensions were significant age effect. After Duncan's multiple range tests, these dimensions could be stratified seven subgroups. However, the sizes of disposable diaper are divided into stages and sold on three distinct groupings in Seaddlers, Cruisers and Baby Dry (e.g. Pampers Diapers). The sizing systems of diapers are only classified according to growing stage and infant's weight. Thus, present study could provide the anthropometic database for infants to redesign the new diaper size. The new sizing system should be considered on waist circumference and stature. The total coverage rate of present standard sizing systems reached 91%, with the total number of size groups being seven categories. Results of present study could provide a systematic approach of identifying anthropometrical characteristics for developing standard sizing systems to diaper production.

REFERENCES

Ashdown, S.P., 1998, An investigation of the structure of sizing systems. *International Journal of Clothing Science Technology*, Vol. 10, pp. 324–341.
Borkowski, S., 2004, Diaper rash care and management. *Pediatric Nursing*, Vol. 30, No. 6, pp. 467–470.
Chung, M.J., Lin, H.F. and Wang, M.J., 2007, The development of sizing systems for Taiwanese elementary- and high-school students. *International Journal of Industrial Ergonomics*, Vol. 37, pp. 707–716.
Emanuel, I., Alexander, M., Churchill, E. and Truett, B., 1959, A height-weight sizing system for flight clothing. In. *WADC Technical Report 56-365*, Wright Air Development Center (Ohio: Wright-Patterson Air Force Base).
Gorgos, D., 2006, Underlying disorder may cause diaper dermatitis. *Dermatology Nursing*, Vol. 18, pp. 501.
Hsu, C.H., Liu, B.S. and Chen, S.C., 2007, Using systematic approach to discover sizing knowledge from anthropometric data for improving production. *WSEAS Transactions on Systems*, Vol. 6, No. 4, pp. 852–857.
Jorden, W.E., Larson, K.D., Berg, R.W., Frandman, J.J. and Marrer, A.M., 1986, Diaper dermatitis: Frequency and severity among a general infant population. *Pediatric Dermatology*, Vol. 3, pp. 198–207.
Kubiak, M., Kressner, B., Raynor, W., Davis, J. and Syverson, R.E., 1993, Comparison of cloth and single use diapers using a simulated infant faces. *Pediatrics*, Vol. 91, pp. 632–636.

Li, Z., 2006, Anthropometric Topography. In *International Encyclopedia of Ergonomics and Human Factors*, edited by Karwowski, W., (London: Taylor & Francis).

Liptak, G.S., 2001, Diaper rash. In *Pediatric primary care*, edited by Heekelman et al., pp. 1225–1228.

Nui, J., Li, Z. and Salvendy, G., 2006, Multi-resolution Description of 3D Anthropometric Data. In *Proceedings of the 16th World Congress on Ergonomics*, Maastricht, International Ergonomics Association.

Ray, G., Ghosh, S. and Aterya, V., 1995, An anthropometric survey of Indian schoolchildren age 3–5 years. *Applied Ergonomics*, Vol. 26, pp. 67–72.

Sanders, M.S. and McCormick, E.J., 1993, *Human Factors in Engineering and Design*, 7th ed. (New York: McGraw-Hill).

Van Onselen, J., 1999, Rash advice. *Nursing Times Skin Care*, p. 12.

Van, R., Wun, C.C. Morrow, A.L. and Pickering, L.K., 1991, The effect of diaper type and overclothing on fecal containment in day care centers. *Journal of American Medical Association*, Vol. 265, pp. 1840–1844.

Wickens, C.D., Lee J.D., Liu, Y. and Gordon-Becker, S.E., 2004, *An Introduction to Human Factors Engineering*, 2nd ed., (New Jersey: Pearson Education).

Wysocki, B.A. and Bryant, A.R., 1992, Skin. In *Acute and chronic wounds nursing management*, edited by Bryant, A.R., (St. Louis: Mosby Yearbook), pp. 1–30.

Ergonomic Trends from the East – Kumashiro (ed)
© 2010 Taylor & Francis Group, London, ISBN 978-0-415-88178-4

A study on fatigue of neck muscle according to the extension degree of the cervical vertebrae while using Neck Support

Chang-Min Lee, Yeon-Ju Oh, Dae-Woong Kim, Jung-Hyeon Yoo & Kwang-Hyeon Shin
Department of Industrial and Management Engineering, Dong-Eui University Eumgwangno, Busanjin-Gu, Busan, South Korea

ABSTRACT: A Neck Support used for driving and office work, can release strain of the cervical vertebrae by holding up the cervical even for short time, and therefore, increase work efficiency. However, the ergonomic analysis and various researches on neck support, except pillows used for sleeping, have been limited to a part of design regarding material of Neck Support and shape of the cervical. The purpose of this study is to analyze muscle fatigue according to the extension angle of the cervical when using a Neck Support. Moreover, this study performs EMG analysis on the sternocleidmastiod and upper trapezius muscles for two variables (extension angles) of the Neck Support. And, the study presents the optimal specification of the Neck Support for correct use.

Keywords: EMG, Neck Support, Cervical Vertebrae

1   INTRODUCTION

As the one of main parts for maintaining stability of body, the cervical vertebrae support the skull and have the function for the posture adjustment with the spinal segments (Kim et al, 2007). Maintaining curvature of cervical vertebra is very important for a correction of body posture and reduction of muscle fatigue in physical activity as well as when someone is sleeping. Because the use of pillow having appropriate shape and height supports function of cervical vertebrae, a little research has been performed so far (Her et al, 1998, Hur et al, 2006, Jeon et al, 2007). These pillows have been used during sleep as well as for a break in various environments, such as airplanes, trains and at the office, to take a short rest.

In 2007, Kim presented that the pain and damage of the cervical vertebrae is a very common complaint that 67% of the U.S.A population have experienced at least once more (Kim et al, 2007).

In particular, computer workers, drivers, and workers who are sitting on chair or doing repetitive tasks over a long period have experienced the pain and damage of the cervical vertebrae (Jin et al, 2006, Kim et al, 2007). These kinds of problems can be solved by using an aid such as a Neck Support for the work break (Shih et al, 1997).

Several researches presented the effect of decreasing muscle fatigue and cervical vertebrae pain showed various results according to the height and degree of the pillow used for sleeping and breaks. However, research on the biomechanical effect of the Neck Support is rare (Jin et al, 2006, Pack et al, 2006). Most of the studies for the pillows have performed with a purpose to obtain the comfortable sleep when lying down on the floor. On the other hand, the Neck Support is used only for short break after work. The function of the pillow and Neck Support is same but the objective is different. Thus, further research into Neck Supports is necessary. The biomechanical effect in the cervical vertebrae can be measured by EMG, the subjective evaluation, and posture analysis (Kwon et al, 1998, Jin et al, 2006, Pack et al, 2006, Pack 2007). Moreover, it can be the confirmed by muscle fatigue and comfort of the neck. Therefore, this paper will analyze data of the Neck Support objectively through biomechanical evaluations not to be limited only design, also considering the materials of the Neck Supports and the type of the cervical vertebrae.

The analyzing factors to find out the effect of the Neck Support on the cervical vertebrae were muscles, break time, the existence or not of the Neck Support, and the angle of the seat. The angle

Table 1.   Statistical specification of subjects (n = 6).

Item	Height (m)	Weight (kg)	Age (yr)
Subjects	174 ± 5.93	71.67 ± 8.29	29.17 ± 3.54

Table 2.   Experimental variables.

	Case A	Case B	Case C
Chair angle	113	113	113
Neck support	Yes	Yes	No
Cervical angle	120°	110°	140°

of chair, used in most offices, trains and airplanes were considered. As for the neck support, one of the most highly used was selected when considering the material. Second, the muscles investigated for the measurement of muscle fatigue while experiments are 'Upper Trapezius', 'Sternocleid', and 'Middle Trapezius Muscle'.

This paper will measure the muscle fatigue on the cervical vertebrae according to the change of the angle of the cervical vertebrae as well as biomechanical effect, with and without Neck Support. Especially, it will find out quantifiable value for the comfort degree of the cervical vertebrae through the ergonomics experiment and biomechanical analysis. Thus, it will indicate appropriate specification of the Neck Support.

2   THE EXPERIMENT METHOD

Six men with no previous history from curvature of the cervical vertebrae or sprain participated in this study. By X-ray, the study confirmed whether all subjects have the related complaint or not. The average height of all subjects was 174 cm (SD = 5.93), the average weight was 71.67 kg (SD = 8.29) and the average age of the subjects was 29.17 years (SD = 3.54) (Table 1).

The Neck Support used in this study is a high quality and the material is Flocking PVC. The shape is '⊏' and the size (angle) can be changed by increasing the air pressure in the Neck Support. The angle of the Neck Support measured through a pilot test can be defined as the extension angle of the cervical vertebrae, that are 110° and 120°. An assistance model the same as defined extension angle (110° and 120°) was produced to minimize the error among subjects and to identify the posture of the cervical vertebrae.

The chair used for this study is easy to adjust and can support the cervical vertebrae in the environments. The range of the angle varies from 113° to 128° as is the standard of the Korea high speed railway. Thus, the chair's angle (the angle of the lumbar) was adopted to 113°. The three cases of experiments were performed for 30 minutes (Table 2).

The measurement positions were the Upper Trapezius, Sternocleid, and Middle Trapezius Muscle to find out the level of fatigue of the cervical vertebrae. The electrode was attached to each muscle and the root mean square (RMS) values of IEMG were analyzed. The data was collected for 10 sec and 5 times at the 10, 20, and 30 minute break times. The measurement equipment is ME6000-T8 and the software is Mega Win Ver 2.3.1. These data were analyzed by Minitab a statistical program.

3   RESULTS

This study has analyzed the muscle fatigue of the neck by measuring EMG to find out the biomechanical effects while wearing the Neck Support during rest.

Figure 1.   The subject's posture and measurement equipment for EMG with the Neck Support.

Table 3.   The RMS and ANOVA results ($p\text{-}value = 0.05**$).

Experiment time (min)	Measurement Parts (Muscle)	Cervical 110°	Cervical 120°	Cervical 140°	p-value
10	Upper Trapezius	39.47 ± 39.2	**10.43 ± 18.69**	27.17 ± 27.19	0.001***
	Sternocleid	19.97 ± 15.14	**13.07 ± 14.71**	13.93 ± 16.96	0.183
	Middle Trapezius	17.97 ± 20.85	**8.5 ± 15.85**	12.1 ± 17.56	0.133
	p-value	0.004***	0.563	0.013***	
20	Upper Trapezius	49.53 ± 71.57	**8.07 ± 14.72**	21.3 ± 21.94	0.002***
	Sternocleid	19.43 ± 17.44	17.9 ± 16.38	**13.4 ± 17.28**	0.366
	Middle Trapezius	21.7 ± 25.83	**6 ± 11.89**	12.6 ± 17.36	0.009***
	p-value	0.019***	0.004***	0.152	
30	Upper Trapezius	68.9 ± 96.33	**9.5 ± 15.35**	22.2 ± 20.32	0.000***
	Sternocleid	15.13 ± 14.24	19 ± 16.52	**14.27 ± 14.46**	0.437
	Middle Trapezius	20 ± 24.3	**7.3 ± 12.63**	13.53 ± 17.16	0.035***
	p-value	0.001***	0.007***	0.110	

The considered factors for the study are the three measurement parts, the break time, and the existence or not of the Neck Support. No use of the Neck Support was defined as a 140° extension angle of the cervical vertebrae. The use of the Neck Support was classified with two types that are 110° and 120° extension angles of the cervical vertebrae.

In the case of a 140° extension angle (no use of the Neck Support) of the cervical vertebrae, the Middle Trapezius Muscle showed the lowest EMG value. Followed the Sternocleid, and the Upper Trapezius was last. It was similar trend at the both 110° and 120° extension angle of the cervical vertebrae (see the shaded cell in Table 3).

There wasn't a significant difference on 95% confidence intervals between the muscle and break time in case of 140° extension angle of the cervical vertebrae. However, in case of the 110° and 120° (after 10 minute break time), there is a significant difference between the muscles.

As the results of the EMG according to extension angles of the cervical vertebrae, the value of the EMG was the lowest at the 120° extension angle. Followed by 140°, and 110° was last. The bold characters in the table 3 showed the results. On the Upper Trapezius Muscle, the EMG of 120° extension angle was 10.43 μVs. The use of the Neck Support shows significantly larger effect for decreasing the muscle fatigue than when there was no Neck Support.

Therefore, the 120° extension angle of the cervical vertebrae is the most effective reduction effect of the muscle fatigue when using a Neck Support.

But, the EMG value at the 110° extension angle of the cervical vertebrae was the highest. It showed that the use of the Neck Support with 110° extension angle could cause bad effect to the neck.

And there are no significant differences among rest time except Upper Trapezius of 110° extension angle.

245

4 CONCLUSION

The purpose of this study is to find out relief effects that Neck Support has in relation to the neck muscles, if any.

First, the use of the Neck Support shows that the muscle fatigue of the Neck decreases significantly. The use of the Neck Support on 120° extension angle is promoted because the EMG value of the 120° extension angle shows the lowest EMG value than any other case while resting. These results are very similar to, both the Upper Trapezius Muscle and the Middle Trapezius Muscle, but not the Sternocleid Muscle. However, there wasn't significant effect between the break times.

These means that a Neck Support needs to be used with a proper posture of the extension angle of the neck, because the use of inappropriate Neck Support could be a possible cause of additional fatigue. In addition, the results show that the EMG value of the Middle Trapezius Muscle is the lowest value at the most parts, but there is no significant difference on 95% confidence intervals. Also, we found that there is no significant difference among the break times. Like long trips (airplanes), have same effects during a few hours use of Neck Support? This could lead to further research.

This study confirmed fatigue relief effects of three kinds of neck muscles according to the type of the Neck Support. And inappropriate use of a Neck Support could increase the muscle fatigue in the neck. The chair used in the experiment was fixed at a 113° lumbar angle and the experiment considered just two types of Neck Support. So, experiments with various types of lumber angles could be performed for the further study to find out the biomechanical effects of Neck Supports.

REFERENCES

Choi, K. S., Koo, J. K., Park, S. J., 2000, Development of Light Seat for High Speed Railroad, *Korean society for railway*, pp. 198–209.

Falla, D., Jull, G., Dall'Alba, P., Rainoldi, A., Merletti, R., An electromyographic analysis of the deep cervical flexor muscles in performance of craniocervical flexion, Physical Therapy 83(10), pp. 899–906.

Her, J. K., Lee, Y. C., Kwon B. A., 1998, A NEW DESIGN OF VERVICAL PILLOW, *Journal of korea institute of sports science*, pp. 109–117.

Hur, J. G., Yang, Y, A., 2006, The Effect of Ergonomic Pillow in Patient with Chronic Neck Pain, *Journal of the Ergonomics Society of Korea*, Vol. 25, No. 1., pp. 17–25.

Jin, G. H., Young, A. Y., 2006, The Effect of Ergonomic Pillow in Patient with Chronic Neck Pain, Journal *of the Ergonomics Society of Korea*, 25(1), pp. 17–25.

Kim, J. C., Jeon, H. S., Yi, C. H., Kwon, O. Y., Oh, D. W., 2007, Strength and Endurance of the Deep Neck Flexors of Industrial Workers With and Without Neck Pain, *Journal of the Ergonomics Society of Korea*, Vol. 26, No. 4 , pp. 25–31.

Kong-King Shieh, Ming-Te Chen, 1997, Effects of screen color combination, work-break schedule, and workpace on VDT viewing distance, *International Journal of Industrial Ergonomics*, 20(1), 11–18.

Kwon, B. A., Lee, Y. C., Her, J. G., 1998, A new design of cervical pillow, *Journal of Korea institute of sports science*, 2(2).

Lin, Y. H., Huang, W. Y., 2007. *Journal of Biomechanics*, 40(2), pp. 414.

Neumann, D. A., 2002, Kinesiology of the musculoskeletal system: Foundation for physical rehabilitation, St. Louis, *Mosby*, pp. 276.

Pack, D. E., Kim, J. Y., 2006, Comparative Evaluation of Ergonomic Pillow made of Korean Traditional Materials, *Journal of the Ergonomics Society of Korea*, pp. 323~326.

Pack, D. E., 2007, Functional Evaluation of Eco- Friendly Pillows by Measuring Cervical Stability, Head Pressure, and Muscle Activity, *Hanyang University*.

Ergonomic Trends from the East – Kumashiro (ed)
© *2010 Taylor & Francis Group, London, ISBN 978-0-415-88178-4*

The influence of continuously intermittent cold immersion on grip strength and associated muscular activity and forearm skin temperature

Wen-Lin Chen
Department of Industrial Management, National University of Science and Technology, Da-an District, Taipei City, Taiwan

Yuh-Chuan Shih
Department of Logistics Management, National Defense University, Beitou District, Taipei City, Taiwan

Cha-Fan Chi
Department of Industrial Management, National University of Science and Technology, Da-an District, Taipei City, Taiwan

ABSTRACT: This paper investigated the effect of a 40-min intermittent immersion in 11°C water on the hand grip strength, muscular activity (in terms of Normalized Root Mean Square of EMG, NRMS), and the corresponding Forearm Skin Temperature (FAST). Twelve male and twelve female subjects took part in this study. A nested-factorial design was employed. The factors were gender, subject (nested within the gender), and immersion duration (8th, 16th, 24th, 32th, and 40th minute after immersion). The ANOVA results indicated that a longer immersion leads to a lower FAST and a less grip strength, and a smaller associated EMG of the muscles of Extensor Digitorums (ED) and Flexor Digitorum Superficialis (FDS). For gender effect, female MVC was about 71% of the male MVC (28.04 kgw/ 39.40 kgw), and females had a greater NRMS.

1 INTRODUCTION

Working in a cold condition can strongly influence the efficacy of task performance; further, low temperature has been strongly correlated to cumulative trauma disorders of the forearm (Georgitis, 1978; Kurppa *et al.*, 1991), wrist (Chiang *et al.*, 1990), and shoulder (Pope *et al.*, 1997). Upon human exposure to extremely cold temperatures, the resulting decrease in peripheral blood flow by vasoconstriction reduces convective heat transfer between the body's core and surface, effectively increasing skin insulation. While vasoconstriction helps to retard heat loss and maintain core temperature, it induces a decline in peripheral tissue temperatures. Thus, cold exposure most certainly affects the hands and feet (Goldman, 1964).

Grip strength is essential for manual operations, and it is affected adversely by cold exposure. For instance, Holewijn and Heus (1992) reported that maximal grip force decreased after 30 min of local cooling (15°C), and Petrofsky and Lind (1980) concluded maximal force was reduced after immersion in a cold water bath (10°C). Hellstrom (1965) observed a fall in grip strength at a lowed forearm muscle temperature (30°C).

On the other hand, muscle activity index could be obtained through analytical examination of the amplitude and frequency components of the EMG. Many studies indicated EMG frequency decreases with reduced muscle temperature (Kadefors *et al.*, 1968; Petrofsky and Lind, 1980). The finding of Petrofsky and Lind (1980) indicated that under the same submaximal exertion level there was a decrease in EMG amplitude as muscle was cooled after 30-min immersion in 10 or 20°C water. Contrarily, air cooling of the skin was found to increase the EMG amplitude during muscular work compared to thermo-neutral conditions (Bell, 1993; Oksa *et al.*, 2002; Winkel and Jørgensen, 1991). Recently, Sormunen *et al.* (2006) evaluated the muscular strain of female

Table 1. Anthropometric data of subjects (Mean ± standard deviation).

Items	Age (yr.)	Height (cm)	Weight (kg)	Hand length* (cm)	Forearm Volume* (cm^3)
Male	25.3 ± 6.17	173.7 ± 5.00	69.2 ± 7.42	18.7 ± 0.63	1431 ± 140.4
Female	24.1 ± 1.68	160.7 ± 6.15	53.6 ± 6.46	16.7 ± 0.84	1000 ± 123.3

* right hand

workers in meatpacking work and indicated the absence of a significant effect of cold exposure on EMG. The reason for this discordance of the cooling effect on EMG is due to perhaps differences not only in methodology but also in activities performed; e.g., dynamic concentric contractions on soleus muscle were evaluated by Winkel and Jørgensen (1991), whereas Petrofsky and Lind (1980) assessed the brief and fatiguing isometric contraction of handgrip muscles.

Occupations requiring manual work in cold immersion include fisherman, fish filleters, and divers. Exposure in such work environments are often characterized by repeated immersion, long exposure time, moderate cold and brief pauses. Previous research has been focused on muscle performance effects after a fixed period of immersion, different temperature and different method of cooling (Suizu and Harada, 2005; Daanen et al., 2003). Thus, a lack of information concerning muscle performance in response to intermittent immersion exists.

Therefore, present study examines the effects of intermittent cold water immersion on MVC, EMG and the corresponding forearm skin temperature.

2 METHODS

2.1 Subjects

Twelve male and 12 female subjects were recruited, and they were healthy and without any musculoskeletal disorders. All were right-handed and anthropometric data are presented in Table 1.

2.2 Materials and apparatus

Two water tanks and a throw-in cooler with a cooling efficiency of 750 Kcal/hr were used for immersion test. One of the tanks called warm tank had the water temperature at 34°C, and it was used both to control Forearm Skin Temperature (FST) at the baseline and to recover FST. The other one called cold tank had the water temperature at 11°C and used to cool the forearm continuously. In addition, a digital thermometer and hygrometer was used to monitor the ambient temperature and humidity at the same time (TECPEL Co.; Model: DTM301). A digital 4-channel thermometer made by TECPEL Co. (Model: DTM319) was used to record forearm skin temperature (FAST). A fabricated grip gauge with a load cell (SENSOTEC, Model: 81, capacity: 250 lbs, tension) was used to measure force output. Force output was calculated using a computer program with a sampling rate of 1000 Hz. The grip span was set at 5 cm, the load cell was connected to a 12-bit A/D converter card, and linearity was calibrated using several known weights. EMG was measured by Biofeedback 2000 (Physio Recorder Model: A-2340) with the default maximal sampling rate of 40 Hz. The latex gloves of three different sizes (7, 7.5, and 8 in) were included.

2.3 Experimental design

A nested-factorial design was employed and independent variables included gender, subject (nested within gender), and immersion duration. Besides grip MVC, dependent variables also included %MVC and muscle activity (in terms of normalized root mean square of EMG, NRMS). There were three stages, namely 8-min baseline, 40-min cooling, and 15-min rewarming. The %MVC is a ratio of the last MVC measured in either cooling or rewarming over the MVC$_{Baseline}$, measured in baseline stage. The immersion duration was marked at every 8 min and every 5 min for cooling and rewarming stage, respectively. The software, STATISTICA 7.0, is used for data analysis, and the level of significance (α) is 0.05.

Table 2. ANOVA Results for all evaluated responses.

Sources of variation	Cooling (11°C)				Rewarming (34°C)			
	MVC	%MVC	Flexor NRMS	Extensor NRMS	MVC	%MVC	Flexor NRMS	Extensor NRMS
Subject(Gender)	**	**	**	**	**	**	**	**
Gender	**	N.S.	**	**	**	N.S.	**	**
Immersion time	**	**	**	**	**	**	**	**
Gender × Immersion time	N.S.	N.S.	N.S.	N.S.	N.S.	N.S.	N.S.	N.S.

**:$p < 0.01$

Table 3. Mean and standard deviation of MVC, Flexor and Extensor NRMS.

Factor		FAST (°C)		MVC (kg)		Flexor NRMS		Extensor NRMS	
Stage	Time	Male	Female	Male	Female	Male	Female	Male	Female
1 (34°C)	5 min	34.8	34.3	40.7 ± 7.93	28.5 ± 4.41	0.52 ± 0.18	0.53 ± 0.27	0.75 ± 0.32	0.76 ± 0.23
2 (11°C)	8 min	23.9	22.7	38.4 ± 8.05	27.2 ± 6.10	0.39 ± 0.18	0.45 ± 0.20	0.56 ± 0.14	0.67 ± 0.19
	16 min	21.1	19.7	39.5 ± 7.09	26.8 ± 6.59	0.33 ± 0.09	0.43 ± 0.15	0.41 ± 0.14	0.52 ± 0.15
	24 min	18.9	18.2	38.4 ± 6.71	27.0 ± 6.02	0.27 ± 0.13	0.39 ± 0.19	0.31 ± 0.11	0.40 ± 0.14
	32 min	17.5	17.1	36.6 ± 5.23	26.0 ± 5.51	0.27 ± 0.10	0.31 ± 0.14	0.23 ± 0.08	0.31 ± 0.11
	40 min	16.7	16.6	36.1 ± 6.69	26.2 ± 5.34	0.22 ± 0.10	0.28 ± 0.12	0.18 ± 0.08	0.31 ± 0.18
3 (34°C)	5 min	26.8	26.5	42.1 ± 6.13	29.9 ± 6.18	0.29 ± 0.11	0.35 ± 0.16	0.44 ± 0.12	0.50 ± 0.18
	10 min	29.2	29.5	41.3 ± 6.39	30.8 ± 5.62	0.41 ± 0.17	0.52 ± 0.26	0.61 ± 0.18	0.73 ± 0.27
	15 min	30.2	31.5	41.5 ± 7.12	30.0 ± 4.67	0.48 ± 0.22	0.60 ± 0.28	0.71 ± 0.25	0.81 ± 0.19

2.4 Procedures

All participants were well informed of the goals and procedures first. The mean ambient temperature (standard deviation, SD) was 23.29°C (2.32°C), and mean relative humidity (SD) was 57.5% (2.40%). The procedure consisted of three stages, namely baseline, immersion, and rewarming stage.

In stage 1, the baseline stage, all participants immersed their right hand into the 34°C normal tank for 8 min. The purpose are (1) controlling the initial skin temperature at 34°C, and (2) measuring grip MVC and the corresponding EMG as a reference for comparison (denoted by $MVC_{Baseline}$ and $EMG_{Baseline}$, respectively). In stage 2, the cold immersion stage (11°C), the immersion duration was 40 minutes, and MVC and EMG were measured every 8 min (8, 16, 24, 32, and 40 min). In stage 3, the rewarming stage (34°C), subjects were asked to immerse their right hands into the normal tank again as soon as completing the Stage 2. MVC and EMG were measured every 5 min (5, 10, and 15 min) during this 15-min rewarming period.

During the formal measurement, subjects sit on a chair without armrests and they immersed their right hands into the water tank up to the elbow joint. While measuring grip MVC subjects removed the hands from the tank. At the same time, the corresponding EMG was recorded as well. In addition to standardizing limb submersion across participants, participants were signaled by the computer when to submerge and when to remove their right hands.

3 RESULTS AND DISCUSSION

The ANOVA results indicate that gender had a significant effect on MVC, and the NRMS for both flexor and extensor, but not on %MVC (see Table 2). Generally, males had greater grip MVC and less NRMS of EMG (Table 3). On the other hand, during the cold immersion period, the MVC and muscular activity (NRMS) decreased as immersing time increased. Cold immersion adversely impacted working muscle capacity and maximum force production has been reported. One possible reason is that decreased temperature slows all muscular metabolic processes (Sesboüé and Guincestre, 2006) and muscle oxygen saturation may then lead to decreased muscle performance (Clifford *et al.*, 2004).

During the rewarming period, a longer exposure led to a more increase in both forearm skin temperature and muscular activity (NRMS). For MVC, a complete recovery seems to occur after 5 minutes. This recovery can be ascribed to an increase in forearm skin temperature and muscle blood flow during immersion in 34°C (warmer) water.

4 CONCLUSIONS

Longer cold exposure led to lower forearm skin temperature, and then to decrease all evaluated responses. After warming, all muscular performance had a gradual recovery.

REFERENCES

Bell, D. G., 1993. The influence of air temperature on the EMG/force relationship of the quadriceps. *European Journal of Applied Physiology*, 67:256–260.

Chiang, H. C., Chen, S. S., Yu, H. S., Ko, and Y. C., 1990. The occurrence of carpal tunnel syndrome in frozen food factory employees. *Kaohsiung Journal of Medical Sciences*. 6:73–89.

Clifford, H., Paul V., and Robert S.P., 2004. Peripheral skin temperature effects on muscle oxygen levels. *Journal of Thermal Biology* , 29:785–789.

Daanen, H. A. M., Vliert, E. V. D., and Huang, X., 2003. Driving performance in cold, warm, and thermoneutral environments. *Applied Ergonomics*, 34:597–602.

Georgitis, J., 1978. Extensor tenosynovitis of the hand from cold exposure. *Journal of Maine Medical Association*, 69(4), 129–131.

Goldman, R. F., 1964. The Arctic soldier: Possible research solutions for his protection. In C. R. Kolb and F. M. G. Holstrom (Eds.) Review of research on military problems in cold regions. *United States Air Forc Arctic Aeromedical Laboratory Technical Documentary Report* No. 64–28.

Hellstrom, B., 1965. *Local effects of acclimatization to cold in man*. Oslo: Universitetsforlaget.

Holewijn, M. and Heus, R., 1992. Effects of temperature on electromyogram and muscle function. *European Journal of Applied Physiology and Occupational Physiology*, 65:541–545.

Kadefors, R., Kaiser, E., and Petersen, I., 1968. Dynamic spectrum analysis of myo-potentials and with special reference to muscle fatigue. *Electromyography*, 8:39–74.

Kurppa, K., Viikari-Juntura, E., Kuosma, E., Huuskonen, M., and Kivi, P., 1991. Incidence of tenosynovitis or peritendinitis and epicondylitis in a meat-processing factory. *Scandinavian Journal of Work Environment and Health*, 17:32–37.

Oksa, J., Ducharme, M., and Rintamäki, H., 2002. Combined effect of repetitive work and cold on muscle function and fatigue. *Journal of Applied Physiology*, 92:354–361.

Petrofsky, J.S. and Lind, A.R., 1980. The influence of temperature on the amplitude and frequency components of the EMG during brief and sustained isometric contraction. *European Journal of Applied Physiology*, 44:189–200.

Pope, D., Croft, P., Pritchard, C., Silman, A., and Macfarlane, G., 1997. Occupational factors related to shoulder pain and disability. *Occupational and Environmental Medicine*, 54:316–321.

Sesboüé, and B., Guincestre, J. Y., 2006. Muscular fatigue. *Annales de réadaptation et de médecine physique*, 49:348–354.

Sormunen, E., Oksa, J., Pienimäki, T., Rissanen, S., and Rintamäki, H., 2006. Muscular and cold strain of female workers in meatpacking work. *International Journal of Industrial Ergonomics*, 36:713–720.

Suizu, K. and Harada, N., 2005. Effects of waterproof covering on hand immersion test using water at 10°C, 12°C and 15°C for diagnosis of hand-arm vibration syndrome. *International Arch Occupational Environment Health*, 78:311–318.

Winkel, J. and Jorgensen, K., 1991. Significance of skin temperature changes in surface electromyography. *European Journal of Applied Physiology*, 63:345–348.

Ergonomic Trends from the East – Kumashiro (ed)
© *2010 Taylor & Francis Group, London, ISBN 978-0-415-88178-4*

Evaluations of the subjective discomfort ratings and EMGs of lower extremity muscles in various knee flexion angles

Yong-Ku Kong & Seong-Tae Sohn
Department of Industrial Engineering, Sungkyunkwan University, Suwon, Korea,
Cheoncheon-dong, Jangan-gu, Suwon, Syeonggi-do, Korea

ABSTRACT: Five muscle groups (the erector spine and four lower-extremity muscles – Biceps Femoris, Vastus Medialis, Gastrocnemus and Tibialis Anterior) were tested to evaluate the effects of knee flexion angles on subjective discomfort ratings and muscle fatigues (median frequency of EMGs – MDF) in a static-sustaining task. The analyses of subjective discomfort (whole body as well as each body part) were evaluated in this study. Overall, participant rated knee flexion angle 60° and 90° as significantly more discomfort than the other knee flexion angles. Sitting height (40 cm) was rated the least discomfort sitting posture. In addition, the analyses of MDF for each muscle group in various knee flexion angles during 15 minute static-sustaining task were also investigated in this study.

Keywords: knee angles, WMSDs, median frequency, subjective discomfort rating

1 INTRODUCTION

Many researchers have reported on the relationship between inadequate working postures of the neck or limbs and incidences of WMSDs of the neck and upper extremities. The risk factors of lower-extremity disorders and lower back disorders have also been investigated by many researchers. Repeated or prolonged knee flexion was reported as a cause of knee injuries, whereas bending or twisting of the trunk was evaluated as one of main causes of lower back disorders. It has been reported that in US and Korea, many workers have carried out tasks with prolonged squatting or kneeling postures at various worksites such as shipbuilding shops, vehicle assembly lines, farms, etc. (Lee and Chung, 1999; NIOSH, 2001). Although many studies have been carried out that mainly targeted the upper-body postures and workloads to prevent WMSDs of upper extremities, fewer studies on the lower parts of the body have been performed, as compared to the upper body (Corlett et al., 1986; Bernard, 1997). Only a few studies have considered the effects of the lower-body postures in the ergonomics literature, in spite of many lower-extremity problems reported in many industries. Thus, the objective of this study is to evaluate the effects of different working postures associated with various knee angles that are commonly required at agricultural and industrial sites on subjective discomfort ratings and fatigue of lower-limb muscles.

2 METHODS

2.1 Subjects

For this study, thirty healthy university students participated. None of the participants had experienced musculoskeletal disorders of the back or lower limbs. The mean age of all participants was 27.3 (±3.1) years, with average height and weights of 170.6 (±1.7) cm and 66.1 (±8.2) kg, respectively.

2.2   Lower-limb postures and muscles

Thirteen lower-limb postures were evaluated in this study. The postures were classified into four groups as follows: standing, squatting, kneeling, and sitting. Standing postures were divided into four sub-postures associated with knee flexion angles such as standing, and knee flexion angles of 150°, 120°, and 90° (KF150, KF120, and KF90), respectively. Squatting postures were also divided into three sub-postures, knee flexion angles of 60°, 30°, and 30° with tiptoes (KF60, KF30, and KF30T), respectively. Kneeling postures were divided into two groups with one-leg and two-leg kneeling (KNL_1 and KNL2). Lastly, sitting postures were classified into four sub-postures based on sitting chair heights and sitting postures: SC0, SC20, SC40, and S_CRS for sitting heights of 0, 20, and 40 cm, and sitting with crossed legs, respectively. Five muscles from the back and lower limbs were selected. The selected muscles were the erector spine, biceps femoris, vastus medialis, gastrocnemius, and tibialis anterior.

2.3   Experimental procedures

Each participant was asked to perform 13 body postures and maintain each posture for 15 minutes. Each participant was also asked to provide subjective ratings of discomfort for his whole body and each body part associated with each body posture at three-minute intervals. To reduce the muscle fatigue effects, each participant had 15 minutes of resting time between trials. The trends of subjective discomfort ratings, and EMG frequency associated with sustaining 13 lower-body postures over a 15-minute time domain were analyzed as dependent variables in this study.

3   RESULTS

3.1   Average subjective discomfort ratings for each posture

A significant effect of the lower-limb posture factor on the average subjective discomfort ratings for the whole body and each body part was found in this study ($p < 0.05$). Table 1 shows the results of the average subjective discomfort ratings associated with various body postures. In the analysis of the discomfort on the whole body, most participants preferred body postures of SC40, S_CRS, SC20, STD and SC0 rather than other postures. KF90 and KF60 were rated as the most uncomfortable postures, followed by KF120 for the whole body.

For the discomfort studies of back, thigh, and leg, the Borg CR10 scale was applied. The findings showed similar trends of discomfort ratings on the whole body assessment except for a few differences. In the results of subjective discomfort ratings on the back, most participants rated the STD, SC40, SC20, S_CRS, KF150 as the least uncomfortable postures, while they rated KF90, SC0 and KF60 as the moderately difficult postures. For the thigh, SC40 and SC20 also were associated with lower discomfort rates than the others, followed by S_CRS, SC0 and STD. The most uncomfortable body postures for the thigh were again KF60 and KF90, followed by KF120. The discomfort for legs showed a similar trend to the thigh.

3.2   Average MDF (median frequency) for each posture

The average MDF of five muscles in working postures over a period of fifteen minutes are listed in Table 2. Regardless of body postures, the VM and GAS had generally higher average MDFs than the other muscle groups (ES, BF, and TA). Comparing the average MDFs for various postures, low values of MDF were observed at the knee angles of 150°, 120°, 90°, and 60° and the kneeling postures, while high values of MDF were obtained while sitting with chair heights of 40, 20, and 0 cm, with crossed legs, and for the standing posture.

4   CONCLUSIONS

In order to investigate the potential risks of WMSDs with respect to knee joint angles, thirteen postures that were commonly required at agricultural and industrial workplaces were evaluated using subjective discomfort ratings, heart rates, and EMG data for five muscle groups.

Table 1. Average subjective discomfort ratings for each body posture.

Posture	Whole body	Posture	Back	Posture	Thigh	Posture	Leg
SC40	7.3^A	STD	1.2^A	SC40	0.5^A	SC40	0.5^A
S_CRS	7.7^A	SC40	1.3^A	SC20	0.6^A	SC20	0.6^A
SC20	7.7^A	SC20	1.5^A	S_CRS	1.0^B	S_CRS	1.1^A
STD	8.5^A	S_CRS	1.8^A	SC0	1.6^B	SC0	1.5^A
SC0	10.0^A	KF150	1.9^A	STD	1.6^B	STD	1.8^A
KF150	12.2^B	KF30T	2.1^B	KF30	4.0^C	KF150	4.0^B
KF30	12.3^B	KNL-1	2.4^C	KNL-1	4.2^C	KF120	4.9^C
KF30T	13.7^B	KF30	2.7^C	KNL-2	4.3^C	KF30	5.3^C
KNL-1	13.9^B	KNL-2	2.8^C	KF30T	4.3^C	KNL-2	5.9^C
KNL-2	14.3^B	KF120	3.2^C	KF150	4.5^C	KNL-1	5.9^C
KF120	15.4^C	KF90	3.3^D	KF120	7.2^D	KF90	6.0^D
KF90	17.3^D	SC0	3.3^D	KF60	8.7^E	KF30T	6.6^D
KF60	17.4^D	KF60	3.6^D	KF90	8.9^E	KF60	7.3^D

*letter indicates significant statistical grouping; **shaded areas were the discomfort ratings for only three minutes' duration

Table 2. Average MDF of each muscle against working posture.

	ES	BF	VM	GAS	TA	Avg. $\pm$ SD
STD	91.5	92.4	135.7	123.5	98.8	108. $\pm$ 20.0
KF150	82.4	67.9	65.4	120.1	89.9	85.1 $\pm$ 22.0
KF120	78	59.1	65	106.7	89.7	79.7 $\pm$ 19.2
KF90	70.6	54.5	70.2	98	100.4	78.7 $\pm$ 19.8
KF60	73.6	57.2	66.4	97.2	110.2	80.9 $\pm$ 22.1
KF30	85.4	86.9	121.7	117.4	99.9	102.3 $\pm$ 16.8
KF30T	78.4	86.3	115	109.2	89.7	95.7 $\pm$ 15.6
SC40	88.7	100.1	154.1	167.6	93.9	120.9 $\pm$ 37.0
SC20	97.8	95.3	146.4	165.9	90.7	119.2 $\pm$ 34.5
SC0	91.4	85.8	137.3	161.1	96.3	114.4 $\pm$ 33.1
S_CRS	89.4	95.5	146.6	160.3	94.5	117.3 $\pm$ 33.5
KNL-one	72.5	76.6	100.6	91	90.4	86.2 $\pm$ 11.5
KNL-two	90.3	73.9	115.8	107.8	91.5	95.9 $\pm$ 16.4
Avg.	83.8 $\pm$ 8.6	79.4 $\pm$ 15.7	110.8 $\pm$ 33.9	125.1 $\pm$ 28.4	95.1 $\pm$ 6.0	

For the whole body discomfort, overall participants felt the greatest discomfort in the postures of knee flexion angles of 60°, 90° and 120°, followed by kneeling postures. Unlike the sitting postures with 20-cm and 40-cm chair heights, the sitting posture with 0-cm chair height showed large discomfort over time. Based on this finding, kneeling and knee flexion postures, as well as sitting on a floor without any planks, were not adequate for a task which required sustaining these postures. Similar to the whole body discomfort analysis, sitting with no chair height was one of the most uncomfortable postures for the back, whereas this posture was generally one of the least uncomfortable postures for the whole body, thigh and leg. It was noted that the average discomfort ratings for the thigh and leg also showed results similar to those for the whole body. Overall, participants rated knee flexion postures as the most uncomfortable postures, whereas they rated sitting and standing postures as the least uncomfortable postures for the thigh and leg. As expected, the postures with knee flexions of 60° and 90° were the most stressful postures, whereas the sitting posture with a 40 cm chair height was rated as the least stressful posture, followed by SC20 and S_CRS.

253

The analysis of differences in the average MDFs between five muscles (ES, BF, VM, GAS, and TA) for each posture showed relatively higher MDF values for the VM (110.8 ± 33.9) and GAS (125.1 ± 28.4) compared to the TA, ES, and BF. According to previous researchers' findings, fiber composition (type I and II fibers) and fiber size have been shown to influence the frequency of the EMG signals: for example, higher values of the MDF of the EMG signals were observed for muscles with a greater percentage of type II fibers, and generally type II fibers have been shown to be larger than type I fibers in men, whereas the opposite can be observed in women (Wretling et al., 1987; Miller et al., 1993; Gerdle, et al., 2000).

REFERENCES

Bernard, B. (Ed.), 1997. Musculoskeletal disorders and workplace factors: A critical review of epidemiological evidence for work-related musculoskeletal disorders of the neck, upper extremity, and lower back. DHHS (NIOSH) Publication No. 97–141, US Department of Health and Human Services.

Corlett, E., Wilson, J., and Manenica, I., 1986, The ergonomics of working postures: Models, methods and cases, London: Taylor & Francis.

Gerdle, B., Karlsson, S., Crenshaw, A., Elert, J., and Fridén, J., 2000, The influences of muscle fibre proportions and areas upon MEG during maximal dynamic knee extensions, European Journal of Applied Physiology, 81, 12–10.

Lee, I., and Chung, M., 1999. Workload evaluation of squatting work postures. In: L. Straker, and C. Pollock, (Eds.), CD-ROM Proceedings of CybErg 1999: The second International Cyberspace Conference on Ergonomics, The International Ergonomics Association Press, Curtin University of Technology, Perth, Austrlia, 597–607.

Miller, A., MacDougall, J., Tarnopolsky, M., Sale, D., 1993, Gender differences in strength and muscle fiber characteristics, European Journal of Applied Physiology, 66, 254–262.

NIOSH Report No. EPHB 229-13a, 2001, Pre-intervention quantitative risk factor analysis for ship construction processes at Bath Iron Works Corporation Shipyard.

Wretling, M., Gerdle, B., and Henriksson-Larsen, K., 1987, EMG: a non-invasive method for determination of fibre type proportion, Acta Physiology Scandinavian, 131, 627–628.

Ergonomic Trends from the East – Kumashiro (ed)
© 2010 Taylor & Francis Group, London, ISBN 978-0-415-88178-4

Evaluation of subjective discomfort ratings on various lower limb postures for prevention of WMSDs

Yu C. Kim
Department of Industrial Management Engineering, Dong-Eui University, Korea

Chang W. Hong
Extension Service Bureau, Rural Development Administration(RDA), Korea

Kwan S. Lee
Department of Industrial Engineering, Hongik University, Korea

1 INTRODUCTION

Work related musculoskeletal disorders (WMSDs) have emerged as a social issue in Korea and possibility of the outbreak of this disease has been high in various professions. Economical damage due to compensation expenses and productivity decrease were found very big. Workers and families who were exposed to WMSDS have been also suffering from financial loss and physical pain as well.

According to industrial accident statistics that was announced by the Ministry of Labor (2005), 62.9% of diseases related to industrial work were WMSDs in Korea in 2004. Among people who had the work loss of more than 1 year, WMSDs accounted for 11.3%. This situation caused problems to companies among which manufacturing industry accounted for 64% of total WMSDs.

In manufacturing industry, awkward and difficult work postures were frequent in manual material handling tasks. Further, workers needed to take different postures since they often lifted various size boxes. They were also exposed to vibrations due to power tools. Workers have been exposed to other ergonomic risk factors. Thus their tasks could impose stress on worker's tendons, muscles, and nerves.

The evaluation of tasks for risk factors was an important process to prevent WMSDs. This is necessary to reduce work load or to secure safety of work. There have been many studies conducted about the load estimation of lumbar and upper limbs because WMSDs have occurred a lot at automobile industry in western countries such as U.S.A. But the researches for lower limbs postures have not been done. Lower limbs were important in causing WMSDs since it influences the stability of postures and the work load of the whole body in case of squatting and kneeling postures.

Therefore, lower limbs' imbalance postures can be a risk factor of WMSDs and thus it is necessary to study on effects of lower limbs on the work load biodynamically and physiologically. In this experiment, after keeping various postures of lower limbs according to the chosen procedure, the work load of each posture was evaluated using a subjective discomfort scale (Borg's CR-10 scale). These results can be utilized as a basic data for a work load of lower limbs postures which can be used to assess the work load of workers.

2 METHOD

2.1 *Subjects*

Twelve healthy male college and graduate students, who were paid volunteers, participated in the experiment. All had been informed of the experimental protocol and risks before the experiment. None had a history of WMSDs in the shoulder, lower-back, or leg regions. The means and standard

Table 1.   Postures used in the experiment.

Postures	Level
Balanced posture by knee angle	1, 2, 3, 4 posture
Imbalanced posture by knee angle	5, 6, 7, 8 posture
One-foot posture by knee angle	9, 10 posture
Kneeling posture	11, 12, 13 posture

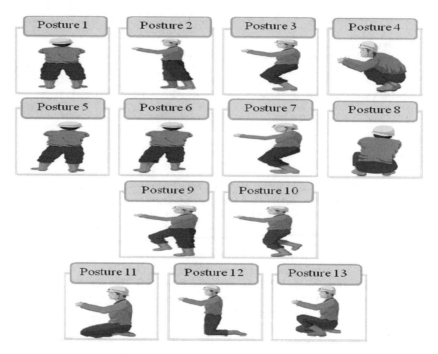

Figure 1.   Leg postures used in the experiment.

deviations of age, height, and body weight for the subjects were 24.5±2.43 years, 175.6±3.28 cm, and 69.2±6.40 kg, respectively.

2.2   Leg postures in the experiment

This study was designed to find effect of body balance, knee angle, one or two feet, and kneeling. These five factors were considered to have a strong influence on the physiological workload.

Thirteen leg postures were tested in the experiment. The postures were grouped into seven categories as follows: standing, knee-flexed posture, squatting, imbalanced posture and squatting, one-foot standing, kneeling. The leg postures used in this study are illustrated in Figure 1 and a short description for each posture is in Table 2.

2.3   Measurement of perceived discomfort and experimental procedure

Assessment of perceived discomfort for lower limb postures has been measured by the Borg's CR-10. The scales CR-10(category ratio) constructed by Borg(1982) has been recommended to be used for measurement of subjective symptoms such as pain of static work. Ratings on these scales give reliable measures of the general perception of fatigue as a consequence of work. The level of work load was classified by the heart rate level. If the Borg's scale was less than 3, it was expected to be a safe work, 3 to 5, a risk work, and more than 5, a high risk work.

Table 2. Leg postures used in the experiment.

Posture No.	Category	Description
1	Standing	Mediolateral distance between heels as shoulder breadth
2		Anterior-posterior distance between heels as foot length
3	Knee-flexed posture	Knee angle at 150 degrees
4	Squatting	Knee angle at less than 90 degrees
5		Posture 1 with loading on the right foot
6	Imbalanced posture	Posture 2 with loading on the right foot
7		Posture 3 with loading on the right foot
8	Imbalanced squatting	Posture 4 with loading on the right foot
9	One-foot standing	Standing on one foot with no knee flexion
10		Standing on one foot with knee flexed by 30 degrees
11		Kneeling with knees fully flexed
12	Kneeling	Kneeling with knees flexed by 90 degrees
13		Kneeling on one knee

In the experiment, the leg posture was used in random order for each subject. Subjects were asked to take designated posture for 5 minutes. They reported their perceived discomfort levels using the Borg's CR-10 scale method on waist, hip, upper leg, lower leg, feet and whole body at 1, 3, and 5 minutes after taking each posture. A 3 kg weight load was selected considering that the weight of hand tools such as a drill, wrench, and hammer which are used a lot in the industrial workplace ranges from 1 kg to 3 kg. All subjects were allowed to take enough rest between consecutive trials.

3 RESULTS

The effects of leg postures on subjective discomfort ratings were analyzed for the positions of the body and working time. Also, an analysis of variance (ANOVA) was performed for all 13 leg postures and working time.

3.1 *Discomfort rating for leg postures classified by the body*

The means of the perceived discomfort for the waist was shown in Table 3. Posture 1 was found to be a safe work for the whole time and the other postures except posture 4, 5 and 8 were found to be risk works if these postures were taken longer than 3 minutes.

Table 4 shows that the hip discomfort ratings for postures 1 and 4 were found to be safe during 5 minutes. Posture 3 and 7 were found to be risk works after 3 minutes of postures and the rest of leg postures were appeared as risk works after 5 minutes.

In case of the upper leg, posture 1 was found to be a safe work during 5 minutes. But Posture 3, 7, 8, 10, 11 and 13 were found to be risk works even with 1 minute of taking the posture and especially posture 7 and 10 were found to be high risk works after 3 minutes. It might be difficult for subjects to maintain the body balance and load in imbalanced postures (Table 5).

With respect to the lower leg, all postures except posture 1, 2, 5 and 12 were classified as risk works after 1 minutes of posture. Postures 7, 8, 10, 11 and 13 were found to be high risk works after 3 minutes and all postures except postures 1, 2 and 12 were also found to be high risk works after 5 minutes of posture (Table 6).

As shown in Table 7 on the feet, all postures except posture 1, 2 and 12 were risk works from 1 minute. Posture 7, 8, 10 and 11 were high risk works from 3 minutes and posture 13 was also high risk work from 1 minute. On the other hand, the posture 12 was appeared to be a safe work during 5 minutes.

Table 8 showed the means of the perceived discomfort for the whole body. All postures except posture 1, 2 and 12 were found to be risk works from 1 minute and postures 7, 8, 10, 11 and 13 were found to be high risk works after 3 minutes. All postures except posture 1 and 12 were high risk works after 5 minutes.

Table 3. Discomfort rating of waist.

Time / Leg Posture	1 Min. (Mean)	3 Min. (Mean)	5 Min. (Mean)
Posture 1	1.0	1.7	2.3
Posture 2	2.1	3.1	3.9
Posture 3	2.4	3.2	4.0
Posture 4	1.7	2.5	3.3
Posture 5	2.0	2.7	3.6
Posture 6	2.3	3.3	4.5
Posture 7	2.2	3.2	4.0
Posture 8	2.0	2.8	3.6
Posture 9	2.1	3.2	4.1
Posture 10	2.0	3.0	4.0
Posture 11	2.2	3.4	4.4
Posture 12	2.4	3.5	4.2
Posture 13	2.8	3.9	4.9

| Safety | | Risk | | High risk |

Table 4. Discomfort rating of hip.

Time / Leg Posture	1 Min. (Mean)	3 Min. (Mean)	5 Min. (Mean)
Posture 1	0.8	1.1	1.4
Posture 2	1.6	2.4	3.0
Posture 3	2.3	3.0	3.7
Posture 4	1.5	2.2	2.8
Posture 5	1.6	2.3	3.0
Posture 6	1.7	2.7	3.8
Posture 7	2.2	3.0	3.8
Posture 8	1.8	2.6	3.4
Posture 9	1.9	2.7	3.6
Posture 10	1.8	2.6	3.6
Posture 11	2.0	2.8	3.9
Posture 12	1.6	2.4	3.3
Posture 13	1.9	2.8	3.6

| Safety | | Risk | | High risk |

Table 5. Discomfort rating of the upper legs.

Time / Leg Posture	1 Min. (Mean)	3 Min. (Mean)	5 Min. (Mean)
Posture 1	1.2	1.6	2.3
Posture 2	2.0	2.9	3.7
Posture 3	3.6	4.6	5.7
Posture 4	2.5	3.5	4.6
Posture 5	2.6	3.8	4.7
Posture 6	2.7	3.8	4.8
Posture 7	3.9	5.3	6.4
Posture 8	3.0	4.3	5.3
Posture 9	2.9	4.0	4.8
Posture 10	3.8	5.0	6.2
Posture 11	3.1	4.4	5.3
Posture 12	2.2	3.0	3.8
Posture 13	3.4	4.5	5.6

| Safety | | Risk | | High risk |

Table 6. Discomfort rating of the lower legs.

Time / Leg Posture	1 Min. (Mean)	3 Min. (Mean)	5 Min. (Mean)
Posture 1	1.7	2.3	3.0
Posture 2	2.4	3.4	4.3
Posture 3	3.7	4.6	5.8
Posture 4	3.3	4.5	5.6
Posture 5	2.9	4.2	5.0
Posture 6	3.3	4.3	5.2
Posture 7	4.2	5.4	6.8
Posture 8	3.7	5.1	6.3
Posture 9	3.2	4.5	5.4
Posture 10	4.0	5.4	6.8
Posture 11	3.9	5.4	6.5
Posture 12	2.0	2.6	3.3
Posture 13	4.5	5.7	6.7

| Safety | | Risk | | High risk |

Table 7. Discomfort rating of feet.

Time / Leg Posture	1 Min. (Mean)	3 Min. (Mean)	5 Min. (Mean)
Posture 1	1.9	2.9	3.5
Posture 2	2.7	3.6	4.5
Posture 3	3.3	4.5	5.6
Posture 4	3.6	4.8	6.1
Posture 5	3.2	4.5	5.6
Posture 6	3.3	4.5	5.6
Posture 7	4.2	5.4	6.3
Posture 8	4.1	5.7	6.7
Posture 9	3.3	4.9	5.9
Posture 10	4.1	5.6	7.0
Posture 11	4.7	6.3	7.2
Posture 12	1.5	2.0	2.6
Posture 13	5.0	6.4	7.3

| Safety | | Risk | | High risk |

Table 8. Discomfort rating of the whole body.

Time / Leg Posture	1 Min. (Mean)	3 Min. (Mean)	5 Min. (Mean)
Posture 1	2.1	3.4	4.6
Posture 2	2.6	3.9	5.0
Posture 3	3.3	4.5	5.7
Posture 4	3.0	4.2	5.2
Posture 5	3.0	4.4	5.5
Posture 6	3.2	4.5	5.8
Posture 7	3.9	5.3	6.6
Posture 8	3.3	5.0	6.4
Posture 9	3.1	4.4	5.5
Posture 10	3.8	5.2	6.7
Posture 11	4.0	5.5	6.9
Posture 12	2.8	4.0	4.8
Posture 13	4.5	5.9	6.9

| Safety | | Risk | | High risk |

3.2 Analysis of variance

An analysis of variance (ANOVA) was conducted on the discomfort rating data between 13 leg postures and working time. The result for all leg postures showed that the posture and time effect were statistically significant respectively ($p < 0.05$) as shown in Table 9.

4 CONCLUSIONS

In this study, 13 leg postures were evaluated on the basis of a subjective discomfort rating for three different working time.

Table 9. ANOVA results for discomfort ratings of all leg postures.

Source of variation	SS	DF	MS	F	p
Leg posture	1128.298	12	94.025	32.977	0.001[a]
Working time	3353.805	2	1676.903	588.128	0.003[a]
Total	4482.103	14			

[a] statistically significant at $\alpha = 0.05$

The results showed that the means of the perceived discomfort rating on leg postures were generally increased as the working time increased. In particular, the biggest mean values were found in the legs and feet. This may be due to the work load of lower limbs in squatting and kneeling postures. Based on the results of ANOVA, leg postures and working time together had significant effects on perceived discomfort and it was found that discomfort ratings increased as the time increased.

To maintain a correct posture at work is an important consideration in the design of work method and workplaces, because it has a strong relationship with the postural stability and mobility of the whole body.

Therefore, the results of this study can be used as a basic data in evaluating the work load of lower limb postures and preventing adverse health effects such as discomfort, fatigue and musculoskeletal disorders from poor working postures.

REFERENCES

Kim, Y.C., Ryu, Y.S., 2005, Ergonomic evaluation of the hazardous jobs in squatting work posture, Journal of the Ergonomics Society of Korea Vol.24, No.1, pp. 37–41.

Ministry of Labor in Korea, 2005, The status report of industrial disaster in 2004.

Lee, I.S., Chung, M.K., 1998, Workload evaluation of squatting work postures, Journal of the Korean Institute of Industrial Engineers Vol.24, No.2, pp. 167–174.

Lee, I.S., Chung, M.K., Kee, D.H., 1998, Evaluation of postural load of varying leg posture using the psychophysical scaling, Journal of the Ergonomics Society of Korea Vol.21, No.4, pp. 47–65.

Armstrong, T.J., 1986, Upper-extremity postures: definition, measurement and control, The Ergonomics of Working Postures: Models, Methods and Cases, Taylor & Francis, London, pp. 59–73.

Borg, G., 1982, Psychological basis of physical exertion, Medicine and Science in Sports and Exercise 14, pp. 377–381.

Corlett, E.N., Wilson, J., Manenica, I., 1986, The Ergonomics of Working Postures: Models, Methods and Cases, Taylor & Francis, London.

Genaidy, A.M., AI-shedi, A.A., Karwowski, W., 1994, Postural stress analysis in industry, Applied Ergonomics 25 (2), 77–87.

Grandjean, E., Hunting, W., 1977, Ergonomics of postures review of various problem of standing and sitting postures, Applied Ergonomics 8 (3), pp. 135–140.

Chapter 13   Ergonomics in occupational health 4

Ergonomic Trends from the East – Kumashiro (ed)
© 2010 Taylor & Francis Group, London, ISBN 978-0-415-88178-4

Assessing the effect of antioxidative with Yoga and Tai-Chi training

Tin-Chang Chang
Department of Business Administration, Asia University, Taichung, Taiwan

Shu-Ling Chang
Overseas Chinese Institute of Technology, Wufeng, Taichung, Taiwan

ABSTRACT: This study examines similarities and differences in Yoga and Tai-Chi exercise. A total of 36 voluntary undergraduate female students were categorized as three treatment groups. The two experimental groups received a regular schedule on Yoga and Tai-Chi training (40 minutes each time, three times a week) for ten weeks. Fasting blood samples were drawn before and after the 10-week training program in order to determine the antioxidant capacity which are superoxide dismutase (SOD) and glutathione peroxidase (GSHPx) and oxidative stress measures which are creatine kinase (CK), lactate dehydrogenase (LDH) and malondialdehyde (MDA). The study suggests that Tai-Chi and Yoga training were found to be an effective way to improve many fitness measures over a ten weeks period. Tai-Chi training was also found to be significantly better than Yoga training in enhancing certain measures of fitness.

Keyword: *Tai-Chi, Yoga, Antioxidant Capacity, Oxidative Stress*

1 INTRODUCTION

In recent years, around the Asia region, health promotion entertainments has gradually become more popular for people who wish to accomplish the goal of keeping in good health. Among all of the forms of entertainment, Tai-Chi is the most well-known traditional Chinese health promotion exercise. Tai-Chi is moderate exercise similar to Yoga. Although it has little movement of the body and no extensive exercise, it leads to health benefits since all of our joints are used. Therefore, to avoid sport injuries by doing entertainment sports, most people choose Tai-Chi or Yoga as options. Moderate exercise can result in greater health benefits than vigorous exercise (Masuda, Tanabe, & Kuno, 2002). Recently, public-health recommendations have emphasized the value of moderate exercise for improving cardiovascular health and reducing the risk of coronary heart disease (Swain & Franklin, 2006). In China, Tai-Chi is a famous exercise that has been practiced for many centuries (Wang, Collet, & Lau., 2004), but Yoga is an increasingly popular therapy. Yoga is used to maintain wellness and assist with the management of a range of health complaints (Smith, Hancock, Blake-Mortimer, & Eckert, 2007; Carson, Carson, Porter, Keefe, Shaw, & Miller, 2007), while Tai-Chi appears to be effective in promoting balance and flexibility, enhancing cardiovascular function, and reducing mental conditions such as depression and anxiety (Chen & Snyder., 1999; Chou, Lee, Yu, Macfarlane, Cheng & Chan., 2004; Wang et al., 2004; Taylor-Piliae, Haskell, Waters, & Froelicher, 2006; Cheng, 2007).Yoga can produce an animate effect on psychological and physical energy (Carson, et al., 2007). It is able to remove the habit of unhealthy nutrition and to establish homeostatic balance (Singh, 2006).

The benefits of regular moderate exercise on the wellbeing and the immune system are well known. However, these benefits are lost with exhaustion and lack of training. It has been reported that exhaustive exercise induces oxidative stress, which occurs when oxidant production overwhelms the antioxidant capacity [1–3]. Moderate exercise significantly decreased the aging-associated development of oxidative stress by preventing but after high-intensity exercise will increase of oxidative stress (Badano, Boveris, Stoppani, & Vidal, 1973; Bailey, Davies, Young, Jackson, Davison, Isaacson, & Richardson, 2003; Navarro, Gomez, Lopez-Cepero, & Boveris,

2004). Yoga and Tai-Chi has been defined as an exercise of moderate intensity (Lord, Russell, & Iran, 1991; Rai, Ram, Kant, Madan, & Sharma, 1994; DiCarlo, Sparling, Hinson, Snow, & Rosskopf, 1995; Wang, et al., 2004). In the study of Subudhi, Fu, Strothkamp, and Murray (2003) subjects were classified as trained and untrained, it points out that aerobic power values of trained subjects were significantly higher than those of untrained subjects. Exercise training not only induces activities of antioxidant enzymes, but also affects the antioxidative capacity of the vascular wall protection (Kojda & Hambrecht, 2005). Some antioxidant system in the human body is for protecting our cells from harm by oxidization or free radical. The main components of the system are small molecules of antioxidant substances such as vitamin C, E and some kinds of enzymes such as superoxide dismutase (SOD) or glutathione peroxidase (GSHPx). These enzymes play an important role in preventing exterior oxidizations(Chung et al, 1999).

Another set of theories suggests that exercise induces oxidative stress, which occurs when oxidant production overwhelms the antioxidant capability (Tauler, Sureda, Cases, Aguiló, Rodríguez-Marroyo, Villa, Tur, & Pons, 2006). This is different to the previous researches. Free Radical has been confirmed by the medical profession that it will cause the oxidation of human cells injury, aging, cancer and cardiovascular disease. Course of the campaign, in addition to several times the muscle will increase oxygen consumption, it would also lead to the formation of free radicals brought oxidative stress (Aruoma, 1994). Malondialdehyde (MDA), a by-product of lipid peroxide, is the most frequently studied marker of oxidative tissue damage during exercise. MDA levels have been found to increase both in different tissues and plasma during exercise.

This aim of the present study was to determine the effect of antioxidative capability and the occurrence of oxidative stress after the Yoga and Tai-Chi training. And antioxidant capacity is measured by SOD and GSHPx. Simultaneously, oxidative stress is measured by CK, LDH and MDA (Terblanche, Masondo, & Nel, 1998; Satoh, Yokozawa, Cho, Okamoto, & Sei, 2004; Hoelzl, Bichler, Ferk, Simic, Nersesyan, Elbling, Chakraborty, & Knasmuller, 2005). This study then compares which is good at enhancing antioxidative capability between Yoga and Tai-Chi.

2 METHOD

Participants in the present study include 36 voluntary undergraduate female students (Tai-Chi group, n = 12; Yoga group, n = 12; control group with no training, n = 12). Table 1 summarizes the baseline characteristics of the subjects in each group. With the means and standard deviations about age, height, and weight for pre-training, there were not significant differences between three groups on age (mean max = 18.83, min = 18.75), height (mean max = 160.58, min = 159.58) and weight (mean max = 51.75, min = 50.33). No participants had previous experience with any exercise training.

There were two different exercise training groups and a control group without any training. In group 1, the Tai-Chi program was a 13-movement styles (Opening form, Crane spreads its wings, Strum the lute, Step back to repulse monkey, Single whip, Wave hands like clouds to both sides, Single whip, High pat on horse, Right heel kick to the corner, Double wind goes through the ears, Turn to kick with the left heel, Cross hands, Close form) taught by an experienced instructor, one-to-two movements were taught each week for 10 weeks (40 minutes each time, three times a week). At the same period, participants with Yoga program taught by an experienced instructor with 9 styles (Savasana, Surya-Namaskara, Majariasana, Ustrasana, Padmasana, Mudra, Natarajasana,

Table 1. Descriptive Statistics of Three Groups.

	Age Mean(SD)	P value	Height Mean(SD)	P value	Weight Mean(SD)	P value
Tai-Chi group	18.75(0.62)	0.914	159.88(5.58)	0.908	51.75(5.45)	0.806
Yoga group	18.83(0.39)		159.58(4.83)		50.33(4.91)	
Control group	18.75(0.62)		160.58(6.60)		51.00(5.44)	
Total	18.78(0.54)		160.01(5.57)		51.03(5.15)	

*The mean difference is significant at the 0.05 level.

Utthita Parsvakonasana, Utthita Hasta Padangusthasana). In group 3, participants were control group without any training.

Thirty-six undergraduate female students volunteered to participate in the study. They were assigned randomly to one of three groups: a Tai-Chi group, a Yoga group and one control group. Tai-Chi and Yoga groups are training at same time and difference places to process the experiment. In phase 1, the objective and procedure of the experiment were clearly described to participants. In phase 2, before the experiment, each participant was required to signed consent and took a check list about the health situation. This study excludes those participants who had hurt, getting hurt, have muscle or skeleton disease, or not fit for this experiment. In Phase 3, the participants' blood biochemical data was also collected before experiment. The data were compared with pretest and posttest. In Phase 4, the two experimental groups received a regular schedule on Yoga and Tai-Chi training for ten weeks. In Phase 5, fasting blood samples were collected after the 10-week training program for measurements of plasmatic indices of antioxidant capacity (SOD and GSHPx) and oxidative stress (CK, LDH and MDA).

3 RESULTS

Table 2 list the results of the analysis of variance (ANOVA) in terms of the pretest data. There is no statistically significant difference among three groups in the pretest. All of the measured five markers show no statistical significant difference among three groups. Table 3 shows the ANOVA of the posttest data. Two measured markers, LDH and MDA, are both significant different

Table 2. ANOVA of Pretest (three groups).

Item	F-value	P-value
CK	0.02	0.984
LDH	0.11	0.894
MDA	0.32	0.728
SOD	0.17	0.847
GSHPx	0.05	0.953

*The mean difference is significant at the 0.05 level.

Table 3. ANOVA of Posttest (three groups).

Item	F-value	P-value
CK	1.71	0.196
LDH	9.16	0.001*
MDA	4.72	0.016*
SOD	1.12	0.338
GSHPx	0.65	0.530

*The mean difference is significant at the 0.05 level.

Table 4. Post Hoc of LDH.

	alpha = .05	
Groups	1	2
Tai-Chi(n = 12)	386.25	
Yoga(n = 12)	397.58	
Control(n = 12)		457.33

Table 5.   Post Hoc of MDA.

| Groups | alpha = .05 | |
	1	2
Tai-Chi(n = 12)	0.33	
Yoga(n = 12)	0.41	0.41
Control(n = 12)		0.43

(LDH: $F = 9.16$, $P = 0.001 < 0.05$; MDA: $F = 4.72$, $P = 0.016 < 0.05$). We further used the post-hoc Scheffe test to perform group comparisons for LDH and MDA. In LDH, result explains that the LDH value in both Tai-Chi and Yoga group are significantly lower than that in control group (Table 4). But in MDA, Yoga group's LDH value is significant higher than the Control group, as well as the Tai-Chi group (Table 5).

4   DISCUSSION

The study shows that a form of Tai-Chi can have significant effect and Yoga can have partial effect on antioxidative capability and oxidative stress in female. It was found that the blood levels of LDH and MDA value decreased significantly ($p < 0.05$), while SOD and GSHPx value increased significantly ($p < 0.05$) for the Tai-Chi group at the end of training, as compared to the onset. It was also noted that LDH value decreased significantly ($p < 0.05$), while the other indexes remained unchanged ($p > 0.05$) during the course of study for the Yoga group. The results demonstrated that Tai-Chi will enhance antioxidative capability and may well reduce oxidative stress in the undergraduate female students. Yoga also may reduce oxidative stress in the undergraduate female students.

There is one of the strongest contributions of Tai-Chi to enhance antioxidative capability and to reduce oxidative stress. Tai-Chi has been defined as an exercise of moderate intensity (Lord, Russell & Iran, 1991; Wang, et al., 2004). The findings of this study confirm those of previous research that moderate exercise significantly decreased the aging-associated development of oxidative stress (Badano, et al., 1973; Bailey, et al., 2003; Navarro, et al., 2004).

The benefits of introducing Yoga are supported by ample literature. In this study, although LDH value has been decreased, the performance of Yoga is not better than Tai-Chi. This non-significant observation may result from a fact that incorporating a training session of 10 weeks is not long enough for this study to have statistically significant results. Although in literature comparatively less discussions has focused on Tai-Chi, this study empirically supports that Tai-Chi has better antioxidative capacity than Yoga. There are statistically significant difference and significant decrease oxidative stress (LDH and MDA) and increase antioxidative capability (SOD and GSHPx). The findings suggest that further research can have more time for training and chase the following development in the future, and compared the difference between Yoga and Tai-Chi.

REFERENCES

Badano, B. N., Boveris, A., Stoppani, A. O., & Vidal, J. C. (1973) The action of Bothrops neuwiedii phospholipase A2 on mitochondrial phospholipids and electron transfer. *Mol Cell Biochem*, 2, 157–167.
Bailey, D. M., Davies, B., Young, I. S., Jackson, M. J., Davison, G. W., Isaacson, R., & Richardson, R. S. (2003) EPR spectroscopic detection of free radical outflow from an isolated muscle bed in exercising humans. *Journal of Applied Physiology*, 94, 1714–1718.
Carson, J. W., Carson, K. M., Porter, L. S., Keefe, F. J., Shaw, H., & Miller, J. M. (2007) Yoga for Women with Metastatic Breast Cancer: Results from a Pilot Study. *Journal of Pain and Symptom Management*, 33, 331–341.
Chen, K. M., & Snyder, M. (1999) A research-based use of Tai Chi/movement therapy as a nursing intervention. *Journal of Holistic Nursing*, 17, 267–279.

Cheng, T. O. (2007) Tai Chi: The Chinese ancient wisdom of an ideal exercise for cardiac patients. *International Journal of Cardiology*, 117, 293–295.

Chou, K. L., Lee, P. W., Yu, E. C., Macfarlane, D., Cheng, Y. H., & Chan, S. S. (2004) Effect of Tai Chi on depressive symptoms amongst Chinese older patients with depressive disorders: a randomized clinical trial. *International journal of geriatric psychiatry*, 19, 1105–1107.

DiCarlo, L. J., Sparling, P. B., Hinson, B.T., Snow, T. K., & Rosskopf, L. B. (1995) Cardiovascular, metabolic, and perceptual responses to hatha yoga standing poses. *Medicine, Exercise, Nutrition and Health*, 4, 107–112.

Herrick, C. M., & Ainsworth, A. D. (2000) Invest in yourself. Yoga as a self-care strategy. *Nursing Forum*, 35 (2), 32–36.

Hoelzl, C., Bichler, J., Ferk, F., Simic, T., Nersesyan, A., Elbling, L., Ehrlich, V., Chakraborty, A., & Knasmuller, S. (2005) Methods for the detection of antioxidants which prevent age related diseases: A critical review with particular emphasis on human intervention studies. *Journal of Physiology and Pharmacology*, 56(2), 49–64.

Kojda, G., & Hambrecht, R. (2005) Molecular mechanisms of vascular adaptations to exercise.Physical activity as an effective antioxidant therapy? *Cardiovascular Research*, 67, 187–197.

Lan, C., Lai, J. S., & Chen, S. Y. (2002) Tai Chi Chuan: An ancient wisdom on exercise and health promotion. *Sports Medicine*, 32(4), 217–224.

Lord, S. R., Russell, D. C., & Iran, W. W. (1991) Physiological factors associated with falls in an elderly population. *Journal of the American Geriatrics Society*, 39, 1194–1200.

Masuda, K., Tanabe, K., & Kuno, S. Y. (2002) Exercise, oxidative stress and health benefit. *Bulletin of Institute of Health and Sport Sciences-University of Tsukuba Ibaraki ken*, 25(1), 1–11.

Navarro, A. M., Gomez, C., Lopez-Cepero, J. M., & Boveris, A. (2004) Beneficial effects of moderate exercise on mice aging: survival, behavior, oxidative stress, and mitochondrial electron transfer. *American Journal of Physiology-Regulatory Integrative and Comparative Physiology*, 286, 505–511.

Rai, L., Ram, K., Kant, U., Madan, S. K., & Sharma, S. K. (1994) Energy expenditure and ventilatory responses during siddhasana—a yogic seated posture. *Indian Journal of Physiology and Pharmacology*, 38 (1), 29–33.

Satoh, A., Yokozawa, T., Cho, E. J., Okamoto T., & Sei, Y. (2004) Antioxidative effects related to the potential anti-aging properties of the Chinese prescription Kangen-karyu and Carthami Flos in senescence-accelerated mice. *Archives of Gerontology and Geriatrics*, 39, 69–82.

Singh, A. N. (2006) Role of yoga therapies in psychosomatic disorders. *International Congress Series*, 1287, 91–96.

Smith, C. A., Hancock, H., Blake-Mortimer, J., & Eckert, K. (2007) A randomised comparative trial of yoga and relaxation to reduce stress and anxiety. *Complementary Therapies in Medicine*, 15, 77–83.

Subudhi, A. W., Fu, M. X., Strothkamp, K. G., & Murray, D. M. (2003) Effect of graded exercise on blood glutathione status in trained and untrained humans. *International sports journal-West-Haven-Conn*, 7(2), 82–90.

Swain, D., & Franklin, B. (2006) Comparison of Cardioprotective Benefits of Vigorous Versus Moderate Intensity Aerobic Exercise. *The American Journal of Cardiology*, 97, 141–147.

Tauler, P., Sureda, A., Cases, N., Aguiló, A., Rodríguez-Marroyo, J. A., Villa, G., Tur, J. A., & Pons, A. (2006) Increased lymphocyte antioxidant defences in response to exhaustive exercise do not prevent oxidative damage. *Journal of Nitritional Biochemistry*, 17, 665–671.

Taylor-Piliae, R. E., Haskell, W. L., Waters, C. M., & Froelicher, E. S. (2006) Change in perceived psychosocial status following a 12-week Tai Chi exercise programme. *Journal of Advanced Nursing*, 54, 313–329.

Terblanche, S. E., Masondo, T. C., & Nel, W. (1998) Effects of cold acclimation on the activity levels of creatine kinase, lactate dehydrogenase and lactate dehydrogenase isoenzymes in various tissues of the rat. *Cell Biology International*, 22, 701–707.

Wang, C. C., Collet, J. P., & Lau, J. (2004) The Effect of Tai Chi on Health Outcomes in Patients With Chronic Conditions. *Archieves of Intermal Medicine*, 164, 8.

Jenkins, R. R. (2000). Exercise and oxidative stress methodology: A critique. American Journal of Clinical Nutrition, 72, 670–674.

Johnson, P. (2002). Antioxidant enzyme expression in health and disease: Effects of exercise and hypertension. Comparative Biochemistry and Physiology, 133, 493–505.

Kasapoglu, M., & Ozben, T. (2001). Alterations of antioxidant enzymes and oxidative stress markers in aging. Experimental Gerontology, 36, 209–220.

Reckelhoff, J. F., & Fortepiani, L. A. (2004). Novel mechanisms responsible for postmenopausal hypertension. Hypertension, 43, 918–923.

Aruoma, O. I., (1994). Free radicals and antioxidant strategies in sports. J Nutr Biochem, 5:370–381.

Ergonomic Trends from the East – Kumashiro (ed)
© 2010 Taylor & Francis Group, London, ISBN 978-0-415-88178-4

Introduction of occupational health care activity in Japan

Yoshika Suzaki
Japanese Red Cross Kyushu International College of Nursing

Naoko Takayama
Anan National College of technology

Minoru Moriguchi
Hiroshima International University

Hiromi Ariyoshi
Saga Medical School Faculty of Medicine, Institute of Nursing

Keywords: Stress, Questionnaire, Occupational health nurses

1 BACKGROUND

As depressive patients or suicide cases are increasing in number, issues on mental health are becoming crucial in Japanese companies. In 2005, the legal guidelines for mental health measures were clarified through the 2005 "Amendment of the Occupational Health and Safety Law." In 2006, "Guidelines for Maintaining and Promoting the Mental Health of Workers" were issued, while the "Obligation to Have Mental Health Measures" was instituted in 2008.

Based on a survey that employed a multifaceted, stress-related questionnaire designed for medium-sized workplaces, where mental health measures can be difficult to implement and working conditions are often harsh, we introduced mental health measures into such a workplace and clarified the problem there.

2 AIMS

This study aimed to reveal the stress conditions of employees who worked for medium-sized workplaces, to clarify the problems in mental health measures in those workplaces, and to find their solutions.

3 RESEARCH METHOD

We conducted the survey at a steel-related enterprise, which had 72 employees: 59 males aged 22–64 (average age: 45.33 ± 13.12) and 13 females aged 22–61 (average age: 47.15 ± 14.24). The survey period was March 2006–December 2008.

Before March 2006, when we intervened, this company had not had the security and health care system, and any health supervisor or industrial physician had not been appointed. At present, a public health nurse is in charge of industrial health care activity as below, cooperating with a part-time industrial physician (once a week) who is also a certified industrial health consultant.

i) Fact-finding
  We used the Multifaceted, Stress-related Questionnaire for this survey.
ii) Health Support according to the actual conditions
  The public heal nurse undertook the health care activity below, based on the survey result of actual conditions.

Table 1. Comparison before and after mental health educating to male employee.

| | | Mental health education | | | | |
| | | Before | | After | | |
		n	%	n	%	The P-values
Life-style	Good	42	72.4	47	87.0	
	Caution is necessary	47.2	27.6	52.8	13.0	0.045*
Degree of fatique	Good	49	84.5	51	94.4	
	Caution is necessary	9	15.5	3	5.6	0.080+
Adaptation to life	Good	38	65.5	42	77.8	
	Caution is necessary	20	34.5	12	22.2	0.110
Depressive tendency	Good	25	43.1	35	64.8	
	Caution is necessary	33	56.9	19	35.2	0.017*
Neurotic condition	Good	33	56.9	40	74.1	
	Caution is necessary	25	43.1	14	25.9	0.043*
Psychosomatic condition	Good	36	62.1	44	81.5	
	Caution is necessary	22	37.9	10	18.5	0.019*
Alcohol dependence	Good	35	60.3	41	75.9	
	Caution is necessary	23	39.7	13	24.1	0.059+
Emotional loss	Good	26	44.8	36	66.7	
	Caution is necessary	32	55.2	18	33.3	0.016*
Overadaptation	Good	39	67.2	42	77.8	
	Caution is necessary	19	32.8	12	22.2	0.151
Aggression	Good	40	69.0	43	79.6	
	Caution is necessary	18	31.0	11	20.4	0.142
Happiness in daily life	Good	21	36.2	34	63.0	
	Caution is necessary	37	63.8	20	37.0	0.004**

* The P-values were calculated by Fisher's direct test +p < .10, * p < .05, ** p < .01

a. Management of working hours
   The public health nurse, cooperating with the part-time industrial physician, established the classification about working hours: (A) "Can work overtime," (B) "Should not work overtime," and (C) "Should not work at all," based on the survey result of stress conditions and the results of the medical examination.
b. Follow-up to the cases with problems in working
   The public health nurse and the industrial physician periodically interviewed with workers who had mental health problems or who were classified into B (should not work overtime) or C (should not work at all). We also interviewed their supervisors with their consent in some cases.
c. Mental health education for managers and general workers
   In August 2006, we started lectures to teach knowledge and skills of mental health care to managers and general workers.
iii) Method of analysis and statistical processing
   Using a statistical software system, SPSS 15.0, we analyzed the actual conditions before and after mental health education by means of a chi-square test or Fisher's exact test.

Ethical conditions
   We explained the research to the managers and general workers both orally and in writing and got their consent to fill in the questionnaire when we received the answer sheets.

Table 2. Comparison before and after mental health educating to female employee.

| | | Mental health education | | | | |
| | | Before | | After | | |
		n	%	n	%	The P-values
Life-style	Good	12	92.3	11	91.7	
	Caution is necessary	1	7.7	1	8.3	0.740
Degree of fatique	Good	11	84.6	10	83.3	
	Caution is necessary	2	15.4	2	16.7	0.672
Adaptation to life	Good	9	69.2	9	75.0	
	Caution is necessary	4	30.8	3	25.0	0.550
Depressive tendency	Good	6	46.2	8	66.7	
	Caution is necessary	7	53.8	4	33.3	0.265
Neurotic condition	Good	10	76.9	10	83.3	
	Caution is necessary	3	23.1	2	16.7	0.541
Psychosomatic condition	Good	7	53.8	9	75.0	
	Caution is necessary	6	46.2	3	25.0	0.248
Alcohol dependence	Good	10	76.9	11	91.7	
	Caution is necessary	3	23.1	1	8.3	0.328
Emotional loss	Good	10	76.9	10	83.3	
	Caution is necessary	3	23.1	2	16.7	0.541
Overadaptation	Good	11	84.6	10	83.3	
	Caution is necessary	2	15.4	2	16.7	0.672
Aggression	Good	9	69.2	9	75.0	
	Caution is necessary	4	30.8	3	25.0	0.550
Happiness in daily life	Good	10	76.9	11	91.7	
	Caution is necessary	3	23.1	1	8.3	0.328

* The P-values were calculated by Fisher's direct test

4 RESULTS

Following education, significant differences were found among the males in six of the questionnaire categories: lifestyle, tendency to depression, tendency to neurosis, tendency to psychosomatic symptoms, degree of apathy, and degree of satisfaction with life. There were no significant differences among the females (Table 1 & Table 2).

The survey results were fed back to the employees through personal interviews. Eleven males and one female had a tendency to illness. Two of these 13 employees were hospitalized in neurology departments and were prohibited from working, namely classified to (C) and 10 were classified to (B), which means that they should not work overtime. Five of these 10 B-classified employees underwent mental examinations, and two were hospitalized in specialized medical institutions. Three of the males are currently outpatients, and three have been reassigned to different posts. A total of four males, including two who were hospitalized, left the company, as did a female employee.

5 REMARKS

According to Hirata et al. (2002), that mental health measures were not active in small- and medium-sized companies, compared with large companies, and our survey illustrated the situation Hirata claimed. One of the reasons might be insufficient human resources for mental health education. The industrial health nurse reported the president and the management that there were a lot of male workers who have various stress-related illness and this situation might cause suicide or death

Table 3. Content and execution frequency of mental health Educating to executive job worker and general worker.

Executive job worker		Frequency
The first	Caring by management supervisor	Once/month
The second	The stress is understood.	Once/month
The third	Self care is experienced.	Once/month
The fourth	Road to good listener	Once/month
The fifth	Recent topic(suicide)	Once/month
Five times in total were executed twice.		

General worker		Frequency
The first	The stress is understood.	Once/month
The second	Self care is experienced.	Once/month
The third	Road to good listener	Once/month
The fourth	Recent topic(suicide)	Once/month
Four times in total were executed three times.		

from overwork and started to improve the workplace environment. About 75% of the employees we interviewed frequently asked such a question, "What I am talking in this interview will be definitely between you and me, right? Could you promise that it will be never reported to the president?" We spent much time letting them know industrial doctors and industrial health nurses preserve the confidentiality of their clients. Through this intervention, we revealed that the company offered health guidance after medical examination for only one day so that some employees could not have the guidance. Therefore, we had to intervene to the health guidance at this company.

After establishing the working hour classification and providing the mental health education, the president or the management had not been oppressive as they were before and had come to listen to employees' opinions. Afterward, the stress conditions of the employees appeared to be improved in all aspects. This was considered to be the results of our industrial health activities for more than two years such as health guidance to all employees.

6 CONCLUSION

We evaluated the industrial health activity mainly done by a public health nurse in a medium-sized company that had not taken any mental health measures. Using the stress-related questionnaires, this study proved that we could grasp workers' psychosomatic conditions including the state of adaptation to the workplace through personal interviews and mental health education.

REFERENCES

European trade Union Congress: European framework agreement on work-related stress. Brussels, Belgium: ETUC, UNICE, UEAPME and CEEP, 2004.

Kumai Mitsuharu, Yoshika Suzaki: Evaluation of the Validity and Utility of the Multi-faceted Life Stress Questionnaire Newly made out for Workers, Japan Health Medicine Association, 16(1), 8–16, 2007.

Levi L: Guidrance on work-related stress. Luxembourg: Office for Official Publications of the European Communities, 2000.

Nobuko Hirata, Yuko-Ohara Hirano: Statistics Study Method on Gender-Based Stressor Affecting Mental health of Working Women in Japan, Memoirs Kyushu Univ, Vol 4, 57–66, 2002.

NORA Organization of Work Team Members, The changing organization of work and the safety and health of working people, Cincinnati, Oh, National Institute for Occupational Safety and Health, 2002.

Sauter Sl, Murphy LR, Hurrel JJ Jr: Prevention of work-related psychogical disorders. A national strategy proposed by the Natinal Institute for Occupational Safety and Health (NIOSH), Am Psychol, 45(10), 1146–1158, 1990.

van der Klink JJ, Blonk RW: Reducing long term sickness absence by an activating intervention in adjustment disorders: a cluster randomized controlled design, Occup Environ Med, 60(6), 429–437, 2003.

Ergonomic considerations related to the globalization of the healthcare workforce in Japan

Nozomi Sato
School of Science & Engineering, Kinki University, Higashi-Osaka, Japan

1 INTRODUCTION

Because of the rapidly increasing aging population and the high retirement rate of nurses and caregivers due to their harsh working conditions, a shortage of healthcare workers is a critical issue in Japan. In 2007, the Ministry of Health, Labour and Welfare estimated that Japan needed to have an additional 400,000–600,000 caregivers in the next ten years (in 2004–2014).

One of the efforts that could contribute to easing this labor_shortage would be recruitment of nurses and caregivers from overseas. In fact, developed countries such as the USA, Canada, the UK, and Australia, and oil-rich Middle East countries (e.g., Saudi Arabia, Kuwait, Libya) have been receiving healthcare workers from many countries in Asian and South African regions (Buchan, et al., 2005). Japan also started to recruit nurses and caregivers from Indonesia under the Economic Partnership Agreement (EPA), and plans to recruit nurses and caregivers from the Philippines as well. This situation is challenging for Japanese healthcare staff, as most of them have no experience working with persons whose racial and cultural backgrounds differ from their own.

At this stage, it is not clear what problems would be expected by receiving foreign nurses and caregivers into the workforce. In order to effectively incorporate and integrate such nurses and caregivers into the Japanese healthcare system, it is essential to discuss potential problems related to this issue. Thus, this paper provides some ergonomic considerations related to the globalization of the Japanese healthcare workforce for maintaining and improving healthcare quality.

2 ERGONOMIC CONSIDERATIONS RLEATED TO GLOBALIZING HEALTHCARE SYSTEMS

2.1 *Paradigms of healthcare delivery systems for patient safety*

The most important goal of the healthcare system is to provide patients with safe and high quality care. To date, much research has been done to achieve this goal. Karch *et al.* (2006) categorized a variety of paradigms related to prevention of adverse patient outcomes into three groups. These paradigms focused on: 1) reducing healthcare professional errors, 2) reducing patient injuries, and 3) improving the use of evidence-based medicine. They pointed out that each paradigm focused on patient-related outcomes, but little attention was paid to the central role of healthcare staff and their activities performed within the healthcare systems.

To complement the limitations of these paradigms, they proposed a framework for improving the quality of healthcare systems from the human factors engineering viewpoint. In this framework, they emphasized the need for designing healthcare systems to support healthcare staff performance, which leads to patient safety, employee safety, and improved quality of care. Elements of performance inputs and transformation processes described in the paradigm are shown in Table 1. According to Karch *et al.* (2006), performance inputs in the paradigm refer to the work system, which are the system pre-conditions, and influence healthcare staff performance. Transformation processes in the paradigm refer to the actual acts of transforming inputs into outputs. It would be

Table 1. Elements of performance inputs and transformation processes in the paradigm proposed by Karsh, *et al.* (2006).

Performance inputs	Transformation processes		
Patient and provider factors	**Physical performance**		
Skills, knowledge, training, education	Carrying	Reaching	Lifting
Size, weight, reach, strength	Lowering	Walking	Running
Age, gender, ethnicity, language	Sitting	Seeing	Manipulating
Needs, biases, beliefs, mood	Injecting	Preparing	Charting
Work system/unit factors	Typing	Writing	Adjusting
Task demands, complexity difficulty	Feeling		
Time and sequence demands	**Cognitive performance**		
Availability of usable technology	Sensing		Perceiving
Technology functions/features	Localizing		Searching
Noise, temperature, lighting	Memory		Attention
Physical layout and geography	Imaging		Forethought
Organization factors	Communicating		Analyzing
Organizational policy/priorities	Problem solving		Vigilance
Organizational structure	Pattern matching		Awareness
Financial resources	Controlling		Learning
Rewards structure	**Social/behavioral performance**		
Management structure	Attributing causality		
Training provided	Self-regulating		
Staffing levels	Social learning		
Social norms and pressures	Motivation		
Social climate/culture	Interpersonal communication		
External environment	Planning behavior		
Extra-organizational rules, standards	Cost-benefit analysis		
Legislation, enforcement	Decision-making		
Industry social influence			
Industry workforce characteristics			

helpful to consider ergonomic issues related to globalization of the Japanese healthcare workforce referring to this paradigm because it describes in detail the kinds of factors that may affect health-care staff performance within the healthcare system and systematically explains how each factor interacts to affect patient safety.

2.2 *Ergonomic considerations associated with language barrier*

For caregivers and nurses working overseas, the biggest concern is to communicate using a non-native language. In Table 1, it is related to the elements of the performance inputs (providers' skills, knowledge, training, education, ethnicity, and language) and acts as a barrier for improving performance. It would considerably affect elements of transformation processes (seeing, preparing, typing, writing as physical performance; sensing, perceiving, searching, communicating, learning as cognitive performance; and interpersonal communication as social/behavioral performance) and will lead to adverse outcomes of healthcare systems. Shearer (2008) suggested that health care providers who were required to provide care in their non-native language might misinterpret a medication printed on the medication administration record and leave the patient with too much or too little medication, and that educating a patient on the wrong medication would continue to promote inaccuracy. Ogawa (2008) stated that various cases of miscommunication happen with the lack of language skills, which leads to discomfort, mistrust, and sometimes abuse if caregivers are unable to communicate well with the care receiver.

To support improvement of language proficiency in migrant caregivers and nurses, recipient countries provide intensive language training courses for them. According to the EPA, Indonesian

Figure 1.   Brief outline for supporting the integration of migrant caregivers and nurses into the healthcare system.

nurses and caregivers receive Japanese language training for 6 months. However, it is very difficult to acquire good knowledge of the Japanese language in such a short period of time. Even though they have good command of Japanese, it is difficult for them to correctly understand what is spoken or written in Japanese in the unfamiliar and complex healthcare system.

One of the most important issues related to integrating foreign staff in the healthcare workforce is prevention of medical accidents induced by poor instructional design of medical materials and miscommunication with staff or patients. Therefore, to lower this language barrier that foreign healthcare staff encounter, in the first place, it is necessary to redesign instructional materials and the way of communication following ergonomic principles. For example, as Rogers *et al.* (2001) suggested, it could be effective to keep the vocabulary simple and explicit, to provide redundancy of information either through multiple modes or to repeat critical information for avoiding medical accidents due to misunderstanding the language.

2.3   *Ergonomic considerations associated with cultural differences*

With the growth in numbers of migrant caregivers and nurses, much literature has focused on their lived experiences to understand the problems they encounter in the workplace or community. Xu (2007) conducted a metasynthesis of the literature and showed that Asian immigrant nurses working in Western countries became targets of marginalization, discrimination, and exploitation. Omeri *et al.* (2002) reported that immigrant nurses living in Australia experienced professional negation, lack of support, otherness, cultural separateness, and silencing. Hunt (2007) reported that immigrant nurses encountered widespread discriminatory practices, including overuse of complaints and grievances against them.

Insensitivity to racial and cultural differences could lead to racism or discrimination in the workplace. Such situations may have negative effects on mental health in migrant caregivers and nurses, which would decrease care quality. In Table 1, insensitivity to racial and cultural differences refers to social climate/culture in the organization factors of performance inputs, and it leads to poor social/behavioral performance (e.g., creating low motivation). Thus, it is necessary to introduce a diversity awareness training system into the workforce for growth and development of both individuals and organizations.

2.4   *Role of ergonomists for incorporating and integrating migrant nurses and caregivers into the healthcare systems*

As mentioned in the first section, it is not clear what ergonomic problems would occur by receiving foreign nurses and caregivers into the workforce, because Japan has just started receiving migrant nurses and caregivers. However, it would be helpful to discuss potential problems related to this issue in advance to prevent medical accidents and to mitigate the difficulties migrant nurses and caregivers encounter. A brief outline of the ergonomists' role for supporting incorporation and integration of migrant nurses and caregivers into healthcare systems is illustrated in Figure 1. Working with employers, managers, healthcare staff, occupational health staff, ergonomists could contribute to designing safer work systems for migrant nurses, caregivers, and Japanese healthcare staff, and design culturally competent organizations. Working with culturally sympathetic social workers or support groups for migrants, ergonomists could contribute to empowering migrant nurses and caregivers and improving their physical and mental health.

3 CONCLUSION

With the globalization of the healthcare workforce, making the healthcare system safe and comfortable is the challenge. Applying the principles of ergonomics could contribute to find solutions to the many aspects of this issue.

REFERENCES

Buchan, J., Kingma, M., and Lorenzo, F. M., 2005, International migration of nurses: trends and policy implications, *www.icn.ch/global/Issue5migration.pdf*.

Hunt, B., 2007, Managing equality and cultural diversity in the health workforce, *Journal of Clinical Nursing*, Vol. 16, pp. 2252–2259.

Karsh B-T., Holden, R. J., Alper, S. J., and Or, C. K. L., 2006, A human factors engineering paradigm for patient safety: designing to support the performance of the healthcare professional, *Quality and Safety in Health Care*, Vol. 15 (Supplement 1), pp. i59–i65.

Ogawa, R., 2008, Caregiving in a cross culture context: The Japanese elderly in the nursing homes in the Philippines, *Presentation paper to the International Symposium, "Globalizing Nursing and Care: Discussions Over Foreign Workers' entry into Japan's Labor market"*, Fukuoka, Japan, organized by Asia Center, Kyushu University.

Omeri, A., and Atkins, K., 2002, Lived experiences of immigrant nurses in New South Wales, Australia: searching for meaning,*International Journal of Nursing Studies*, Vol. 39, pp. 495–505.

Rogers, W. A., Mykityshyn, A. L., Campbell, R. H., and Fisk, A. D., 2001, Analysis of a "Simple" medical device, *Ergonomics in design*, Vol. 9 (1), pp. 6–14.

Shearer, J. B., 2008, English as a second language among licensed practical nurses: Implications and relevance in practice, *The Journal of Practical Nursing*, Vol. 58(1), pp. 26–27.

Xu, Y., 2007, Strangers in strange lands: A metasynthesis of lived experiences of immigrant Asian nurses working in Western countries, *Advances in Nursing Science*, Vol. 30 (3), pp. 246–265.

Ergonomic Trends from the East – Kumashiro (ed)
© *2010 Taylor & Francis Group, London, ISBN 978-0-415-88178-4*

KAIZEN research on skill transmission and increase of productivity using CAD/CAM against aged society

Koki Mikami
Hokkaido Institute of Technology, Sapporo, Japan

Kenichi Iida & Kenichi Hatazawa
Hokkaido Industrial Research Institute, Sapporo, Japan

Masaharu Kumashiro
University of Occupational and Environmental Health, Kitakyushu, Japan

ABSTRACT: In "super-advanced age and fewer children" society, what is of urgent necessity is establishment of an efficient production system including skill transmission and solution of Q (quality improvement), C (cost reduction) and D (lead time shortening) for Japan's manufacturing industry to survive.

In order to contribute to the above issues and activation, KAIZEN research was conducted on skill transmission and increase of productivity using a CAD/CAM system with an agricultural machinery manufacturing factory in Hokkaido as an object.

The results obtained is as follows:

1) New BEND CAM software and an NC bender were introduced into the existing CAD/CAM system as a transmission method of manufacturing skills.
2) The new CAD/CAM system enabled the beginner to master the bending work in 30 minutes. Mastering the work was once said to take 3 years.
3) Safety from the danger of the coming off of metal molds and incised wounds in the bending work was established.
4) In addition, 10 kinds of improvements such as "Making of a movable work table", "Arrangement of a left over steel plate space", "Arrangement of a scrap steel plate space" and "Setting of a new jig space" were embodied.

Keywords: Aging, skill transmission, KAIZEN, QCD, CAD/CAM

1 INTRODUCTION

Japan's economy is now said to be in boom though most people can hardly feel that boom. On the other hand, the northern main island of Japan Hokkaido is still suffering from slow business. In this "super-advanced age, fewer children" society, the government has enforced laws of deferred annuities and continued employment (up to age 65). The manufacturing companies in this district need to be the-only-one-type companies with original skills and products. For that, it is of urgent necessity in the production field to construct an efficient production system including handing down of skill against "super-advanced age, fewer children" society and solve years' problems of quality improvement (Q), cost reduction (C) and lead time shortening (D).

In order to contribute to the above issues and activation, KAIZEN research was conducted on skill transmission and increase of productivity using a CAD/CAM system with an agricultural machinery manufacturing factory in Hokkaido as an object.

2 METHODS

2.1 *The approach*

The research took the form of the cooperation of companies and external research institute. It was thought important for smaller companies in Hokkaido to participate in KAIZEN research along with third parties (universities or Hokkaido Industrial Research Institute).

The research was performed as in the following, based on the Ergoma Approach[1,2]

Ergoma is a coined compound term composed of "ergo" from ergonomics and "ma" from management.

2.2 *Ergoma approach*

The 1st step: Aim {Pick out problems for the company to solve}

The 2nd step: Whole-company-tackling {Organize an improvement project team consisting of employers, managers and workers}

The 3rd step: Preparatory investigation {Hold a hearing to listen to workers' opinion, and conduct a preparatory observation in the workplace}

The 4th step: Discovery of problem worksites and items {Seize target worksites or tasks awaiting solution, and clear up the causes}

The 5th step: Analysis of the present conditions {Conduct research in the target worksites taking into account the indication matters of Step 6 from the viewpoints of IE, Ergonomics and industrial psychology. Choose approaches appropriate for the target sites. }

The 6th step: Indication matters {Classify the results and indicate the direction of KAIZEN}

The influence of work on humans & The Influence of humans on work

(1) Unsafe operations, (2) Workload, (3) Unsafe conditions, (4) Work content, (5) Uncomfortable working postures, (6) Health, (7) Unsafe actions, (8) Job consciousness, (9) Degree of concern for work, (10) 5 S's, (11) Management conditions, (12) Workers' background, (13) Labor productivity, (14) Job satisfaction, etc.

The 7th step: Decide what should be priority matters {The KAIZEN project team discovers the true cause from step 6, makes KAIZEN plans, and examines them. The use of work improvement support tools such as Work Posture Burden Evaluation System, Support Information System for creating new ideas and Virtual Simulation enables effective improvement.}

The 8th step: Practice of KAIZEN {Incorporate the improvement plans. Ensure that all the workers practice the improved work}

The 9th: After-improvement evaluation {Measure the after-improvement effects. Especially hearing of workers is important for the next KAIZEN}

2.3 *The target factories*

The target factory manufactured the onion picker machines shown the one in Fig.1.

Figure 1.   The onion picker machines.

3 RESULT AND DISCUSSION

The Ergoma approach into this manufacturing as many as 60 problems clear. First of all, the improvement committee established through this research gave priority to how to hand down aging worker's skills. We made a new process/work manual to make workers' unwritten knowledge visible, and then began to examine how a New BEND CAM software and a new NC vender usable for the existing CAD/CAM system would work because older workers complained about the heavy workload of the bending work the most (Fig. 2).

It was thought that the effect of the introduction would be the following.

<Introduction of a new NC vender>

1) CAD / CAM system would enable the control of the process & work procedure and blueprint / Material information. This would make worker's unwritten knowledge visible, and lead to skill handing down.
2) Using the information input by CAD/CAM, anybody would be able to make highly precise components without any experience or instincts.
3) The bending condition of a material would appear on the screen, which would lead to no work errors.
4) No need to hold the heavy punch would lead to little holding burden and safety.

The NC vender introduced is shown in Fig 3. A New BEND CAM software is shown in Fig 4.

Table 1. shows the results of the effect measurement of the NC vendor and New BEND CAM software.

Figure 2.　The workload of the bending work.

Figure 3.　The NC vender.

Figure 4.　Parts drawing using a New BEND CAM software.

Table 1. Comparison with before and after.

	Experienced		Beginner	
	Evaluation index	Time (min)	Evaluation index	Time (min)
The former vender	98.7	18.0	88.9	23.7
The NC vender	73.4	10.2	75.9	16.5
Ratio of variation	−24.2%	−43.0%	−14.6%	−30.0%

Figure 5. Parts on the monitor screen of NC vender and operation in the workplace.

Figure 6. A movable storing table for punches.

Figure 7. A movable work table.

The component bending process for onion pickers used to take the skilled worker 18.0 minutes, and it was shortened to 10.2 minutes. The beginner's time was shortened to 16.6 minutes from 23.7 minutes, and the working time showed increases in productivity. And the monitoring of the bending procedure led to no process mistakes.

Fig 5. shows Parts on the monitor screen of NC vender and operation in the workplace.

Besides, the work posture burden evaluation indices[3] of both and the beginner showed decreases in workload.

Figure 8.   Arrangement of a left over steel plate space.

Figure 9.   Arrangement of a scrap steel plate space.

We also measured the time an older worker with no experience of the NC vender needed to be able to do the works by himself. The result was that he mastered the work in about 30 minutes. About 3 years was needed to master it before.

In addition, 10 kinds of improvements such as "development of a movable storing table for punches (Fig. 6)", "Making of a movable work table (Fig. 7)", "Arrangement of a left over steel plate space (Fig. 8)", "Arrangement of a scrap steel plate space (Fig. 9)" and "Setting of a new jig space" were embodied.

4   CONCLUSION

The agricultural machinery manufacturing were incorporated 10 important items about QCD and workload reduction were incorporated here. The enlightenment activities with the Ergoma by the outside researchers led to the employees' own KAIZEN activities, and then to fostering of the company KAIZEN climate for a sustainable production field construction.

REFERENCES

M. Kumashiro,1987. Work load, Postures and Job Redesign – An Ergonomic and Industrial Management (Ergoma) Approach -, NEW METHODS IN APPLIED ERGONOMICS, Taylor & Francis. London. 247–252.

K. Mikami 2002, The Theory and Practice of Work-design in Small and Medium-sized Manufacturing Enterprises in an Aging Society:, AGING AND WORK, Taylor & Francis. London: 233–244.

K. Mikami, M. Shibuya, G. Sasaki, N. Funada, T. Hasegawa, and M. Kumashiro, 1998. A supporting system for work improvement to create a high-productivity workplace -A System forevaluation the burden of work postures and virtual simulation: Global Ergonomics, ELSEVIER, pp 497–500.

Chapter 14   Ergonomics in occupational health 5

Ergonomic Trends from the East – Kumashiro (ed)
© *2010 Taylor & Francis Group, London, ISBN 978-0-415-88178-4*

Sculptors' workstation modification to reduce muscular fatigue and discomfort

Sara Arphorn, Yuparat Limmongkol & Vichai Pruktharathikul
Department of Occupational Health and Safety, Faculty of Public Health, Mahidol University

Watana Jalayondeja
Faculty of Physical Therapy and Applied Movement Science, Mahidol University

Chaikittiporn Chalermchai
Department of Occupational Health and Safety, Faculty of Public Health, Mahidol University

ABSTRACT: This study aims to improve the sculptors' workstation in pottery handicraft to reduce muscular fatigue and discomfort. This study was conducted in 24 healthy carving subjects who no illness history, muscular injury and bone diseases. Demographic data showed average age of 33.58 ± 6.40 years old. Also, their average experience in carving work was 8.88 ± 5.36 years. The improvements of the workstation were redesigned of the banding wheel, storage of carving equipment and adjusted height of seat to 2 levels, 28 and 42 centimeters.

The results found that discomfort of general body, left and right low back muscles and right shoulder muscle for operating at the modified workstation were significantly less than operating at the traditional workstation (p-value < 0.05) at 12.00 p.m., 1.05 p.m. and 4.00 p.m.. In addition, the ΔMF in right upper trapezius in the afternoon for working at the modified workstation was significantly reduce compared to the traditional workstation (p-value < 0.05). Also, the reduction of average time spent per one piece and the increased score of workstation satisfaction with the change of the posture for operating at the modified workstation were statistically significantly different (p-value < 0.05). However, no significant differences of the ΔMF in the left and right sides of erector spinae L4-L5 and the left upper trapezius at all carving period and the total score of workstation satisfaction between performing at the modified and the traditional workstations were found. Additionally, no correlation of muscular fatigue indicators between the subjective measurement by using questionnaire and the objective measurement by using EMG was found.

It could be summarized that the modified workstation could clearly reduce discomfort in low back muscles and right shoulder muscle with significant difference. Moreover, the average time spent per one piece of carving work was reduced and the score of workstation satisfaction with the change of the posture was increased. Therefore, these results confirm increased productivity and comfort for the sculptors using this modified workstation.

Keywords: Sculptor's Workstation/Muscular fatigue/Discomfort/Pottery handicraft/Median frequency

1 INTRODUCTION

The carving process is an important process of making the pottery handicraft. The sculptors always cope with ergonomics hazards such as workstation including table, chair and all equipments. Natapintu K. *et al.*, 2002 and Kroemer KHE. *et al.*, 1994 found that muscular fatigue and eyestrain will be developed and sometimes the sculptors might have to work in awkward posture. They also reported that pottery handicraft's operators always exert static muscle loading for long working duration. Moreover, the unsuitable workstation may affect user's comfort and productivity (Pulat BM., 1991 and Ayoub MA .,1990). The primary unpublished survey showed the fatigue and pains of low back (70.70 percent), right shoulder (67.20 percent), neck (41.40 percent) and left shoulder

(39.70 percent) among these sculptors. There are two main causes of these complaints. Firstly, the characteristic of the carving is repetitive work because the sculptors have to sit all day and use the same muscular. Most of them always sit in leg crossed posture on the floor. This posture might affect the cardiovascular system, reduce quantity of work items, and cause the muscular fatigue. Secondly, the carving workstation is unsuitable causing unsafe working posture such as reaching above the shoulder, bending or twisting at the wrist. There are three main reasons why the carving workstation was complained by the operators. Firstly, the height of banding wheel can not be adjusted to appropriate level of eye sight and several sizes of the products. According to Das B., 1996 , the operators work at too height workstation that can effect on trapezius and neck muscular fatigue. If the workstation is too low level, it will have fatigue problem at back muscle. Moore B. et al., 1992 also reported that intervertebral disc can be degenerated because of awkward posture, awkward sit and long working duration. Secondary, trimming, carving and piercing tools, and a water tank with a sponge were putted on long reaches. Accordingly, the long reaches can increase waste time, make work more difficult and strain the sculptors' body (MacLead D., 1994). Finally, the banding wheel, and seat were designed without taking the operators' anthropometry. In these situations, work-related muscular skeletal disorders (WMSDs) may be developed among these operators (Pulat BM., 1991). Therefore, it is necessary to have well design workstation. That can decrease discomfort and muscular fatigue. Moreover, the sculptors could achieve more productivity and better score of the workstation satisfaction. It is clear that the sculptors had been suffered from muscular fatigue and reducing work performance due to carving process is repetitive work and has unsuitable workstation. This study aimed to improve the sculptors' workstation in the pottery handicraft for reducing muscular fatigue and discomfort to increase productivity. The improvement was mainly based on ergonomics principles, and anthropometry data from the operators to design the workstation that can be adjusted individually.

2   METHODS

A survey research and a quasi-experimental research by pre-post test for one group were performed in this study. The general information, the incidence of muscle fatigue using the interviewed forms is collected. Quasi-experimental research was used for the comparison of discomfort by using Visual Analogue Scale (VAS) and muscular fatigue by using electromyography (EMG) which was represented as median frequency shift (ΔMF) in low back muscles and upper shoulder muscles. The average time spent per one piece and the score of workstation satisfaction between operating at a traditional workstation and operating at a modified workstation were determined. The improvements of the workstation were redesigned of the banding wheel, storage of carving equipment and adjusted height of seat to 2 levels, 28 and 42 centimeters. This was studied in 24 healthy carving subjects who had not suffered from any illness history, muscular injury and bone diseases. The populations for this experiment were the sculptors who worked at cottage industries of Danchai village (Moo.7), Dankwien sub-district, Chokechai district, Nakhonratchaseema province. The subjects had been acclimatized for two days by operating at the modified workstation before the measurement. They were interviewed by using questionnaire to show theirs feeling of discomfort. At the same time, the electromyography is used to determine their muscular fatigue. After the subjects worked at the traditional and the modified workstations, they were inquired about the workstation satisfaction by using questionnaire. The subjects were measured the muscular fatigue as the median frequency which were calculated as the median frequency shift (ΔMF) in the low back muscle, erector spinae; L4-L5 and the upper shoulder muscle, upper trapezius by using electromyography. All of the measurement and the interview were performed when the subjects carved the same size and the similar design of the pottery handicraft while they worked at the traditional workstation and at the modified workstation at the particular time.

3   RESULTS

Most of the subjects were female (83.33%), but few of them were male (16.67%). Their average age was 33.58 ± 6.40 years old. All of them have used their dominant hand by right-hand. Their average work experience of sculpture was 8.88 ± 5.36 years. Also, mean of working duration per day, and

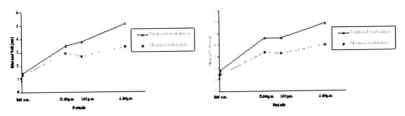

Figure 1. Level of the mean score of discomfort of the left (left) and right (right) low back muscle of working between the modified and the traditional workstations at various working-hours.

average time of a lunch break were 7.87 ± 1.31 hours and 42.50 ± 17.51 minutes, respectively. Approximately, 83.33% of them worked seven days a week. All of them (100%) had worked as sculptor at these occupations for more than one year. Analyzed normal distribution was used to test all parameters (the data of discomfort, the median frequency shift (ΔMF) in upper trapezius, the median frequency shift (ΔMF) in erector spinae L4-L5, workstation satisfaction and average time spent per one piece) by Shapiro-Wilks Test. The results shown that some parameters were normal distribution: the data of discomfort, the median frequency shift (ΔMF) in upper trapezius, the median frequency shift (ΔMF) in erector spinae L4-L5 and average time spent per one piece. Thus, Independent – samples t – test was determined the significant difference of these parameters. However, the results of the workstation satisfaction were not normal distribution therefore Mann – Whitney U test were used to examine the statistical significant difference between working at the traditional and the modified workstations. All of the subjects worked from 9.00 a.m. to 4.00 p.m. (6 working- hours). The subjects were inquired for discomfort scale determination for 5 times: before work (9.00 a.m.), start work for 5th minutes in morning (9.05 a.m.), 3rd hours (12.00 p.m.), start work for 5th minutes in afternoon (1.05 p.m.), and 6th hours (4.00 p.m.) of carving period. Independent – samples t – test with 95% of confidence interval (P-value < 0.05) was determined for the significant difference of the discomfort of left low back muscle. The results were illustrated in Figure 1. The result demonstrated that the discomfort of the left low back muscle of the subjects worked at the modified workstation was significantly different less than they operated at the traditional workstation at start work for 5th minutes in afternoon (1.05 p.m.) and 6th hours (4.00 p.m.) of carving period. In contrast, there were no significant change of the discomfort between operating at the modified and the traditional workstations at before work (9.00 a.m.), start work for 5th minutes in morning (9.05 a.m.) and 3rd hours (12.00 p.m.) of carving period. The decreasing trend of discomfort of the left low back muscle at start work for 5th minutes in morning (9.05 a.m.) and 3rd hours (12.00 p.m.) of carving period were observed. While, the discomfort of the right low back muscle were significantly lower in working at the modified workstation than working at the traditional workstation in 3 periods: 3rd hours (12.00 p.m.), start work for 5th minutes in afternoon (1.05 p.m.) and 6th hours (4.00 p.m.) of carving period (Figure 1 right). There were no differences in discomfort in the right low back muscle between carving at the modified workstation and operating at the traditional workstation in 2 periods: before work (9.00 a.m.) and start work for 5th minutes in morning (9.05 a.m.) as well as for the left low back muscle.

The results of muscular fatigue as the median frequency shift (ΔMF) by using electromyography were collected during 9.00 a.m. – 4.00 p.m. (6 hours of carving task). Beside, the left and right sides of the upper trapezius and the erector spinae L4–L5 were recorded for electrical activities in raw electromyography. Furthermore, all of these subjects were measured for 4 times: before work, 3rd, 4th, and 6th hours of carving period. The subjects took lunch during 12.00 p.m. to 1.00 p.m. The analyzed results in Figure 2 indicated that the ΔMF in the right upper trapezius was significantly less reduction in working at the modified workstation than working at the traditional workstation in the afternoon. While, non-significantly different reports of the ΔMF in the right upper trapezius in the morning were found. However, the ΔMF in this muscle in the morning for working at the modified workstation was less reduction than working at the traditional workstation.

However, no significant differences of the ΔMF in of the left upper trapezius and the left and right sides of erector spinae all carving period and the total score of workstation satisfaction between performing at the modified and the traditional workstations were found.

287

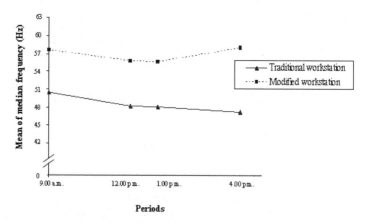

Figure 2.   Comparison of the median frequency in the right upper trapezius of working between the modified and the traditional workstations at various working-hours.

4   DISCUSSION

The subjective feeling of discomfort intensity was applied to evaluate subjective feeling of fatigue by inquiring the subjects to report the location of discomfort and the severity before and after examination with the workstation. The severity of VAS scores were estimated by measuring the distance in centimeters between no discomfort and very discomfort. The results from using this questionnaire might show slightly error because of the misunderstanding of the content of ques-tionnaires. This error was prevented by implementing the training on how to use and answer the questionnaire. Furthermore, this study and previous of Mekhora K. *et al.*, 2000, stated that the response of discomfort intensity may vary in each subject because of differences in discomfort threshold. Thus, the objective measurement by using electromyography (EMG) was applied in this study resulted in the better representative of the overall muscular fatigue. Moreover, the subjects were informed about the time for answer discomfort form no more than 10 seconds. This method is also used in the previous study of McLean L. et al. The VAS results of the general body, the left and right low back muscle (erector spinae L4–L5), and the right shoulder muscle (upper trapezius) at 3rd hours (12.00 p.m.), start work for 5th minutes in afternoon (1.05 p.m.) and 6th hours (4.00 p.m.) of carving period indicated significant lower in working at the modified workstation. From the study outcome in morning time, it was found that the discomfort of the left and right low back muscle (erector spinae L4–L5) and the right shoulder muscle (upper trapezius) showed the same profile. There were not significantly different discomfort between working at the traditional workstation and the modified workstation. One possible reason was found that the subjects might not be acclimatized with the modified workstation because of short period for training and testing. According to Mekhora K. *et al.*, 2000, they studied the long-term effects of ergonomics intervention on neck and shoulder discomfort. Therefore, comparison between the results from this previous study and results from the current study found that there was clearly significant difference of less discomfort after intervention when used long period for working at the intervention. This study is also consistent with the findings from other studies (Mekhora K. *et al.*, 2000, Dieen JHV. *et al.*, 1997 and Henk F. *et al.*, 2004) using ergonomics recommendations for individual anthropometry to improve the workstation could lead the reduction of discomfort. It was clearly found that most of the subjects reduced in discomfort level of low back, hip, thigh, neck, knee and calf. Of the findings in this study, this was also consistent with the findings of Babski-Reeves K. *et al.*, 2005. found that the interaction of monitor height and chair type significantly affects the loads placed on human body of minimizing static loading and reducing in discomfort. Related to the ΔMF, the less reducing of ΔMF in the right upper trapezius of carving periods was observed when the subjects operated at the modified workstation. From Figure 2 the decreasing of the ΔMF was observed at the traditional workstation possibly resulted in the muscular fatigues in the right upper trapezius (Schulte E. *et al.*, 2004). This study found that the reducing of this muscular fatigue was similar to the study of Arne A *et al.*, 1998, which the muscular fatigue in trapezius was a significant reduction after improved the workplace which was composed of new table, new chair and support their forearms on the table top. However, the much reduction ΔMF in the left upper trapezius muscle had also been observed when the banding wheel and seat were higher level at the modified workstation than the traditional workstation. When it was compared to the study of Moffet HA. et al., 2002, it was found that this present result was consistent with the recommendation which the high level of a seat could affect to raise shoulder and increased muscular activities in trapezius and deltoideus muscles. Similar results have been reported in the study of Schulte E. *et al.*, 2004. They showed a fatigued-related increase of muscular activity and a decrease of the median frequency in trapezius muscle. This study was also consistent with the findings of Madeleine P *et al.*, 2002. They found that decrease in the median frequency was observed for localized fatigue in the upper trapezius muscle in various arm positions.

5 CONCLUSION

It could be summarized that the improvement of the workstation could reduce discomfort in low back muscles and right shoulder muscle with significant difference. From EMG analysis, the results were that there was a reduction trend of ΔMF in the left and right sides of erector spinae L4–L5 while working at the modified workstation, but there were no significant differences. Moreover, the average time spent per one piece of carving work was reduced and the score of workstation satisfaction with the change of the posture was increased. Therefore, these results confirm increased productivity and comfort for the sculptors using this workstation.

REFERENCES

Arne A, Gunnar H, Bjorset HH, Ola R, Magne T., 1998, Musculoskeletal, visual and psychosocial stress in VDU operators before and after multidisciplinary ergonomic interventions. In *Applied Ergonomics*, Vol. 29(5), pp. 335–54.

Ayoub MA., 1990, Ergonomic deficiencies: II. probable causes. In *Occupational Medicine*, Vol. 32(2), pp. 131–6.

Babski-Reeves K, Stanfield J, Hughes L., 2005, Assessment of video display workstation set up on risk factors associated with the development of low back and neck discomfort. In *International Journal of Industrial Ergonomics*, Vol. 35, pp. 593–604.

Das B, Arijit KS., 1996, Industrial workstation design: a systematic ergonomics approach. In *Applied Ergonomics*, Vol. 27(3), pp. 157–63.

Dieen JHV, Jansen SMA, Housheer AF., 1997, Differences in low back load between kneeling and seated working at ground level. In *Applied Ergonomic*. Vol. 28(56), pp. 355–63.

Henk F. Van Der Molen et al., 2004, Efficacy of adjusting working height and mechanizing of transport on physical work demands and local discomfort in construction work. In *Ergonomics*, Vol. 47(7), pp. 772–83.

Kroemer KHE, Kroemer HB, Kroemer-Elbert KE. 1994, Ergonomics how to design for ease and efficiency, (New Jersey: Prentice-Hall).

MacLead D., 1994, The ergonomics edge improving safety, quality, and productivity, (New York: Van Nostrand Reinhold).

McLean L.,Tingley M., Scott R.N., Richards J. , 2001, Computer terminal work and the benefit to microbreaks. In *Applied Ergonomics*, Vol. 32, pp. 225–37.

Madeleine P, Farina D, Merletti R, Arendt-Nielsen L., 2002 , Upper trapezius muscle mechanomyographic and electromyographic activity in humans during low force fatiguing and non-fatiguing contractions. In *European Journal of Applied Physiology.*, Vol. 87(4–5), pp. 327–36.

Mekhora K, Liston CB, Nanthavanij S, Cole JH., 2000, The effect of ergonomic intervention on discomfort in computer users with tension neck syndrome. In *International Journal of Industrial Ergonomics*, Vol. 26, pp. 367–79.

Moffet HA, Hagberg MB, Hansson-Risberg CE, Karlqvist LC., 2002, Influence of laptop computer design and working position on physical exposure variables. In *Clinical. Biomechanics*, Vol. 17, pp. 368–75.

Moore B, et al., 1992, The ergonomics evaluation of several chairs: a case study; advances in industrial ergonomics and safety IV, (London: Taylor&Francis).

Natapintu K. et al., 2002, The study and development of the learning model to improve occupational health and safety problem in home-based industry in North-East of Thailand Konkhaen : Research Promoting Fund

Pulat BM, Alexander DC., 1991, Industrial work station and work space design: industrial ergonomics case studies, (New York: Industrial Engineering and Management Press).

Schulte E, Ciubotariu A, Arendt-Nielsen L, Disselhourst-Klug C, Rau G, Graven-Nielsen T., 2004, Experimental muscle pain increase trapezius muscle activity during sustained isometric contractions of arm muscles. In *Clinical Neuroscience*, Vol. 115, pp. 1767–78.

Ergonomic Trends from the East – Kumashiro (ed)
© 2010 Taylor & Francis Group, London, ISBN 978-0-415-88178-4

Evaluating plantar pressures in workers wearing safety shoes

Kohei Nasu & Takao Tsutsui
Department of Health Policy and Management, University of Occupational and Environmental Health, Japan

Yasuhiro Tsutsui
Hitachi Metals, Ltd.

Hiro Oguni
Department of Computer Science, Faculty of Electro-Communications, University of Electro-Communications

Shoko Kawanami & Seichi Horie
Department of Health Policy and Management, University of Occupational and Environmental Health, Japan

1 INTRODUCTION

Mechanical irritation of the plantar region is one of the primary causes of calluses and clavuses (Hideoki, 1998). Dishan *et al.* (1996) and Denise (2002) note that chronic excessive plantar pressure is one of the most likely causes of hyperkeratosis. Wearing safety shoes is believed to cause pain and hyperkeratosis due to the redistribution of plantar pressures. However, no published articles to date have reported measurements of plantar pressures in workers wearing safety shoes. This paper reports a study in which we sought to evaluate whether safety shoes are associated with higher risk of hyperkeratosis than other shoes. We also sought to evaluate whether insoles or sneaker-type safety shoes may prevent such adverse health effects. We proceeded by performing quantitative measurements to compare plantar pressures in ordinary safety shoes (condition *Or*); in those with insoles (condition *Is*); and in sneaker-type safety shoes (condition *Sn*).

2 METHODS

Four males free of known orthopedic conditions (34.5 ± 9.0 years, 169.2 ± 2.4 cm in height, weighing 67.7 ± 7.9 kg) volunteered to participate in our study. We measured plantar pressures with an F-SCAN system (Tekscan Inc.) composed of 60×21 pressure sensors. Before measurement, disposable sensors were cut and shaped to fit the shoes. Pressure in kilograms per square centimeter (kg/cm^2) was determined by F-scan (software version 4.213F). We sampled plantar pressures at a rate of 5 Hz over a period of one minute as the subjects stood or walked. The protocol was repeated three times for each of conditions *Or*, *Is*, and *Sn*.

We calculated the average value of the plantar pressures as the subjects stood or walked. Since plantar forms and sizes differed among the subjects, we mathematically converted all results into a standard form and generated standardized results. Hereafter, we defined the average of these values as 'average pressure.' We sorted plantar pressures into three groups: Group 1 (0.01–$0.19 \, kg/cm^2$); Group 2 (0.20–$0.39 \, kg/cm^2$); and Group 3 (0.40- kg/cm^2). We then compared the average distribution of plantar pressures for each shoe type. Peak pressure was defined to be the highest pressure exerted while walking. We averaged peak pressures for each shoe type and defined these values as the peak pressure for that particular shoe type.

We used SPSS software version 16.0.2J (SPSS, USA) for the Kruskal-Wallis test.

Condition *Or*   Condition *Is*   Condition *Sn*

Figure 1.   Shoes and insoles assessed in this study.

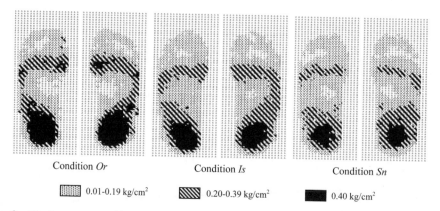

Condition *Or*   Condition *Is*   Condition *Sn*

0.01-0.19 kg/cm²   0.20-0.39 kg/cm²   0.40 kg/cm²

Figure 2.   Plantar pressures with subjects standing.

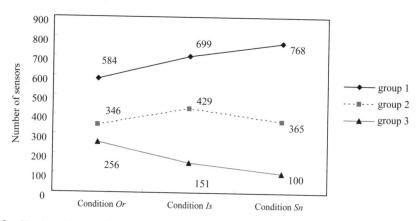

Figure 3.   Number of sensors in each group while standing.

2.1 *Average pressure results*

The average area of the lowest plantar pressures (Group 1) among the four subjects was significantly larger for conditions *Is* and *Sn* than for condition *Or*. The proportions of Group 1 sensors among all contact sensors were 49.2%, 54.7%, and 62.3% while standing and 64.9%, 76.6%, and 84.6% while walking, for condition *Or*, *Is*, *Sn*, respectively. These values differed significantly when the subjects stood (P = 0.037) or walked (P = 0.017).

The proportion corresponding to Group 1 was the largest for condition *Sn*, followed by condition *Is*. Average plantar pressures were highest for condition *Or*, followed by condition *Is*.

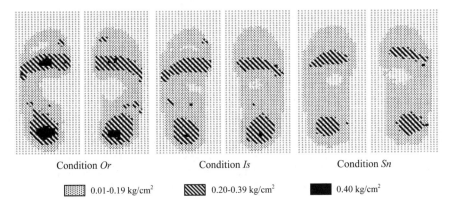

Condition *Or*   Condition *Is*   Condition *Sn*

▨ 0.01-0.19 kg/cm^2 ▨ 0.20-0.39 kg/cm^2 ■ 0.40 kg/cm^2

Figure 4. Plantar pressures with subjects walking.

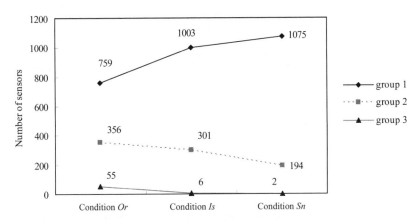

Figure 5. Number of sensors in each group while walking.

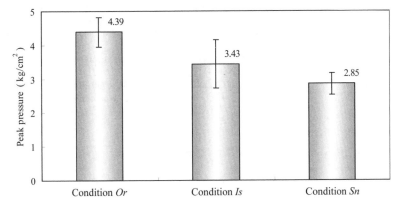

Figure 6. Peak plantar pressure in each condition while walking.

2.2 *Peak pressure results*

Peak plantar pressures (mean ± SE) while walking differed significantly (P = <0.001) among the three conditions: 4.4 ± 0.4 kg/cm^2 for condition *Or*, 3.4 ± 0.7 kg/cm^2 for condition *Is*, and 2.9 ± 0.3 kg/cm^2 for condition *Sn*. Relative to peak plantar pressures for condition *Or*, peak plantar pressures were reduced significantly (by about 22%) for condition *Is* and (by about 35%) for condition *Sn*.

3 DISCUSSION

Previous articles have extensively addressed the relationship between peak plantar pressures and callus formation for ordinary shoes. The peak plantar pressures at patient calluses were $3.82 \pm 0.71 \, \text{kg/cm}^2$, $2.80 \pm 0.70 \, \text{kg/cm}^2$, and $3.47 \pm 0.34 \, \text{kg/cm}^2$, respectively, in studies by Piet *et al.* (1999), Slater *et al.* (2006), and Pataky *et al.* (2002). Following removal of the calluses, peak plantar pressures fell to $2.56 \pm 0.57 \, \text{kg/cm}^2$, $1.99 \pm 0.76 \, \text{kg/cm}^2$, and $1.44 \pm 0.23 \, \text{kg/cm}^2$, respectively. Anthony *et al.* (2003) found peak plantar pressures of $4.74 \, \text{kg/cm}^2$ for healthy subjects with calluses and $4.39 \, \text{kg/cm}^2$ for diabetics with calluses but $3.98 \, \text{kg/cm}^2$ for subjects in either group without calluses.

The peak plantar pressures measured in our study indicate a clear high risk for the development of calluses for condition *Or* and a potentially high risk for condition *Sn*. Our study suggests that insoles and sneaker-type safety shoes may help prevent hyperkeratosis by reducing peak plantar pressures from those experienced with ordinary safety shoes.

The distribution of plantar pressures observed suggests that the wider contact area with condition *Is* and condition *Sn* compared to condition *Or* reduces the area of highest plantar pressure (Group 3). Thus, efforts should focus not only on reducing the area of Group 3, but on increasing contact areas. Since the total weight applied by the feet is equal to body weight, reducing plantar pressures in one area will simply increase pressures in other areas. Ideally, to prevent hyperkeratosis, we should seek to disperse plantar pressures over a wider area of the feet by inserting inner soles into shoes or by wearing shoes made of soft materials.

REFERENCES

Anthony, C.D., Robert, K., Albert, C. and Kim, C.D., 2003, High plantar pressure and callus in diabetic adolescents. *Jornal of the American Podiatric Medical Association*, Vol. 93, No. 3.
Denise, B.F., 2002, Corns and calluses resulting from mechanical hyperkeratosis. *American Family Physician*, Vol. 65, No. 11.
Dishan, S., George, B. and Saul, G. T., 1996, Callosities, corns, and calluses. *British Medical Journal*, Vol. 312.
Ogawa, H., *et al.*, 1998, *TEXT Hifukagaku*, edited by Ogawa, H., *et al.* (Nanzando press), pp. 60.
Pataky, Z., Golay, A., Faravel, L., Silva, J.D., Makoundou, V., Peter-Riesch, B. and Assal, J.P., 2002. The impact of callosities on the magnitude and duration of plantar pressure in patients with diabetes mellitus. *Diabetes and Metabolism*, Nov; 28(5): 356–61.
Pitei, DL, Foster, A. and Edmonds, M., 1999, The effect of regular callus removal on foot pressures. *The Journal of Foot and Ankle Surgery*, Jul–Aug; 38(4): 251–5.
Slater, R.A., Hershkowitz, I., Ramot, Y., Buchs, A. and Rapoport, M.J., 2006, Reduction of digital plantar pressure by debridement and silicone orthosis. *Diabetes Research and Clinical Practice*, Dec; 74(3): 263–6.

Ergonomic Trends from the East – Kumashiro (ed)
© *2010 Taylor & Francis Group, London, ISBN 978-0-415-88178-4*

Injuries for Ice Hockey Skating

Tin-Chang Chang[1]
Department of Business Administration, Asia University, Taichung, Taiwan

ABSTRACT: The purpose of the study is concerning the ergonomics of ice hockey skates design to reduce injuries. We intend to identify with: what concept of hockey skate to be designed for the performance of ice hockey and how to improve the product components to reduce the injuries. All injury data were collected from NHL injury reports of the 30 teams in 18th, 2007~Mar. 26th, 2008, this resource provides detailed of injury location as well as percentage. Moreover, review the industry documents for the questionnaire design, and send to ice hockey skate designer experts by mail interview. In total, 4 questionnaires were completed. According to 744 injuries of NHL players during hockey games, showed 32.93% were lower limp injury. Also the feedback from experts regarding the significant points for the ice hockey design showed quarter package is important for protecting ankle and side wall area of the foot (Mean = 4.5). Lining material is the element of comfort, and it is directly affects the side wall area of the foot (Mean = 4). Heel support is crucial to the heel area of the foot (Mean = 5), Ankle padding is protecting the ankle area (Mean = 4.75). Tongue construction is important for the instep area of the foot (Mean = 4.75). Foot bed proven slippery and better control for the toe area (Mean = 4.25). Thermo formable provide the suitability, it is important for side wall area (Mean = 4.00). Outsole is a foundation of the skate, it is protect the side wall area of the foot (Mean = 3.25). Blade Holder and runner for the performance aspect, it is important for the heel area (Mean = 2.25). The analyses demonstrated a significant point. The nine influence factors of ice hockey skate design were able to increase the safety and high performance. However, according to the injury report, there are still some injuries appeared on ankle and feet location. It is showed other factor might need to be discussed, such as violent and aggressive behaviors, training, characteristics of ice hockey, new construction of skate, different materials sourcing and new technology innovation.

Keywords: Ice Hockey Skate, NHL Injury report.

1 INTRODUCTION

1.1 Ice Hockey

Ice Hockey is a popular professional sport in North America. It is one of the four major North American professional sports and is represented by the National Hockey League (NHL) at the highest level. From 1967 to 2000, the NHL has expanded from 6 to 30 teams (Diamond, 2003). By 2005, almost 1.3 million Canadians were hockey players. In the United States, there are more then 539,000 registered players, including 103,533 adult players, 353,505 youths and 57,549 female players (USA Hockey, 2007). Ice Hockey has been depicted as a game played with "clubs-hockey sticks, knives- skates, and bullets- pucks (Chao 1978, Mcfaull 2001). It is a fast and dynamic sport (Federolf et al, 2007). The players travel up to 30 mph and pucks exceeding 100 mph (Stuart, Smith, 1995), thus, ice hockey is a kind of sport considered with high risk of getting injured. Most athletic injuries can be divided into orthopedic injuries (e.g. bruises, strains, sprains, fractures, muscle ligaments and tears) and concussive injuries (Schneider , 2006). These injuries are caused

---

[1] Address of correspondence and requests for reprints to Yi-Ling Li, Department of Business Administration, Asia University, 500, Lioufeng Rd., Wufeng, Taichung 41354, Taiwan or e-mail (ervine@asia.edu.tw)

Table 1. Injury location.

Injury location	Injury sampling	percentage
Head	47	6.32%
Face	35	4.70%
Neck	27	3.63%
Shoulder	85	11.42%
Hand	87	11.69%
Chest	16	2.15%
Back	55	7.39%
Abdomen	13	1.75%
Hip	48	6.45%
Groin	86	11.56%
Thigh	9	1.21%
Leg/Feet	78	10.48%
Knee	97	13.04%
Ankle	61	8.20%

by checking, collision, falls, and contact with the puck, high sticks, and occasionally, skate blades. Studies of ice hockey injuries began to appear regularly in sports literature during the early 1970s. According to the Center for Disease Control, it is the second leading cause of winter sports injury among children. A study of 9 to 15 years old hockey players found that body checking caused 86% of all injuries during games. It is estimated that direct trauma (a sudden forceful injury) accounts for 80% of all injuries (Hughston Sport Medicine Foundation). In research from the American Medical Association by Dr. Barry J. Maron, sudden death due to neck blows among amateur hockey players, awareness of such risks to young athletes is crucial for developing an informed public and formulating protective measures to enhance the safety of sports activities. This research is to study the ergonomics of ice hockey skates designed to reduce player injures.

1.2 Ice Hockey skate

A good understanding of the mechanism of injury is important to develop preventative strategies. (Gibbs, 1994). Hockey skates are not only constructed to allow skaters to move their feet alternately, but also designed to protect them from getting injured during games and practices. Moreover, it is also designed to meet the performance and comfort needs of the players. Although innovations have been developed and marketed as devices which were designed to prevent injuries during skating, there are still some design concepts that need to be concerned.

1.3 Injury locations

The most common injuries include lacerations (cuts) to the head scalp, concussion, scalp, bruises, strains, sprains, fractures, muscle ligaments and tears. The injury locations are in the head, face, neck, shoulder, hand, chest, back, abdomen, hip, groin, thigh, leg/feet, knee, and ankles. According to Nordic Musculoskeletal Questionnaire (NMQ), NMQ was to develop and test a standardized questionnaire methodology allowing comparison of neck, shoulder, back, elbow, waist, hand, hip/thigh, keen and ankle/feet complaints for use in epidemiological studies (figure 2). The data were collected from injury reports of 30 NHL teams from Sep. 18th, 2007~Mar. 26th, 2008 (Table 1). Data shows 13.04% (n = 97) were knee injury, 11.69% (n = 87) were hand injury, 11.56% (n = 86) were groin injury, 11.42% (n = 85) were shoulder injury, 10.48% (n = 78) were feet injury, 8.20% (n = 61) were ankle injury. According to the above data, the lower limb is the most commonly injured area accounting for a significantly higher proportion of injuries than any other body parts, indicating that the protection of ice hockey skate is significant.

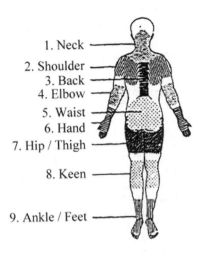

1. Neck
2. Shoulder
3. Back
4. Elbow
5. Waist
6. Hand
7. Hip / Thigh
8. Keen
9. Ankle / Feet

Figure 1.   Nordic Musculoskeletal Questionnaire (NMQ).

A area: instep (1&2)
B area: ankle (3&4)
C area: heel (5&6)
D area: side wall (7&8)
E area: toe (9)

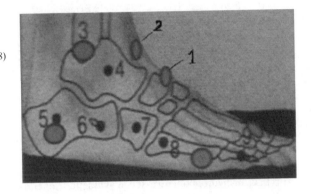

Figure 2.

2   METHOD

This research is conducted by interviewing design experts of ice hockey skates. Nine factors were found to influence the concepts of ice hockey skate design and development (Nike Bauer, 2007). Nine factors included quarter package, lining material, heel support, ankle padding, tongue construction, foot bed, thermo formable as well as outsole and blade holder with runner. Total four skate design experts are interviewed with questionnaires. The questionnaire was designed in the following manner. The importance is each of 9 factors regarding safety and high performance on each part of foot area on ice hockey skate design. Note: 5 parts of foot area contact: A. instep, B. ankle, C. heel, D. side wall, E. toe (figure 2).

3   RESULTS

The data analyses were conducted by SPSS statistical software. The results are showed in Table 2.
    These data showed that designers have a common view on quarter package designing. It is important for protecting ankle and side wall area of the foot (Mean=4.50), which defines the structure of the skates and therefore, increase the performance of the players. The stiffness is the key point for quarter package design, which provides exceptional resistance to abrasions/cuts. Lining material is the element of comfort, and its performance is measured by durability, moisture management, and its behavior with water. It directly affects the side wall area of the foot (Mean = 4),

Table 2. Influence factors.

Factor	Item	Questions	Mean	Std. Deviation	Factor	Item	Questions	Mean	Std. Deviation
a. Quarter	a1	a. to A area	3.00	1.414	f. Foot Bed	f1	f. to A area	2.00	2.000
Package	a2	a. to B area	4.50	0.577		f2	f. to B area	1.25	0.500
	a3	a. to C area	3.25	1.500		f3	f. to C area	3.25	0.957
	a4	a. to D area	4.50	0.577		f4	f. to D area	2.75	1.258
	a5	a. to E area	3.00	1.414		f5	f. to E area	4.25	0.500
b. Lining	b1	b. to A area	1.75	0.957	g. Thermo	g1	g. to A area	2.50	1.291
Material	b2	b. to B area	2.25	0.957	Formable	g2	g. to B area	4.25	0.957
	b3	b. to C area	3.25	0.500		g3	g. to C area	3.75	0.957
	b4	b. to D area	4.00	0.816		g4	g. to D area	4.00	0.816
	b5	b. to E area	1.00	0.000		g5	g. to E area	2.25	1.258
c. Heel	c1	c. to A area	1.75	0.957	h. Outsole	h1	h. to A area	2.00	2.000
Support	c2	c. to B area	2.75	1.500		h2	h. to B area	1.75	0.500
	c3	c. to C area	5.00	0.000		h3	h. to C area	2.25	1.500
	c4	c. to D area	1.75	0.957		h4	h. to D area	3.25	1.708
	c5	c. to E area	1.00	0.000		h5	h. to E area	2.75	1.500
d. Ankle	d1	d. to A area	1.75	0.957	i. Blade Holder	i1	i. to A area	1.50	1.000
Padding	d2	d. to B area	4.75	0.500	and Runner	i2	i. to B area	1.00	0.000
	d3	d. to C area	3.00	1.826		i3	i. to C area	2.25	1.500
	d4	d. to D area	2.25	0.957		i4	i. to D area	2.00	1.414
	d5	d. to E area	1.00	0.000		i5	i. to E area	1.00	0.000
e. Tongue	e1	e. to A area	4.75	0.500					
Construction	e2	e. to B area	2.25	1.258					
	e3	e. to C area	2.00	1.155					
	e4	e. to D area	2.00	1.414					
	e5	e. to E area	1.75	1.500					

quick-dry grip liner or hydrophobic tru-suede synthetic leather and brushed nylon usually apply on the lining area. According to standard deviation of lining material, most designers agree that lining material is more important for the side wall area. Heel support relates directly to skating performance, which obviously is crucial to the heel area of the foot (Mean = 5), injection heel support and integrated anatomical heel support provide skaters better control when skating. As for ankle padding, the major aspect of comfort and ankle area protection (Mean = 4.75), there is a lot of pressure in the ankle area between the skate quarter and the foot, so a good padding is necessary. In addition, ankle padding locks the heel in the right position. Tongue construction needs certain flexibility since most of the forward flex happens on the tongue when skating. The tongue needs to have enough padding to protect against laces that could create painful pressure on the instep (Mean = 4.75). Anatomically shaped felt tongue remains anatomically positioned and naturally flexes forward. The current foot bed is extremely thin and has no extra cushion because of very millimeter of foam that could compress during a push-off is a reduction of energy transferring to the ice. Therefore, this is the area where players get the most pressure while skating. Dry-grip moisture management is design on foot bed to proven slippery and better control (Mean = 4.25). Most high end ice skates use full upper thermo formable to bring the comfort to the players. Players with suitable, accommodated skates usually have a better control when skating. Correct fit of skates is also vital to the players' comfort. If the boot is too tight, the players could get cramps, on the other hand, if the boot is too loose, the player's feet could blister (Choice, 1992). The side wall area of the foot is also important (Mean = 4.00). Outsole is the foundation of the skate, which is used to assemble the skate boot with runner, beveled and injected outsole, and allows players to have tighter turns and maximum push-off power. Outsole also transfers the energy to the blade, which protects the side wall area of the foot (Mean = 3.25). The blade holder and Runner are designed on the performance aspect, which is to help players accelerate, maximize their speed and turning ability. Blades are generally made of tempered steel and are coated with a high-quality chrome, which are important for the heel area of the foot (Mean = 2.25).

4 CONCLUSION

This study demonstrated the significant points on ice hockey skates design. The nine influence factors of ice hockey skate design are able to increase the players' safety and achieve higher performance. However, according to the injury report, there are still some injuries on ankle and feet location. This shows that other factors still might need to be discussed, such as violent and aggressive behaviors, training, characteristics of ice hockey, new construction of skate, different materials sourcing and new technology innovation.

REFERENCES

Diamond D, Bontje P, Dinger R, *et al*. Total NHL. Chicago, IL: Triumph Books; 2003.

Gibbs N. Common rugby league injuries. Recommendations for treatment and preventative measures. Sports Med 1994;18(6):438–50.

Hughston Health Alert, Ice Hockey Injuries in Hughston Sports Medicine Foundation http://www.hughston.com/hha/a.hockey.htm

Jillian Claire Schneider, 2006, Emotional Sequelae of Sports-Related Injuries: Concussive and Orthopedic Injuries in Drexel University.

McFaull S. Contact injuries in minor hockey: a review of the CHIRPP database for the 1998/1999 hockey season. CHIRPP News. 2001;19:1–9.

Nike Bauer Hockey 2007 product catalog.

P. Federolf, R. Mills and B. Nigg 2007, Agility characteristics of ice hockey players depend on the skate blade design in *Journal of Biomechanics* 40(S2). 006-2007 Season Final Registration Reports. USA Hockey. http://www.usahockey.com/uploadedFiles/USAHockey/Menu_Membership/Menu_Membership_Statistics/0607%20final.pdf

Sharah Grim Hostetler et al. 2004, Characteristics of Ice Hockey – Related Injuries Treated in US Emergency Departments, 2001-2002 in *Pediatrics* Vol. 114 No. 6 December 2004.

Sim FH, Chao EY. Injury potential in modern ice hockey. *Am J Sports Med.* 1978;6:378–384.

S.T. Stevens *et al.* 2006, The effect of visors on head and facial injury in National Hockey League players in *Journal of Science and Medicine* in Sport 9, 238–242.

Stuart MJ, Smith A. Injuries in Junior A ice hockey. A three-year prospective study. *Am J Sports Med.* 1995; 23:458–461.

Chapter 15   Ergonomics modeling and usability evaluation

Ergonomic Trends from the East – Kumashiro (ed)
© 2010 Taylor & Francis Group, London, ISBN 978-0-415-88178-4

The relationship between two dimension of hand and three dimension of grip forms

He-Yi Lin & Eric Min-yang Wang
Department of Industrial Engineering and Engineering Management
National Tsinghua University, Hsinchu, Taiwan, ROC

ABSTRACT: This study describes the relationship between people's palm position with anthropometric and natural grip. The main purpose is to find the correlation for superior modeling rules for product's grip form design. An anthropometric survey was carried out with an optical scanner to measure and perform analysis the hand by inputting the data into a CAD system to calculate the accurate dimensions.

A test system has been developed to generate a test table to get the 2D rendering of the hand. An aluminum test cylindrical rod was developed to measure the diameter of the grip, a suite of circular knives with diameters ranging from 2.8 millimeter to 4.6 millimeter to cut the oil clay grip samples become standard test oil clay to catch model of grip. Twenty subjects (10 males and 10 females) participated in the experiment. First, participants were tested using the aluminum test cylindrical rod to determine the most comfortable diameter of grip. Second, choosing adapting oil clay and griping it to capture the different grip shape according with each user's finger length. Third, using the COMET 3D digital scanner system to scan the model for get the finger's affected section forms on oil clay.

The collected data was statistically analyzed and also compared with each finger's data. Based on this data, the values of the different sections of the model of hand grip can be determined. This study finds the relationship between the 2D table and the 3D model. A proportion value can be generated when dividing the two dimension of hand by the circumference of the finger for each section in the grip. In the male, as the grip diameter increases, the proportion decreases. For example, the lengths of the Indexfinger from 3.8 centimeters to 4.4 centimeters have optimal proportions trending from 1.54 to 1.50 then 1.39. The other fingers have the same relationship but the values of proportion are different. The same trend is observed for the data of the female hand circumferences and dimensions.

The data as obtained is intended to be used for the design optimizations of the product grip interface with a view to understand the optimal dimension to achieve the customization and at the same time increase efficiency, accuracy of dimension in grip interface design. Hence an attempt was made to describe the relevance of the data in the design of a hand grip product interface from ergonomic considerations.

Keywords: Grip interface; Anthropometric; Optimal grasp form; Ergonomics

1 INTRODUCTION

Among the various products grip interface design in anthropometric or ergonomics research, grip diameter has been studied extensively because it is an important factor to determine user grip diameter, and it can influence force exertion in manual work (Khalil 1973, Grant et al. 1992, Blackwell et al. 1999). So far, many investigators have tried to obtain the optimal handle diameter of a cylindrical handle sharp with subject comfort rating (Hall and Bennett 1956, Yakou et al. 1997, Yong et al. 2005). The grip force applied to a cylindrical handle is one of the most important hand force components. So far, however, no attempt has been made to fully describe the exact form of this function and then they used a 30 mm handle to examine some fundamental characteristics of grip forces and to explore the basic pattern of the grip force function (Ren et al. 2007). Grip strength

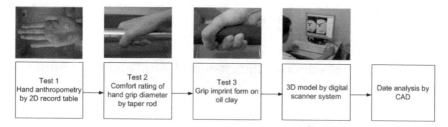

| Test 1
Hand anthropometry
by 2D record table | Test 2
Comfort rating of
hand grip diameter
by taper rod | Test 3
Grip imprint form on
oil clay | 3D model by digital
scanner system | Date analysis by
CAD |

Figure 1. Block diagram of test system.

is usually measured using an apparatus such as EMG, dynamometer. They have geometries that are not representative of cylindrical handle feature used on many manual and powered hand tools. This situation will then cause geometric graphic differences.

Most of the researchers are using the force strength to describe the finger condition. For measuring forces exerted by all fingers, Amis (1987) and Radhakrishnan and Nagaravindra (1993) designed a cylindrical handle instrumented with individual cantilever beams and strain gauges. Lee and Rim (1991) and Yong et al. (2005) were using a flexible sensor glove to record and to distribute the grip pressure by wrapping these sensors around a cylinder for evaluating power tools, hand tools and meat hook handles (Fellows and Freivalds 1991, Yun 1993, Kong et al. 2003). Ren et al. (2007) applied a Jamar dynamometer to measure the grip strength. However, this research will find the relationship between the two dimensional measurements of the hand and three dimension measurements of the grip is unclear. In the above research methods, all of these systems have limitations in terms of evaluating forces on handles with non-cylindrical shape and in precise positioning of individual fingers and phalanges.

It is stated that "when the operator is gripping a cylindrical handle, the direction of the main grip force is generally parallel to the Z-axis defined in ISO". Edgren et al. (2004) used a vector pose with hand as a measure of grip force. The proposed vector was formed using values of two forces measured on two specific orthogonal axes on the hand.

In this present study would like to understand the relationship between the two dimension model of hand and the three dimension model of the grip interface. The form of the grip measurement system was developed by using a cylindrical taper rod and then obtaining the finger shapes using oil clay which is shaped by the user's hand size when gripped. This experiment system (Fig. 1) can be used to measure the grip shape of individual fingers on cylindrical or non-cylindrical grip handles as it captures the precise positioning by all fingers and phalanges.

The objectives of this study were:

(1) To understand and evaluate the relationship between the two dimensional model of the hand and three dimensional model of the grip interface based on the different user's hand size.
(2) To apply normalized three dimensional proportion circumferences recommended gripping handle design base on the user's hand two dimension anthropometry.

2 METHOD

2.1 Subjects

Twenty subjects (10 males and 10 females) between the ages of 18 and 49 years with a mean of 33 years. There were 9 experts randomly chosen in woodworking shop and others recruited through advertisements within Tsinghua University and Hsinchu High School. Participants were all healthy volunteers and had hand tool experience.

2.2 Instrumentation and apparatus

For the data accuracy that designed a non-fixture test instrument to perform measurements of grip, measure the finger length to determine the 2D dimension of the hand, evaluate the adaptable grip

Figure 2.   Grip diameter taper rod.

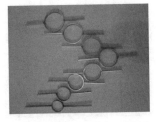

Figure 3.   Circular knife for cutting oil clay cylinders.

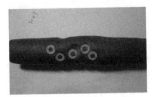

Figure 4.   Grip imprints form on oil clay.

Figure 5.   3D model by digital scanner system.

diameter based on the user's grasp comfort rating and to choose an oil clay cylinder size according to their sign size to get a real grip form to define grip circumferences for each finger. They were asked to grasp the aluminum taper test rod. The rod consisted of two cylinders from 28 mm (small) to 46 mm (large) diameter and there were sign places each 2 cm (Fig. 2).

Based on the hand length of each participant they grasped and moved their hand to rate the comfort diameter with the tapered aluminum test cylindrical rod. With the comfort level established an appropriate oil clay diameter was determined to be used to capture the user's grip. There were 10 diameters of oil clay available based on the circular knives (Fig. 3) and to make an evaluated model (Fig. 4). Using the COMET 3D digital scanner system (Fig. 5) to scan the model for get the finger's affect section forms in each fingers on oil clay. The subject was considered as a random effects variable. This was a balanced design and all oil clay cylinders were made to order for each participants hand size based on the comfort level of the tapered aluminum cylinder.

Figure 6. Measure the distance from the radiocarpal joint of the right hand to the tip of each finger.

Figure 7. The rating of comfort diameter with a handle.

3 EXPERIMENTAL DESIGNS

At the beginning of the experiment, informed consent was obtained and anthropometric measurements of hand length were made of each participant. Hand and fingers size were determined using a table measure to measure the distance from the radiocarpal joint of the right hand to the tip of each finger (Fig. 6). A unique 2D model in table form was recorded for each participant. The participants were then seated according to their hand size, to consider the comfort position and to find the adapting grip diameter with hand (Fig. 7).

4 RESULTS

4.1 *The anthropometric of hand*

The average lengths mean from the radiocarpal joint of the right hand

(1) to the tip of thumb was 9.98 cm,
(2) to Indexfinger was 16.27 cm,
(3) to middle finger was 17.33 cm,
(4) to ring finger was 16.24 cm, and
(5) to litter finger was 13.44 cm.

The average diameter mean of grip was 3.82 cm with the maximum was 4.4 cm and the minimum was 3.2 cm (Table 1).

4.2 *The section circumference of each finger except the thumb*

The mean of section circumference of each finger, which was defined by the following equation:
   $L = 2\pi r$,

Where L is the circumference (cm), π value is 3.1415; r is half-diameter (cm).
That there were
to Indexfinger was 11.1 cm,
to middle finger was 11 cm,
to ring finger was 11.1 cm,
to litter finger was 11.35 cm.

Table 1. The mean length of hand anthropometric and the mean diameter in comfort grip rating. Numbers are in units of centimeters.

Variable	Mean	Median
Thumb	9.98	9.90
Indexfinger	16.27	16.50
Middle finger	17.32	17.60
Ring finger	16.23	16.30
Little finger	13.44	13.20
Mean of Diameter	3.82	3.80

Table 2. The mean circumference of each affect section plane. Numbers are in units of centimeters.

Variable	Mean	Median
Indexfinger	11.11	11.05
Middle finger	11.00	11.00
Ring finger	11.08	11.15
Little finger	11.34	11.45
Mean of Circumference	11.99	11.90

Table 3. The relative proportion on section plane of male.

Male Participant	Grip diameter	Relative proportion			
		Index finger	Middle finger	Ring finger	Little finger
1	4.4	1.39	1.56	1.50	1.19
5	4	1.50	1.61	1.51	1.21
4	3.8	1.54	1.65	1.57	1.29

The average mean of circumference was 12 cm with the maximum was 13.8 cm and the minimum was 11.0 cm (Table 2).

5 DISCUSSION

This objective of this study was to understand the relationship between the two dimensional model of the hand and the three dimensional model of the grip when using the measurement instruments. A body of researcher describes the relationship between the handle form and the grip strength such as Bassey and Harries (1993) found a positive correlation between skeletal size and grip strength. Filbert et al. (1998), Lunde et al. (1970) and Nwuga (1975) showed grip strength to be positively correlated with body height and weight. Unlike Schmidt and Toews (1970) they did not find a significant correlation between age and grip strength (Filbert et al. 1998).

In this study, our main purpose is to show how the shape of grip handle can be adapted within individual hand size, then to compare the numerical between grip diameter and circumference of the shape section of each finger. Some results showed that the male's section proportion which is the finger length divided by the circumference of the shape section, there is a decreasing trend. For example, when the user chooses the adapting handle and the diameter is 4 cm, thus calculate its circumference is 12.57 cm, also applied the Table 3 and 4, which presents the relationship between

Table 4. The relative proportion on section plane of female.

Female Participant	Grid diameter	Relative proportion			
		Index finger	Middle finger	Ring finger	Little finger
1	4	1.38	1.50	1.38	1.10
4	3.8	1.41	1.51	1.39	1.10
4	3.6	1.42	1.58	1.46	1.16
1	3.2	1.49	1.59	1.43	1.17

Table 5. The full proportion table on hand anthropometry of male.

Male R/P	Grid diameter	Relative proportion			
		Index finger	Middle finger	Ring finger	Little finger
real	4.4	1.39	1.56	1.50	1.19
predict	4.2	1.42	1.60	1.50	1.20
real	4.0	1.50	1.61	1.51	1.21
real	3.8	1.54	1.65	1.57	1.29
predict	3.6	1.67	1.71	1.58	1.31
predict	3.4	1.71	1.76	1.65	1.39
predict	3.2	1.85	1.82	1.66	1.41

finger length and circumference of shape section, to predict a better circumference for each finger in the grip form

to Indexfinger was $12.57 \times 1.39 = 17.47$ cm,
to middle finger was $12.57 \times 1.56 = 19.45$ cm,
to ring finger was $12.57 \times 1.5 = 18.7$ cm, and
to litter finger was $12.57 \times 1.19 = 14.96$ cm.

The strong correlation parameter was the natural grip diameter. In other words, the handle form positions as determined when an oil clay cylinder was gripped by the participant, it followed the hand's 2D size. When any user decides the grip diameter and uses the translation table a better shape circumference can be calculated. This study suggests that this advantage is consistent. The underlying reason for this has yet to be proven, as these relationship parameters are still for reference in product interface design.

6 CONCLUSIONS

Traditionally, studies of handle interface for human factors have primarily focused on the control of the arm, hand, and strength. However, most natural human actions such as exercise, work, health aid tool, car control handle, etc., the data could support to them.

There are two conclusions that can be drawn from our data.

(1) The study confirmed that the relationship between the two dimensional model of the hand and the three dimensional model of grip interface showed a decreasing trend.
(2) According to the relationship data shown in the proportion table, also the user could calculate the approximate value using a mathematical methodology. Tables 5 and 6 shows predict values based on this approach.

Table 6. The full proportion table on hand anthropometry of female.

Female R/P	Grip diameter	Relative proportion			
		Index finger	Middle finger	Ring finger	Little finger
predict	4.4	1.38	1.44	1.34	1.05
predict	4.2	1.38	1.49	1.35	1.06
real	4.0	1.39	1.50	1.39	1.10
real	3.8	1.41	1.51	1.40	1.10
real	3.6	1.42	1.58	1.46	1.16
predict	3.4	1.45	1.59	1.47	1.16
real	3.2	1.49	1.59	1.43	1.17

However, with regard to the handle form dimension and position used, it is paramount that the user or designer, maintain consistency in usability, especially in the handle section circumference for individual fingers. For the relationship between 2D of hand and 3D of grip interface, some aspect of the different grip section form was not observed in our data.

REFERENCES

Fiebert, I.M., Roach, K.E., Fromdahl, J.W., Moyer, J.D., Pfeiffer, F.F., Relationship between hand size, grip strength and dynamometer position in women, Journal of Back and Musculoskeletal Rehabilitation 10, 137–142, 1998.

Khalil, T.M, An electromyographic methodology for the evaluation of industrial design. Human Factors 15 (3), 257–264, 1973.

Blackwell, J.R., Kornatz, K.W., Heath, E.M., Effect of grip span on maximal grip force and fatigue of flexor digitorum superficialis. Applied Ergonomics 30, 401–405, 1999.

Hall, N.B., Bennett, E.M., Empirical assessment of handrail diameters. Journal of Applied Psychology 40, 381–382, 1956.

Ren, G. D., John, Z. W., Daniel, E. W., Thomas, W. M., A new approach to characterize grip force applied to a cylindrical handle, Medical Engineering and Physics, 30, 20–33, 2008.

Kong, Y.K., Brian, D. L., Optimal cylindrical handle diameter for grip force tasks, International Journal of Industrial Ergonomics 35, 495–507, 2005.

ISO/DIS 15230. Mechanical vibration and shock—coupling forces at the machine–man interface for hand-transmitted vibration. Geneva, Switzerland: International Organization for Standardization, 2005.

Amis, A.A., Variation of finger forces in maximal isometric grasp tests on a range of cylindrical diameters. Journal of Biomedical Engineering 9, 313–320, 1987.

Lee, J.W., Rim, K., Measurement of finger joint angles and maximum finger forces during cylinder grip activity. Journal of Biomedical Engineering 13, 152–162, 1991.

Yakou, T., Yamamoto, K., Koyama, M., Hyodo, K., Sensory evaluation of grip using cylindrical objects. JSME International Journal, Series C 40 (4), 730–735, 1997.

Edgren, C.S., Radwin, R.G, Irwin, C.B., Grip force vectors for varying handle diameters and hand sizes. Human Factors 46, 244–51, 2004.

Jaruwan, K., Angoon, S., Nantakrit, Y., Patrick, E. P., Anthropometry of the southern Thai population, International Journal of Industrial Ergonomics 38, 111–118, 2008.

Dewangana, K.N., Owarya, C., Dattab, R.K., Anthropometric data of female farm workers from north eastern India and design of hand tools of the hilly region International Journal of Industrial Ergonomics 38, 90–100, 2008.

Anthropometric Source Book Volume III: Annotated Bibliography of Anthropometry, NASA Reference Publication 1024.

Kim, B.H., Yi, B.J., Oh, S.R., Suh, I. H., Non-dimensionalized performance indices based optimal grasping for multi-fingered hands Mechatronics 14, 255–280, 2004.

Jonathan, R.R., Jose, L.M., Gutierrez, A., Hand Size Influences Optimal Grip Span in Women but not in Men, The Journal of Hand Surgery, Vol. 27A No 5 September, 897–901, 2002.

Satoru, K., Summers, V. A., Mackenzie, C. L., Ivens, C. J., Takashi, Y., Grasping an augmented object to analyze manipulative force control, 2002.

Anakwe, R. E., Huntley, J. S., Mceachan, J. E., Grip strength and forearm circumference in a healthy population, Journal of Hand Surgery, European Volume, 32E: 2: 203–209, 2007.

Saling, M., Stelmach, G.E., Mescheriakov, S., Berger, M., Prehension with trunk assisted reaching, Behavioural Brain Research 80, 153–160, 1996.

Mangialardi, L., Mantriota, G., Trentadue, A., A three-dimensional criterion for the determination of optimal grip point, Robotics and computer-Integrated Manufacturing, Vol. 12, No.2, 157–167, 1996.

Ergonomic Trends from the East – Kumashiro (ed)
© *2010 Taylor & Francis Group, London, ISBN 978-0-415-88178-4*

An usability study in developing a mobile flight case learning system in ATC miscommunications

Kuo-Wei Su
Department of Information Management, National Kaohsiung First University of Science and Technology, Kaohsiung City, Taiwan (R.O.C)

Jau-Wen Wang
Department of Information Management, Fortune Institute of Technology, Kaohsiung County, Taiwan (R.O.C)

Jyun-Hua Lin
Department of Information Management, National Kaohsiung First University of Science and Technology, Kaohsiung City, Taiwan (R.O.C)

1 INTRODUCTION

How to provide flight services based on "safety" is the most important key for aviation enterprises. From the statistic data of aircraft accidents, there existed 55 percent of accidents attributed to the flight crew, followed by the airplane at 17 percent, the weather at 13 percent etc. (Darby, 2006). So obviously, most occurrences were resulted from "human." Pilots and ATCS are the principle members responsible for voice radio communications in ATC system. Cushing (1988) averred that the primary cause of most aviation occurrences is from communication errors between pilots and controllers for a long time.

In consideration of convenience of mobile devices, the study would develop a ubiquitous learning system by M-learning (Fischer, 2001) and could be as the start of cross-platform learning system on multi-devices. Due to the constraints of mobile devices, such as screen display size (Zhao et al., 2006), the conception of M-HCI would be employed to develop easy-to-use UIs and evaluate acceptability and utility of the system through a usability testing.

2 METHODOLOGY

2.1 Development of MFCLS

The study confirmed human-centered design process (ISO, 1999) to develop our mobile flight case learning system (MFCLS). The focus of human-centered design process is on user requirements.

2.2 Prototype of MFCLS

The main function of MFCLS is the search by data entry with pull-down menus. Why did we utilize pull-down menus as the data entry? Because pull-down menus are always available to the user by making selection, and a pull-down menus can be used to add frequently used items (Shneiderman and Plaisant, 2005). Moreover, a PDA or a smart-phone is interacted with users by touch-screen, in order to reduce time of entering words, so the study determined to utilize pull-down menus as the data entry.

2.3 Interface design of MFCLS

The design conception of the User Interface (UI) is based on guidelines of the small-screen interface design, which is majorly Eight Golden Rules (Shneiderman and Plaisant, 2005). As figure 3, the

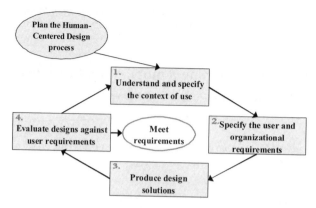

Figure 1.   Human-Centered Design Activities (ISO, 1999).

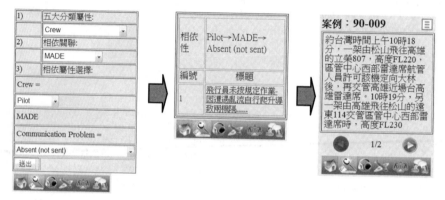

Figure 2.   Case retrieval procedures.

Figure 3.   User interface design.

system provides users' personal learning progress, uses iconic symbols to reduce users' short-term memory loading, and offers informative buttons and feedback to satisfy users' internal locus of control etc.

3   EXPERIMENTAL DESIGN

The experiment was in laboratory. The equipment involved a Dopod P100 pocket PC, a designed experimental scenario, and a Questionnaire for User Interaction Satisfaction (QUIS) (Shneiderman and Plaisant, 2005). There were two experimental groups: for twelve undergraduate students without any experience of operating PDA and for two flight experts with some experience of operating PDA.

4 EXPERIMENTAL RESULTS

4.1 *Questionnaire analysis*

The statistical analysis tool was SPSS software. In normality test, the one-sample test of kolmogorov-smirnov of non-parametrics (>0.05) was used testing by the variable the total time (seconds) of operating the tasks per participant: p-values $= 0.948$, so the experimental sample is normally distributed.

From the result of the reliability analysis, all the values of Cronbach's Alpha were over 0.6 which was the suggesting value of Nunnally and Bernstein (1994). This meant our questionnaire owned adequate internal consistency, stability and homogeneity.

Furthermore, the summarizing statement was very positive and the result represented high satisfaction for our MFCLS. And the study believed adopting the participants' suggestions and the recommendations of two flying experts, which would be presented in the next section, will keep MFCLS better and more satisfactory in the future.

4.2 *Open-ended questions*

Open-ended questions are designed to encourage full and meaningful answers using the participants' own knowledge. The study categories the users' suggestions into two groups:

i. Interface Design:
  - Highlights on the screen may be increased.
  - Hints or instructions for commands or functions may be not enough and need to be enhanced.
  - Colors on the screen may be so similar that main contents are not apparent enough.
ii. Functions of System:
  - Maybe could join voice mode to present the original record of controller-pilot communications by cases.
  - To add a search function, like keyword search, can help users find knowledge more quickly, especially for experts.

4.3 *Discussion*

In this section, the experimental results would be reviewed as the research findings as follows:

  - From the subjective assessment, the participants gave good approval and positive responses.
  - The symbols of "⌗ (Case-based Learning)", "☏ (Voice Communications)" and "☰ (Back to the Table of Contents)" may be redesigned.
  - Besides above, one pilot had mentioned that maybe because of the limitations of small screen size, a mobile system does not suit to long-time and uninterrupted learning.

5 CONCLUSIONS

5.1 *Conclusions*

Through the experiment of users' satisfaction investigation and the expert reviews, it was proved that there are good acceptability and utility of the system. In other words, it was confirmed that MFCLS is a well-designed mobile learning system. Therefore, it was helpful pilots and controllers to learn the knowledge by themselves at any time and any place through an intellectual mobile device, such as a PDA.

5.2 Contributions

The emergence of M-learning provides a shift opportunity for education training environment in aviation. The study aims to offer a full and ubiquitous self-learning environment for pilots and controllers. The following is the contributions for the academic and practical fields:

i. The contributions for the academic field:
   - The iconic symbols in MFCLS were inspected to be efficient for reducing operational memory and promoting learnability for users.
   - The user interfaces of MFCLS based on small-screen interface design were examined to be friendly.
   - The operational procedure of MFCLS based on HCD was evaluated to be easy to use and easy to learn.
ii. The contributions for the practical field:
   - The content of MFCLS is helpful for pilots and controllers to learn by themselves.
   - Pilots or controllers can learn by themselves anytime and anywhere, so that promote their situation alertness and emergency to drop faults.

5.3 Limitations and future works

Limitations of experiment which may affect the completeness of the experimental results:

- The experiment was carried out in the laboratory, the real situation was difficult to be simulated, and the study substituted usual undergraduates for flight crews.
- There were fewer participants invited to attend the experiment.

Some topics worth being probed further:

- This system could be extended for PC or other types of screens by redesigning the interfaces.
- It can be considered to join voice recognition and voice conversation for interactions with users.
- The domain could be extended for laws in aviation, so that to develop a full knowledge base for flight crews or scholars.

REFERENCES

Cushing, S. (1988). *Language and Communication-Related Problems of Aviation Safety.* Proceedings of the 1987 Annual Meeting of the American Association for Applied Linguistics. San Francisco. ED296595.

Darby, R. (2006). Commercial Jet Hull Losses, Fatalities Rose Sharply in 2005. Aviation Safety World.

Fischer, G. (2001). User Modeling in Human-Computer Interaction. Journal of User Modeling and User-Adapted Interaction (UMUAI), 11 (1/2), 65–86.

ISO. (1999). ISO 13407: Human-centred Design Processes for Interactive Systems. Geneva: International Standards Organisation. Also available from the British Standards Institute, London.

Nunnally, J. C., and Bernstein, I. H. (1994). Psychometric Theory (3rd ed.), New York: McGraw-Hill.

Shneiderman, B. and Plaisant, C. (2005). Designing the user interface : strategies for effective human-computer interaction, 4th ed., Boston , MA: Addison Wesley.

Zhao, D., Grundy J., and Hosking J. (2006). Generating mobile device user interfaces for diagram-based modelling tools. ACM International Conference Proceeding Series, Proceedings of the 7th Australasian User interface conference, 169 (50), 101–108.

Ergonomic Trends from the East – Kumashiro (ed)
© 2010 Taylor & Francis Group, London, ISBN 978-0-415-88178-4

Influence of seat angles for sitting comfort of wheelchairs

Toshio Matsuoka
Mie Prefecture Industrial Research Institute, Mie, Japan

Hiroyuki Kanai & Toyonori Nishimatsu
Faculty of Textile Science & Technology, Shinshu University, Nagano, Japan

1 INTRODUCTION

Recently, Japanese society has been aging rapidly. Nursing homes for elderly persons are increasing year by year. So it is important to enrich the quality of life for elderly persons. They use various types of wheelchairs in the nursing homes.

The main functions of a wheelchair are to provide mobility, transport and a place to sit. Many studies (e.g. Yoneda *et al.*, 2003) have reported on "riding comfort" of wheelchairs, however little attention has been paid to "sitting comfort." We (Matsuoka *et al.*, 2005) evaluated sitting comfort of the wheelchair and investigated how backrest angles influenced sitting comfort. Therefore, paying attention to sitting comfort of the wheelchair, we have developed a wheelchair on which elderly persons can comfortably spend considerable time. In this paper, we investigated how cushion angles of the wheelchair influence sitting comfort of the wheelchair for elderly and young persons.

2 METHOD

The wheelchair used in our experiment is a modular-type wheelchair consisting of a seat cushion, a backrest, armrests, leg rests and foot plates. As shown in Figure 1, the backrest angle was 111 degrees and only the seat cushion angle was changed. The backrest angle of 111 degrees was evaluated the most comfortable by elderly persons (Matsuoka *et al.*, 2005). The cushion angles were 3, 11 and 19 degrees.

Subjects were a group of six elderly persons (age 75.2 ± 6.7, height 154.2 ± 10.5 cm, weight 50.1 ± 7.3 kg) in a nursing home, who used wheelchairs in their daily live, and a group of 10 young persons (age 23.1 ± 1.0, height 168.2 ± 7.3, weight 68.8 ± 6.4 kg), who took a pre-test to receive a full explanation on the contents of the experiment before they gave informed consent. Each subject sat on the wheelchair for 30 minutes and evaluated sitting comfort twice – once immediately after sitting on the wheelchair and then after sitting on it for 30 minutes. Eight adjectives, namely "soft at

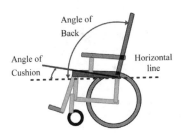

Figure 1.   Definition of the angle.

back", "soft at cushion", "painful at back", "painful at cushion", "easy at back", "easy at cushion", "comfortable" and "tiring" were judged using the semantic differential (SD) method (Masuyama and Kobayashi, 1989).

Contact area, positions of gravity and pressure distribution between their bodies and the cushion or the backrest of the wheelchair were also measured using the tactile sensor system (BIGMAT-2000, NITTA Co., Ltd.) while evaluating sitting comfort. The system consists of mats that are made up of thin flexible sensors, and the sensors were set on the cushion and the backrest. Each measuring area was 430 × 480 mm (2,064 points). The pressure distributions were compared with the subjective interpretation of sitting comfort.

3 RESULTS AND DISCUSSION

3.1 Subjective Measurements

According to results of the sensory evaluation conducted immediately after the elderly sat on the wheelchair, the cushion angle of 11 degrees was evaluated as being "easy", "comfortable" and "not tiring." The results of the initial evaluation by the young showed that when they sat on the wheelchair at these cushion angles, it was "not tiring." Mean preference scores of each adjective, as evaluated by the elderly after 30 minutes are shown in Figure 2, and those by the young are shown in Figure 3. After sitting in the wheelchair for 30 minutes, the elderly persons evaluated that the angle of 11 degrees was "easy", "comfortable" and "not tiring," while the angle of 19 degrees was evaluated as being "hard", "painful", "tight" and "not comfortable." The elderly evaluated that they were less tired after 30 minutes at these cushion angles. Both groups evaluated that the angle of 11 degrees was most comfortable when initially sitting on the wheelchair and also after sitting on it for 30 minutes. The angle between the cushion and the backrest was 100 degrees when the cushion angle was 11 degrees. But we (Matsuoka et al., 2005) reported that the angle of 108 degrees – cushion angle of 3 degrees and backrest angle of 111 degrees – was evaluated as being comfortable. So we suggest that the backrest angle should adequately be linked to the cushion angle.

The sensory evaluation results were examined by factor analysis to define sitting comfort from calculated mean preference scores. We obtained the factor matrices using the principal factor

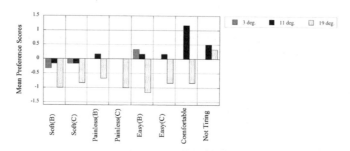

Figure 2. Mean preference scores evaluated by the elderly.

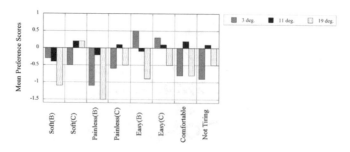

Figure 3. Mean preference scores evaluated by the young.

solutions without repeated assumption of communality, which is the proportion of the test variance ascribed to the action of the common factors. We rotated the factor axes so that the structure of the sensory test relations was indicated more clearly. The results of factor matrices were rotated using the varimax method (Okuno *et al.*, 1980). From the results of factor analysis for the elderly after 30 minutes, two common factors were obtained. Factor 1 is related to "softness", "pain", "ease" and "comfort" and factor 2 is related to "tiredness". It was found that the elderly evaluated sitting comfort of the wheelchair using two adjectives, "easy" and "not tiring" after 30 minutes. From the results of factor analysis for the young after 30 minutes, two common factors were obtained. Factor 1 is related to "pain", "comfort" and "tiredness" and factor 2 is related to "softness" and "ease". It was found that the young evaluated sitting comfort of the wheelchair using two adjectives, "not tiring" and "easy" after 30 minutes. Therefore, for both groups, two factors – "easy" and "not tiring" – were common and significant in evaluating sitting comfort of the wheelchair, whose cushion angle only was changed.

Figure 4 shows the factor loading scores of each cushion angle evaluated by the elderly, while Figure 5 shows those evaluated by the young. It can be seen in Figure 4 that the elderly evaluated the angle of 3 degrees as being "easy" and "tiring", 11 degrees as being "easy" and "not tiring", and 19 degrees as being "not easy." It can be seen in Figure 5 that the young evaluated the angle of 3 degrees as being "easy" and "tiring", 11 degrees as being "easy" and "not tiring", and 19 degrees as being "not easy" and "tiring". The results of both groups were very similar.

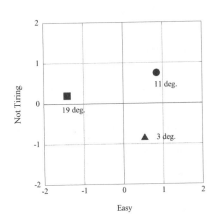

Figure 4.   Factor loading scores evaluate by the elderly after 30 minutes.

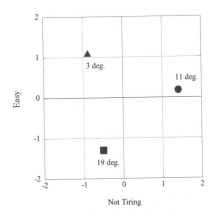

Figure 5.   Factor loading scores evaluated by the young after 30 minutes.

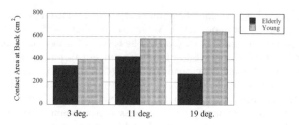

Figure 6.　Contact area of backrest after 30 minutes.

Table 1.　Correlation coefficients between body pressures and sensory values after 30 minute. (Elderly persons) (*P < 0.05 , **P < 0.01).

	Soft at Back	Soft at Cushion	Painless at Back	Painless at Cushion	Easy at Back	Easy at Cushion	not Comfortable	tiring
Contact area of back	0.94	0.86	0.94	0.86	0.80	0.93	1.00**	0.35
Contact area of cushion	−0.07	−0.25	−0.07	−0.25	−0.35	−0.10	0.35	1.00**
Body pressure at Back	−0.63	−0.47	−0.63	−0.47	−0.38	−0.60	−0.90	−0.78
Body pressure at Cushion	0.34	0.16	0.34	0.16	0.05	0.31	0.70	0.94
MvGr at Back (H)	−0.96	−0.89	−0.96	−0.89	−0.83	−0.95	−0.99	−0.29
MvGr at Back (V)	0.71	0.83	0.71	0.83	0.88	0.74	0.36	−0.70
MvGr at Cushion (H)	−1.00**	−0.99**	−1.00**	−0.99	−0.97	−1.00*	−0.89	0.04
MvGr at Cushion (V)	0.11	−0.08	0.11	−0.08	−0.18	0.07	0.51	0.99

***MvGr Moving distance for center of gravity
****H: Horizontal, V: Vertical

Table 2.　Correlation coefficients between body pressures and sensory values after 30 minutes. (Young persons) (*P < 0.05, **P < 0.01).

	Soft at Back	Soft at Cushion	Painless at Back	Painless at Cushion	Easy at Back	Easy at Cushion	not Comfortable	tiring
Contact area of back	−0.75	0.98	0.01	0.43	−0.92	−0.83	0.31	0.66
Contact area of cushion	−0.54	1.00**	0.28	0.66	−0.78	−0.64	0.56	0.84
Body pressure at Back	−0.97	0.77	−0.45	−0.03	−1.00	−0.99	−0.17	0.24
Body pressure at Cushion	−0.73	0.98	0.04	0.46	−0.91	−0.81	0.34	0.69
MvGr at Back (H)	0.55	−1.00**	−0.27	−0.66	0.79	0.65	−0.55	−0.84
MvGr at Back (V)	0.72	−0.99	−0.05	−0.47	0.91	0.81	−0.34	−0.69
MvGr at Cushion (H)	0.81	−0.95	0.09	−0.34	0.96	0.88	−0.21	−0.58
MvGr at Cushion (V)	0.66	−1.00	−0.14	−0.54	0.86	0.75	−0.43	−0.75

3.2　Body pressure measurements

Figure 6 shows the result of the contact area between the backrest and the human back after 30 minutes. As shown in Figure 6, the contact area of the elderly was largest when the angle was 11 degrees and that was smallest when the angle was 19 degrees. That of the young was largest when the angle was 19 degrees and that was smallest when the angle was 3 degrees. When the contact area with the backrest was large, the upper part of the body was supported by the backrest. So the angle providing the largest contact area for the elderly was different from that for the young. Therefore, as the seat angle was bigger, the sitting posture of both groups was different.

3.3　Correlation between sensory evaluations and body pressure distribution

The relation between sensory values and body pressure distributions was examined, and correlation coefficients between body pressure distributions and sensory values by the elderly and the young groups after 30 minutes are shown in Tables 1 and 2, respectively. As shown in Table 1, the adjective "comfort" was closely related to the contact area at the backrest, "not tiring" was closely related to the contact area at the cushion. "Soft at cushion" and "easy at cushion" were closely related to the moving distance in a horizontal direction for the center of gravity at the cushion. As shown in Table 2, "soft at cushion" was closely related to the contact area at the cushion.

4 CONCLUSION

We investigated how cushion angles of the wheelchair influenced sitting comfort, and the relations between sitting comfort and body pressure distribution. The results were as follows.

(1) The cushion angle of the wheelchair influenced sitting comfort for both elderly and young groups.
(2) The cushion angle of 11 degrees was the most comfortable for both groups.
(3) Two adjectives "easy" and "not tiring" were common and significant in evaluating sitting comfort of the wheelchair for both groups.
(4) For the elderly, the adjective "comfort" was determined by the contact area at the backrest, "not tiring" by the contact area at the cushion, while the determining factor for "soft at cushion" and "easy at cushion" was the moving distance in a horizontal direction for the center of gravity at the cushion.
(5) If the elderly felt easy to move when sitting on the wheelchair, they felt comfortable.

REFERENCES

Okuno T., Kume H., Haga T. and Yoshizawa T., 1980, Multiple Regression Analysis, (Tokyo: JUSE Press), pp. 323–372
Masuyama E. and Kobayashi S., 1989, Sensory evaluation, (Tokyo: Kakiuchi Press), pp. 16–23
Matsuoka T., Nishimatsu T. and Toba E., 2005, Information theoretic anakysis on sitting comfort of wheelchair. *Journal of Japan Society of Kansei Engineering*, Vol. 6, No. 1 pp. 1–6
Yoneda I., Kasuya S., Bando M., Sueda O. and Oku H., 2003, Quantitative evaluation and analysis of influence of inclined road surface on maneuvering the wheelchair. *The JSME Symposium on Welfare Engineering*, Vol. 2003, No. 3 pp. 47–50

Ergonomic Trends from the East – Kumashiro (ed)
© 2010 Taylor & Francis Group, London, ISBN 978-0-415-88178-4

Evaluation of bed comfort by body pressure distribution and sensory test

Toshio Matsuoka
Mie Prefecture Industrial Research Institute, Mie, Japan

Hirokazu Kimura & Takanori Yamamoto
Technology Research Institute of Osaka Prefecture, Osaka, Japan

Hiroyuki Kanai & Toyonori Nishimatsu
Faculty of Textile Science & Technology, Shinshu University, Nagano, Japan

Shin Saito
Mie Prefecture College of Nursing Mie, Japan

1 INTRODUCTION

Dialysis patients in Japan are increasing every year with the total number of patients now over 20 million. Dialytic treatment is conducted three times a week, and each treatment session lasts about four hours. Therefore, the patients need a comfortable treatment environment in the hospital. A chair-type bed which saves space has been developed and used in the hospitals. As the patients feel tight and uncomfortable on them, they prefer conventional beds.

Yamazaki *et al.* (2003) reported on the comfort of beds. This study investigated the environment to obtain comfortable sleep on beds. The comfort of a bed for the artificial dialysis treatment is different from those in our daily life. Therefore, we have developed a new bed on which dialysis patients feel comfortable for a long time. In this paper, we investigate how a bed mat influences lying comfort based on sensory evaluations and by measuring body pressure distributions.

2 METHOD

We used three urethane mats (No.1, No.2, No.3) consisting of two layers whose materials were a low repulsion urethane and a high elasticity urethane. The length of one mat was 2,000 mm, divided into three parts with lengths of 650 mm (Part A), 640 mm (Part B) and 710 mm (Part C). The width was 700 mm and the thickness was 60 mm. Details of the samples are shown in Tables 1 and 2, and

Table 1. Summary of mat. "Part A" is the head of the mat.

Sample	Thickness	A part	B part	C part
No.1	20 mm upper 40 mm lower	Urethane A	Urethane B Urethane C	Urethane A
No.2	30 mm upper 30 mm lower	Urethane A	Urethane B Urethane C	Urethane A
No.3	20 mm upper 40 mm lower	Urethane A	Urethane B Urethane D	Urethane A

Table 2. Thickness at a pressure of 9.8 kPa.

Sample	A part (mm)	B part (mm)	C part (mm)
No.1	15.99	23.29	15.99
No.2	15.36	21.85	15.36
No.3	35.16	37.39	35.16

Table 3. Summary of each urethane.

Sample	Density (kg/cm$^{3)}$)	Hardness (N/314 cm^2)	Tensile strength (kPa)	Elongation (%)
Urethane A	60 ± 5	49.0 ± 19.6	29.4	100
Urethane B	50 ± 5	70 ± 30	100	100
Urethane C	45 ± 3	156.9 ± 19.6	69	80
Urethane D	80 ± 15	235.4 ± 39.2	39	40

details of each type of urethane are shown in Table 3. The materials of No.1 and No.2 were the same, but the thickness of each layer was different. The materials used in the lower layers of No.1 and No.3 were different.

Bed comfort when lying on each mat was judged through tactile sensation using the Scheffe-Nakaya's paired comparison method (Research Committee of Sensory Evaluation, 1999). The evaluation adjectives were "soft (at back, hip, leg)", "fitted (at back, hip, leg)", "elastic (at back, hip, leg)", "easy (at back, hip, leg)", "tiring", "relaxing", "like", "peaceful", "high-class" and "comfortable". Three pairs of test samples were randomized and presented to subjects. The samples were covered with a cotton sheet and the subjects could not distinguish the difference among the samples. Subjects were five males in their twenties and thirties.

We statistically calculated the mean preference scores of the three mats for each adjective and used factor analysis to study bed comfort, and used the technique of information theory to study the evaluation structure of bed comfort from the calculated mean preference scores. The information theory to clarify the structure of the response for the stimulation was used for analyzing the hand evaluation of fabrics (Nishimatsu and Sakai, 1988) and the evaluation of sitting comfort (Matsuoka et al., 2005).

Body pressure and contact area between the human body and the mat were measured using the tactile sensor system (BIGMAT-2000, NITTA Co., Ltd.) for five minutes. Measuring area was $1,720 \times 480$ mm (8,256 points). The system consists of mats that are made up of thin flexible sensors, and the sensors were set on the mat with the cotton sheet. Subjects were nine males in their twenties. The pressure distributions were compared with the subjective interpretation of bed comfort.

In clinical evaluations, body movements and body pressure distributions were measured while lying on sample B and a usual bed for dialytic treatment. Subjects were four female dialysis patients.

3 RESULTS AND DISCUSSION

3.1 Subjective measurements

Based on the results of the statistical significance test for the mean preference, there were no significant differences among 10 adjectives. The 10 adjectives, namely "soft at back", "fitted at back", "fitted at leg", "elastic at leg", "easy at back", "easy at hip", "tiring", "relaxing", "peaceful" and "high-class", were excluded in analysis. Mean preference scores for each adjective are shown in Figure 1. Mat No.2 was judged as being easy at hip and leg, fitted at hip, easy at leg, comfortable,

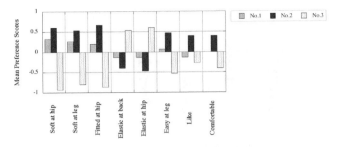

Figure 1.   Mean preference scores.

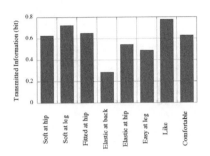

Figure 2.   Transmitted Information.

and it was liked. Mat No.3 was judged as being hard at hip and at leg, not fitted at hip, not easy at leg and uncomfortable. It was found that the mat influenced each adjective of bed comfort.

The mean preference scores of eight adjectives were examined by factor analysis to investigate the subjective words of bed comfort. We obtained the factor matrices using the principal factor solutions without repeated assumption of communality, which is the proportion of the test variance ascribed to the action of common factors. We rotated the factor axes so that the structure of the sensory test relations was indicated more clearly. The results of the factor matrices were rotated using the varimax method (Okuno *et al.*, 1980). From the results of factor analysis, two common factors were obtained. Factor 1 is related to "soft at hip", "soft at leg" and "fitted at hip" while factor 2 is related to "like" and "comfortable". Therefore, two factors – "soft" and "preference" – were common and significant in evaluating lying comfort of bed mats, whose mat material only was changed.

3.2   *Transmitted information*

We applied the information theory (Masuyama and Kobayashi, 1989) to study the evaluation structure of bed comfort from the mean preference scores. Bed comfort was evaluated by human tactile sensation. The input stimulus is the bed material and the response (human output) is the evaluation result. We analyzed how much the stimulus information (namely lying comfort) transmitted to the sensation of subjects. Figure 2 shows the transmitted information of each adjective evaluated by the subjects. The amount of information per stimulus was 1.59 bit. The average of the transmitted information of all adjectives was 0.59 bit and 37% of the stimulus information was transmitted to the human response. The amount of transmitted information of "like" was 0.78 bit and that of "comfortable" was 0.63 bit. These adjectives concerning the judgement of bed comfort were transmitted more information than others.

3.3   *Body pressure measurements*

Figure 3 shows the result of body pressure between the bed mat and the human body. The body pressure of mat No.2 was lower than others at each part and for the entire mat. Also, there were

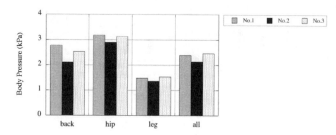

Figure 3. Body pressure between body and mat.

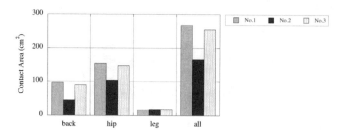

Figure 4. Contact area over 4.3 kPa.

Table 4. Correlation coefficients between body pressures and sensory values. (*P < 0.05, **P < 0.01).

	Soft at hip	Soft at leg	Fitted at hip	Elastic at back	Elastic at hip	Easy at leg	Like	Comfortable
Contact area of back	0.99	0.99	1.00*	−1.00*	−1.00*	1.00*	0.86	0.98
Contact area of hip	0.80	0.81	0.87	−0.86	−0.88	0.92	1.00*	0.96
Contact area of all body	0.98	0.99	1.00*	−1.00*	−1.00*	1.00*	0.88	0.99
HP area at back	−0.41	−0.44	−0.53	0.52	0.54	−0.62	−0.90	−0.71
HP area at hip	−0.62	−0.64	−0.72	0.71	0.73	−0.79	−0.98	−0.86
HP area at all body	−0.53	−0.55	−0.64	0.63	0.65	−0.72	−0.95	−0.80
Body pressure at back	−0.82	−0.84	−0.89	0.88	0.90	−0.93	−1.00*	−0.97
Body pressure at hip	−0.49	−0.51	−0.60	0.59	0.61	−0.69	−0.93	−0.77

less contact areas with pressures over 4.3 kPa, an indicator for preventing bedsores (Landis, 1930), compared to others at the back, the hip and the entire body. The thickness of mat No.2 at a pressure of 9.8 kPa was thin and it was easily compressed compare to the other mats. As the upper layer of mat No.2 was made of low repulsion urethane, it was thicker than No.1 or No.3. Body pressure distributions were influenced by the thickness of the low repulsion urethane and the compression properties of the urethane.

3.4 Correlation between sensory evaluations and body pressure distribution

The relation between sensory values and body pressure distributions was examined and correlation coefficients between body pressure and sensory values were shown in Table 4. As shown in Table 4, the adjectives "elastic at back and at hip" was closely related to the contact area at the back, "like" was to the contact area at hip, and "fitted at the hip" was to the contact area of the entire body.

3.5 Clinical evaluations

In the clinical evaluation, we measured body pressure distribution and the number of body movements while dialytic treatment was conducted. The number of body movements on mat No.2 and

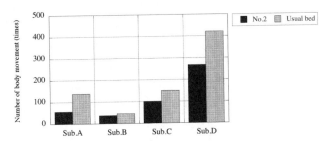

Figure 5.   Number of body movements.

on a usual bed for each subject is shown in Figure 5. As shown in Figure 5, the average number of movements on mat No.2 decreased 36% compared with that on the usual bed. From the result of body pressure distribution, the contact area of mat No.2 was larger than that of the usual bed while body pressure of No.2 was smaller than that of the usual bed. It was found that the body movements of each patient decreased as body pressure distribution improved.

4   CONCLUSION

We investigated how the bed mat influenced lying comfort using sensory evaluations and by measuring body pressure distributions. The results were as follows:

(1) The material of the mat influenced lying comfort of the bed.
(2) Two factors, "soft" and "preference", were common and significant in evaluating lying comfort of the bed, whose mat material only was changed.
(3) The adjectives "like" and "comfort" related to the total judgement of bed comfort were transmitted more information than other adjectives.
(4) The adjectives "elastic at back and hip" was judged by the contact area at the back, "like" was by the contact area at the hip, and "fitted at the hip" was by the contact area of the entire body.

ACKNOWLEDGEMENT

The authors express thanks to TANAKA K Co., LTD and to Prof. Shingo Takezawa at Suzuka University of Medical Science for their sincere support.

REFERENCES

Okuno T., Kume H., Haga T. and Yoshizawa T., 1980, Multiple regression analysis, (Tokyo: JUSE Press), pp. 323–372.
Landis EM., 1930, Micro-injection studies of capillary blood pressure in human skin. *Heart*, 15, pp. 209–228.
Masuyama E. and Kobayashi S., 1989, Sensory evaluation, (Tokyo: Kakiuchi Press), pp. 108–113.
Matsuoka T., Nishimatsu T. and Toba E., 2005, Information theoretic anakysis on sitting comfort of wheelchair. *Journal of Japan Society of Kansei Engineering*, Vol. 6, No. 1 pp. 1–6.
Nishimatsu T. and Sakai T., 1988, Application of the information theory to hand evaluation and proposal of the design equation of the visual and tactual sense values of pile fabrics. *Sen-i Gakkaishi*, Vol. 44, No. 2 pp. 88–95.
Research Committee of Sensory Evaluation, 1999, Sensory Evaluation Handbook, (Tokyo: JUSE Press), pp. 349–393.
Yamazaki N., Satoh S., Tachikawa R., 1994, Biomechanical fitting of bed mattresses. *Biomechanisms*, Vol. 12, pp. 61–71.

Ergonomic Trends from the East – Kumashiro (ed)
© 2010 Taylor & Francis Group, London, ISBN 978-0-415-88178-4

The improvement of the grip span of handbrake of motorcycle

Tsun-Yu Woo, Bor-Shong Liu & Hsien-Yu Tseng
Department of Industrial Engineering and Management, St. John's University, Taipei, Taiwan

1 INTRODUCTION

The motorcycle is an indispensable transportation vehicle in Taiwan, and the efficiency of hand-brake is a noticeable factor which causes traffic accident of motorcycle driving. Handbrake is a vital structure for the whole motorcycle; therefore, this topic aims at the research of handbrake component. According to the measurements of grip span of handbrake of most domestic motorcycle brand, this survey found out that the data exceeds the effectively male and female span of power grip. For suddenly traffic events, large span between handle and handbrake lever would disturb handbrake operation. Based on the reason, the improvement of handbrake lever was chosen to be the study approach.

2 LITERATURE REVIEW

The sum of grip strength is produced by the power grip of palm and fingers which grasps the objective tightly with certain situation. Based on Caldwell (Caldwell et al, 1974) suggestions, the time of instantly static power grip needed to sustain at least 4 to 6 seconds at the biggest force, and the average strength value of the first 3 second was taken to be the biggest muscular strength, also some scholars took maximum value (peak force) in process as a biggest muscular strength (Dempsey and Ayoub, 1994). In the meanwhile, for increasing the reliability of data, the measurement of muscular strength must be carried on repeatedly. The static power grip will be influenced by many factors, such as the age, gender, wrist position, arm posture, height of force generated, posture of force produced, handle diameter, glove as well as subjects' motivation and so on (Chang, 2002).

The posture is one of the primary factors affected the grip force, and the posture or position of hands will change the joint angle between muscle and phalanx, furthermore they influence the relation between muscle length and tension. Therefore any wrist flexure, extension, radial deviation, ulna deviation, or knuckle angle will affect the strength of power grip (Chung, 2002). Human will feel extremely uncomfortable as the angle of wrist flexure is the biggest (Carey et al, 2002), then the grip force will be the largest when the wrist position is neutral. Flexion and extension of wrist will make the grip strength reduce, and the decline rate of grip force of flexure will be bigger than the extension (Linscheid and Chao, 1973; Hallbeck and Mcmullin, 1994; Duque, 1995; Jung and Hallbeck, 2002) (Shih, 1995).

Grip span is a quite important factor for power grip. Ching-Chan Lin and Guan-Gin Shie (2000) suggested that the best grip span for female is 50 mm, the best grip span for male is 60 mm, and 40 mm grip span is disadvantaged for power produced of male, relatively 70mm grip span is not privileged for female. The institute of labor safety and health research and Professor Zhi-Yun You (1998) suggested that the diameter of handle should set up to be situated between 40mm and 50mm (Guo, 2002).

The grip strength of male is obviously bigger than the female. Ching-Chan Lin et al. (2000) pointed out that the strength of feminine power grip is approximately 52.7% of masculine power grip. Fransson and Winkel (1991) thought that the female grip force is 60% of male. Hallbeck and McMullin (1993) discovered that the grip strength is influenced by gender significantly, and the strength of feminine power grip is approximately 74% of masculine (Guo, 2002).

3 METHODS

Experimental contents of this research include the measurement of grip span between handbrake lever and handle of motorcycle, palm size, and wrist angle, the measurement of muscular volt and grip force of hand concerned with the grip span simulated from the handbrake of real motorcycle and the measurement of muscular volt of grasping handbrake by riding a motorcycle. The experimental Factors include three levels of grip span (5.5, 7.83 and 10.17 cm), three different wrist angles (0°, 10° and 30°) of grasping handbrake lever and the gender (male and female), then muscular data was collected on four different muscles (Flexor carpi ulnaris muscle, Brachioradialis muscle, Flexor carpi radialis muscle, Flexor digitorum sublimis muscle).

3.1 *Subjects*

The trials of measurement and the experiment were supported by 5 male and 5 female college students who are 20–22 years old.

3.2 *Independent experimental factors*

1. Gender: Male and female.
2. Grip span of handbrake: The grip span of the biggest mean value of power grip is 5.5 cm, and the grip span of 7.83 cm and 10.17 cm were measured by inside and middle grip span between brake lever and handle, and those positions are usually grasped by most people for all of 125 cc. motorcycles.
3. Wrist angle: The wrist angle of 0° is the neutral posture, as well as the stature of subjects is situated between 150 cm and 183 cm, and their wrist angles are situated between 10° and 30° when riding a motorcycle, therefore the research took these three angles to be the levels of wrist angle.

3.3 *Dependent experimental items*

The voltage value of muscles: Using Electromyography to collect muscular volt as grasping the grip strength dynamometer and actually grasping the handbrake of motorcycle, the experiment observed the movement of power grip of left-hand muscles which include (1) Flexor carpi ulnaris muscle (2) Brachioradialis muscle (3) Flexor carpi radialis muscle (4) Flexor digitorum sublimis muscle. The movement of muscles initiated the change of electric signal, and then the magnitude of grip force was calculated by the amount of muscular volt.

3.4 Experimental instrument

Vernier caliper (2) Goniometer (3) Grip strength dynamometer (4) Electromyography (EMG).

3.5 *Experimental procedure*

The experiments were respectively proceeded to measure different grip width of motorcycle handbrake in the parking lot of campus, and then those grip spans were used to measure the strength and muscular volt of hand muscles by grasping handbrake lever with the different wrist angles in laboratory. The subjects should be constrained by three grip spans, three wrist angles as well as four muscles of left forearm to process the trials.

3.5.1 *Experiment I*

The measurement of grip span of handbrake, subjects' stature, and the wrist angle of riding a motorcycle: (1) Investigating major brands of motorcycles, and then the measurement was taken place at the parking lot of campus. The measuring position of handbrake included inside, middle and outside perimeter and grip span, and three trials was taken for each measurement item. (2) Using the fine lace, vernier caliper (Figure 1) and ruler to be tools, and obtaining each grip span data of handbrake lever. (3) Five male and five female students volunteered to be the subjects to

Figure 1.   Vernier Caliper.

Figure 2.   Goniometer.

Figure 3.   Grip Strength Dynamometer (Reformed).

Figure 4.   Electromyography.

carry on the measurement of palm and fingers (thumb, index finger, middle finger, ring finger, little finger, hand length, hand thickness, palm length, hand width (4 fingers) and palm width by vernier caliper, and the average data is calculated by three trials. (4) The wrist angles were measured by goniometer (Figure 2) when the subjects sat on the 125cc.motorcycle of YAMAHA and grasped handle and handbrake lever simultaneously.

3.5.2   Experiment II
The power-grip simulation of grasping handbrake lever of motorcycle: (1) Reforming the grip strength dynamometer (Figure 3), because the original handle of grip strength dynamometer is different size apart from the handle of motorcycle and the size is insufficient, it must be transformed into the size approaching the real motorcycle. (2) Measuring the muscular volt, the subjects grasped the grip strength dynamometer with three wrist angles and three grip spans, and then the muscular volt was measured by EMG on the muscles of forearm. (3) The subjects grasped grip strength dynamometer by neutral posture ($0°$) and 5.5 cm grip span, and four trials within 10 seconds, then resting 2 minutes in order to avoid the fatigue of muscle. Next, the subject's wrist was fastened by curved plastic tube, and each subject extended the wrist angle to $10°$ and $30°$ and proceeded on the trials. (4) Adjusting the grip span of grip strength dynamometer to 7.83 cm and 10.17 cm, and repeating step (2) to (3).

3.5.3   Experiment III-A
The experiment of grip span of original handbrake lever of real motorcycle: (1) YAMAHA 125cc. motorcycle which is popular for male riders was chosen to be experimental tool. (2) The experiment was proceeded on by four male subjects who had participated in second experiment. (3) The measuring time was set up to be 10 seconds, the subjects rode on the motorcycle, and then stopped moving the vehicle immediately after twisting the accelerator and moving a second, then, there were 4 trials within 10 seconds for each subject (Figure 7 and Figure 8). (4) Then, calculating the strength of grasping the handbrake of motorcycle according to the muscular volt and grip force of last experiment.

3.5.4   Experiment III-B
The experiment of grasping reformed handbrake lever of real motorcycle: (1) Transforming the original design of handbrake lever of YAMAHA 125cc (Figure 5 and Figure 6) through the mechanic

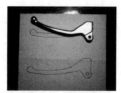

Figure 5. Original Handbrake Lever.

Figure 6. Reformed Handbrake Lever.

Figure 7. Experiment III.

Figure 8. Experiment III.

factory of campus. (2) Replacing the reformed handbrake lever to the original one. (3) Carrying on the step (3) of the experiment III A again.

4 RESULTS AND DISCUSSIONS

After the measurement of handbrake of motorcycles, the research discovered that the biggest size of grip span on the meddle segment of handbrake is 10.17 cm, and the grip span of inside segment of handbrake lever is 7.83 cm. According to previous research, the up limit of grasping span for feminine power grip is 5.5 cm (Gwoa, 2002). Therefore the three grip spans of handbrake were simulated by the experimental data aforementioned for second experiment. The masculine power of grip span at 5.5 cm could be up to 32.8 kg apart from the other data, and the feminine subjects could generate 15.4 kg, but at 10.17 cm the biggest male grip-power was only 14.2 kg and 8.5 kg for female. These results demonstrate that the grasping power is unable to increase for larger grip spans; therefore the present grip spans of the motorcycles manufactured are not suitable for all riders. (Table 1)

The table 2 shows that grip span significantly affects the grip force of male and female ($p < 0.05$), and the empirical datum identify that the grip span of 5.5 cm can save lots of effort (The grip power is the biggest). Table 3 presents that wrist angle is a significant factor for all three grip spans ($p < 0.05$), therefore the wrist angle can notably influence the force of grasping handbrake of motorcycle, and the experimental datum shows the power grip of hand can generate the biggest value for neutral posture. Table 4 demonstrates that the grip span and gender significantly influence the grip force; therefore the research suggests that the manufacturer of motorcycle must consider the gender difference, the magnitude of grip span and wrist angle when designing the products. Especially there are disparities between men and women, females should be suitable to more exquisite handbrake lever than males.

Table 5 expresses that the factors significantly influence the muscular volt at the wrist angle of 0 degree ($p < 0.05$), especially on Flexor Carpi radialis muscle, and also the factors significantly affect the muscular volt of Brachioradialis muscle at 10 degree ($p < 0.05$), but synthetically the factor of grip span dose not influence the muscular volt very much. Table 6 demonstrates that the factors significantly influence the muscular volt of Brachioradialis muscle regarding as feminine ($p < 0.05$), and the wrist angle is a significant factor on Flexor Carpi radialis muscle.

Table 7 demonstrates that the magnitude of grip span of reformed handbrake lever is smaller the original one (Figure 5 and Figure 6). Table 8 shows the average muscular volt of manipulating the original and reformed handbrake lever, and the research discovered that the reformed handbrake lever generates smaller muscular volt than the original one on Flexor carpi radialis and Flexor

Table 1. Value of power grip (kg).

Subjects	Wrist Angle	Grip Span 5.5 cm	7.83 cm	10.17 cm
MS1	0°	32.78	14.18	14.18
	10°	26.95	13.4	13.4
	30°	24.98	12.93	12.93
MS2	0°	19.93	11.93	11.93
	10°	17.85	12.63	12.63
	30°	19.08	9.35	9.35
MS3	0°	17.4	10.43	10.43
	10°	12.35	9.28	9.28
	30°	13.1	8.35	8.35
MS4	0°	22.05	6.03	6.03
	10°	20.95	7.78	7.78
	30°	21.8	8.25	8.25
MS5	0°	17.95	6.15	6.15
	10°	11.83	3.2	3.2
	30°	11.4	4.58	4.58
FS1	0°	12.45	10.43	8.48
	10°	8.6	8.63	2.7
	30°	9.93	2.88	6.3
FS2	0°	9.93	6.85	0
	10°	6.28	4.48	0
	30°	5.85	0	0
FS3	0°	10.35	0	0
	10°	0	0	0
	30°	2.85	0	0
FS4	0°	15.35	8.45	6.73
	10°	14.05	6.85	7.32
	30°	13.8	6	6.83
FS5	0°	12.58	8.2	7.03
	10°	15.1	8.68	4
	30°	11.18	9.18	6.65

Table 2. ANOVA of wrist angle and grip span.

Factors	Male P-value	Female P-value
Wrist Angle	0.3563	0.2368
Grip Span	**1.7E-06**	**0.0005**
Wrist Angle× Grip Space	0.8610	0.8448

Table 3. ANOVA of wrist angle and gender.

Factors	5.5 cm P-value	7.83 cm P-value	10.17 cm P-value
Wrist Angle	**7E-05**	**0.0020**	**0.0003**
Gender	0.2306	0.2654	0.7942
Wrist Angle× Gender	0.9882	0.9602	0.8636

Table 4. ANOVA of grip span and gender.

Factors	0° P-value	10° P-value	30° P-value
Grip Span	**0.0003**	**0.0005**	**0.0003**
Gender	**5E-05**	**0.0074**	**0.0009**
Grip Span × Gender	0.3374	0.7002	0.4206

Table 5. ANOVA of muscular volt of grip span and gender.

Muscles of Forearm	Factors	0° P-value	10° P-value	30° P-value
Flexor Carpi Ulnaris	Grip Span	**0.0076**	0.1159	**1E-04**
	Gender	0.1610	0.8491	0.2296
	Grip Span×Gender	0.9545	0.5998	0.9562
Brachioradialis	Grip Span	0.2707	**0.0241**	*
	Gender	**0.0354**	**4E-05**	*
	Grip Span×Gender	0.9505	0.0817	*
Flexor Carpi Radialis	Grip Span	**0.0138**	0.3142	0.5826
	Gender	**0.0057**	0.2383	0.0581
	Grip Span×Gender	0.9143	0.9484	0.9096
Flexor Digitorum Sublimis	Grip Span	*	*	0.5371
	Gender	*	*	**0.0222**
	Grip Span×Gender	*	*	0.9917

Table 6. ANOVA of muscular volt of grip span and wrist angle.

Muscles of Forearm	Factors	Male P-value	Female P-value
Flexor Carpi Ulnaris	Grip Span	0.5998	*
	Wrist Angle	0.2955	*
	Grip Span×Wrist Angle	0.6970	*
Brachioradialis	Grip Span	*	**0.0349**
	Wrist Angle	*	**2E-05**
	Grip Span×Wrist Angle	*	0.0536
Flexor Carpi Radialis	Grip Span	0.1620	0.1102
	Wrist Angle	**0.0103**	**0.0190**
	Grip Span×Wrist Angle	0.4097	0.4602
Flexor Digitorum Sublimis	Grip Span	0.5469	*
	Wrist Angle	**9E-05**	*
	Grip Span×Wrist Angle	0.9654	*

digitorum sublimis muscles. The results demonstrate that the improved handbrake lever just can decrease the demand of grip force generated by the part of forearm muscles.

5   CONCLUSION AND RECOMMENDATION

This research identified that the hand needs to produce more grip force while the grip span and wrist angle are increased, and also the grip span and wrist angle are decreased and the grip force is

Table 7. The grip space of handbrake of YAMAHA 125cc. (cm).

Handbrake Lever	Grip Space of Medium Segment	Grip Space of Posterior Segment
Original	9.33	11.33
Reformed	8.2	8.75

Table 8. Muscular Volt of Original and Reforming Handbrake (mV)

Subjects	Muscular Volt	Flexor Carpi Ulnaris	Brachio-Radialis	Flexor Carpi Radialis	Flexor Digitorum Sublimis
MS1	Original	2.797	1.174	1.361	3.245
	Reformed	**2.080**	**0.552**	**0.756**	**2.275**
MS2	Original	3.255	0.757	1.762	4.462
	Reformed	**2.907**	0.787	**0.714**	**3.303**
MS3	Original	2.984	0.834	1.091	4.206
	Reformed	3.523	1.188	**1.037**	4.462
MS4	Original	1.441	0.583	1.636	3.631
	Reformed	2.410	0.778	**0.485**	**3.488**

reduced. Therefore the range of grip span should be set up around 5.5 cm, and the wrist angle should be maintained 0° (neutral posture), then the grip power of grasping handbrake will be the smallest. The factor of gender is significant to the grip force, therefore the design of handbrake for men and women should probably be considered differently. Concerned with the forearm muscles, the factors influence the muscular volt of power grip significantly on Brachioradialis and Flexor Carpi radialis muscle, and these muscles should be trained frequently in order to manipulate handbrake lever easily. After installing the reformed handbrake lever to the motorcycle, the muscular volts were decreased on Flexor carpi radialis and Flexor digitorum sublimis muscle, and the research identified again that the strength of grasping the handbrake lever can be saved by small grip span when riding a motorcycle.

According to the results discovered, the research suggests that the grip span of handbrake around 5.5 cm is suitable for male and female riders, and the wrist posture should be neutral (0°). Furthermore, the height of handle bar of motorcycle should be adjustable regarding as the stature of rider, and then the wrist of rider can be maintained neutral posture. Finally, considering the differences of gender to the design of motorcycle, that will be more comfortable and safe than other considerations, and it is also the purpose of this research.

REFERENCES

Caldwell, L. S., Chaffin, D. D., Dukes-Dobos, F. N., Kroemer, K. H. E., Laubach, L. L., Snook, S. H. and Wasserman, D. E., 1974. A proposed standard procedure for static muscle strength testing. *American industrial Hygiene Association Journal*, Vol. 35, 4th, pp. 201–206.

Carey, Eilis J. and Gallwey, Timothy J., 2002. Evaluation of wrist posture, pace and exertion on discomfort. *International Journal of Industrial Ergonomics*, Vol. 29, pp. 85–94.

Chang, Shien-Young, 2002. Grip and key pinch strength: norms for 7 to 22 year-old students in Taiwan. *Tzu Chi Medical Journal*, Vol. 14, 4th, Department of Public Hygiene of Tzu Chi University, Hualien, pp. 241–252.

Chen, Jyh-Jian and Mao, Yan-Jie, 1998. The practice of ablock brake system of motorcycle. *Dayeh Journal*, 7(1), Dayeh University, Changhua.

Chen, Jyh-yong, 2002. Hand measurement data and grip force of civilian. Institute of occupational safety and health, http://www.iosh.gov.tw/seasnpap.htm.

Dempsey, P. G. and Ayoub, M. M., 1994. The influence of gender, grasp type, pinch width and wrist position on sustained pinch strength. *International Journal of Industrial Ergonomics,* Vol. 17, pp. 259–273.

Duque, J., Masset, D. and Malchaire, J., 1995. Evaluation hand grip force from EMG measurements. Applied Ergonomics, Vol. 26, 1st, pp. 61–66.

Guo, Shiao-Yuan, 2002. The effects of grip span, maximal wrist extension/flexion, and gloves on grip strength and time needed to reach different levels of exertion. Thesis, Graduate School of Logistic management of Defense Management College, Tauyang.

Hallbeck, M. S. and McMullin, D. L., 1993. Maximal power grasp and three-jaw chuck pinch force as a function of wrist position, age and glove type. *International Journal of Industrial Ergonomics,* Vol. 11, pp. 195–206.

Jung, M. C. and Hallbeck, M. S., 2002. The effect of wrist position, angular velocity, and exertion direction on simultaneous maximal grip force and wrist torque under the isokinetic conditions. *International Journal of Industrial Ergonomics,* Vol. 29, pp. 133–134.

Lai, Shao-Yu, Cheng, Yi-Lin, Tsai, Fong-He, Tsai, Li-Ching and Chen, Jyun-Ching, (2002), The measurement of grip strength. Project, Department of Industrial Engineering and Management of Chao Yang University of Technology, Taichung.

Lin, Ching-Chiuan and Liu, De-Tsai, 2000. The study of factors influencing the grip force. *Technology Journal,* Department of Industrial Engineering and Management of Kun Shan College of Technology, Tainain, pp 511–516.

Linscheid, R. L. and Chao, E. Y., 1973. Biomechanical assessment of finger function in prosthetic joint design. *Orthop Clinics North America,* Vol. 14, 2nd, pp. 317–330.

McArdle, W. D., Katch, F. I. and Katch, V. L., 1986. Exercise Physiology: Energy. *Nutrition and Human Performance* (2ed). Lea and Febiger, Philadelphia, PA.

Shih, Yu-Chuan, 1995. The study of factors of grip interface to the influence of difference threshold of hand exertion and weight. Dissertation, Graduate School of Industrial Engineering of National Tsing Hua University, Shintsu.

Wang, Shian-Ling, 1999. The effects of muscular load of hand when carrying on gripping and twisting movement for the grip diameter. Thesis, Graduate School of Management of National Taiwan University Technology, Taipei.

Ergonomic Trends from the East – Kumashiro (ed)
© *2010 Taylor & Francis Group, London, ISBN 978-0-415-88178-4*

Design of mobile phone interfaces for taiwanese older adults: Needs, performance and preference

Dyi-Yih Michael Lin, Chi-Nan Cheng & Chun-Jung Chen
Department of Industrial Engineering and Management, I-Shou University, Kaohsiung, Taiwan

1 INTRODUCTION

Population aging has been a major trend in many societies. Considering the prevalent use of digital technology nowadays, the increasing number of older users has given rise to a new challenge for ergonomics researchers. This concern mainly comes from the fact that aging is normally associated with reduced cognitive abilities (Park, 2000) and therefore older adults are generally a slower information processor (Salthouse, 1996). The literature has also empirically demonstrates that older adults are disadvantaged with respect to perceptual acuity (Lindenberger & Baltes, 1994), working memory (Craik & Byrd, 1982) and focused attention (Hasher *et al.*, 1991).

These disadvantages have prevented the elderly from successful interaction with modern information devices. For example, older adults have been shown to require longer processing time but to achieve less hit rates when interacting with hypermedia-based systems (Lin & Hsieh, 2006; Lin, 2004; Lin, 2003). While the majority of the aging and human-computer interface studies have devoted to the issues related to computer-based activities, design of digital mobile devices that incorporate the unique needs of the older population has yet reached a consensus database. As digital mobile devices have become a necessity in our daily life, the present study aimed at investigating the elements that would make mobile phones a friendly and effective interaction media from the older adult's point of view.

According to the aforementioned purpose, the present study first conducted a needs analysis, followed by a performance experiment in which simulated mobile interfaces developed from the results of the needs analysis were tested. Finally, subjective preference over the interfaces was assessed to cross examine the consistency between the user experience and associated performance.

2 NEEDS ANALYSIS

2.1 *Methodology*

The needs analysis was carried out in the form of questionnaire, in which 29 individuals aged over 65 were randomly sampled from a local Grey Hair Club to participate in the survey. The local government-run club was chosen as the subject pool for the present study because its membership was demographically representative in terms of education, economy, and social status. The questionnaire consisted of three parts which respectively collected personal profile, user needs in terms of mobile phones functions and user needs in terms of interaction modes. The functional analysis comprised 11 questions spelled to identify those system management functions older users consider indispensable given their life style. The interaction analysis comprised 8 questions phrased to characterize those interface features older users consider most facilitating given their physical and psychological difficulty (e.g., visual acuity and memory).

2.2 *Results*

The functional data showed that among the various system functions, Menu Settings, Phone Directory, and Communication Records are the most needed items (20%, 19% and 18%, respectively), followed by General Application Tools (12%; e.g., notepad, calendar, calculator, etc.),

Entertainment tools (10%; e.g., album, internet, etc.), Email (8%), Camera (6%), Music (4%) and Games (3%). It appears that despite the prevalence of the Internet and multimedia technology, the older Taiwanese adults seem to consider mobile phones more as simply a telecommunication tool than as an all-purpose move-around platform where emails and audio-visual entertainment are readily available.

As far as the needs with respect to interaction modes, 25 out of the 29 participants agreed that the majority of the mobile phones in the market are not older user friendly ($\chi^2 = 15.21, p < 0.05$). 17 participant considered the display size should have been larger for better visual acuity ($\chi^2 = 27.69$, $p < 0.05$). 21 participant suggested that incorporation of auditory outputs would enhance the ability to recall ($\chi^2 = 5.99, p < 0.05$). 17 out of the 29 participants preferred rectangular-shape keypad to those in round shape or other ($\chi^2 = 13.52, p < 0.05$). Surprisingly, none of the surveyed participants considered icon-based interface an effective interaction mode, with a significant portion ($\chi^2 = 17.03, p < 0.05$) chose either text alone display (18) or text-plus-graph (11) display. Also surprisingly is that an overwhelming 26 preferred traditional key pressing to the PDA-like full display touch screen when the participants were asked to select the easiest way to control the interaction ($\chi^2 = 18.24, p < 0.05$).

3 PERFORMANCE AND PREFERENCE ANALYSIS

3.1 Experimental design

The present experiment was designed to objectively assess how the older adult could fare over different interfaces with respect to a variety of mobile phone tasks derived from the results of the needs analysis. A single-factor experiment was conducted in which interface was the independent variable with full-display touch screen (TS), full-display touch screen plus auditory output (TSA), and key-pad display (KP) being the three levels of the treatment. A within-subject design was adopted in order to minimize the potential confounding from individual differences. Four different tasks were designed for each of the three interfaces and the elapsed time to successfully complete each task was defined as the performance measure. To counterbalance possible carry-over effect from the repeated measurement, the order in which the interface treatment was presented to the subject was randomized.

3.2 Subjects

A new set of 20 subjects were randomly recruited from the same club. The subjects were aged from 60 to 78 with a mean of 67.9 and standard deviation of 4.60. All the subjects were familiar with basic computer usage such as Windows systems and mouse operations. Some of the subjects reported vision difficulty prior to the experiment but these problems were resolved by providing them with corrected eye glasses.

3.3 Experimental material/systems

Three simulated mobile phone interfaces according to the treatment definitions were developed by an interface prototyping system named RAPID Plus (Version 6.6). The three simulated interfaces allow the subject to perform the designated tasks by clicking the mouse on the interface components where appropriate. Figure 1 and figure 2 illustrate snapshots of the TS interaction and the TSA interaction respectively. Figure 3 illustrates the interaction carried on the KP display.

3.4 Procedures

The 20 subjects were first briefed and given warm-up exercises. Then each subject was instructed to undergo on the three interfaces four different mobile phone tasks, including dialling, tow functional setting, and emergency call that were presented in random order. A 3 three-minute break was offered to each subject in the transition of the three interfaces. The time spent in successfully completing the tasks was recorded for subsequent analysis. Upon finishing the performance session, the subjects were asked to rank their preference over the three interfaces on a 1 to 3 scale with 1 representing the most preferred one.

Figure 1. Task interaction on the full-display touch screen.

Figure 2. Task interaction on the full-display touch screen plus auditory output.

Figure 3. Task interaction on the key-pad display.

3.5 *Results*

3.5.1 *Performance results*

Table 1 shows the means and (deviations) of the performance time. ANOVA results indicated that the time differences among the three interfaces are significant ($F[2, 38] = 129.44, p < 0.001$). The posy-hoc analysis employing the Tukey test revealed that the three pair-wise contrasts are all the sources leading to the significant difference. The TSA interface resulted in the shortest completion of the tasks as compare to the TS counterpart (29.97 vs. 36.37, $T = -6.44, p < 0.001$). So did the TSA interface when it was contrasted to the KP interface (29.97 vs. 42.01, $T = -15.99, p < 0.001$). The TS interface also significantly outperformed the KP interface (36.37 vs. 42.01, $T = -9.55$, $p < 0.001$).

3.5.2 *Preference results*

The preference data were analyzed by the rank sum Friedman test in terms of the extent to which the subject enjoy the interaction. The results indicated a pattern consistent to the performance data where the touch screen plus auditory output interface was significantly preferred to the touch screen alone interface and the key pad interface with the latter one being the least desired.

Table 1. Means and (deviations) of the completion time with respect to the three interfaces (in seconds).

Interface	Touch Screen	Touch Screen Plus Audio	Key Pad
	36.37	29.97	42.01
	(5.87)	(4.38)	(5.50)

4 CONCLUSIONS

Population aging has been a major phenomenon in Taiwan where the use of digital technology has been pervasive. As a daily necessity nowadays, mobile phones are of no exception that imposes the usage challenge for older adults. The present study thus conducted a series of investigation including needs analysis, performance assessment and preference ranking in order to design mobile phone interfaces friendly for the Taiwanese elderly. A sample of 29 Taiwanese aged over 65 was surveyed in terms of the types of functions and services they actually need with respect to the current mainstream mobile phone providers. The results indicated that only the function setting, communication records, and phone directly are the most cited needs in mobile phone use. Three simulated mobile systems were developed according to the functionality identified from the needs analysis. The systems enabled these functions to be interacted by touch screen, touch-screen plus auditory output, and key pad display respectively. A single within-group factor experiment with the three interaction modes was conducted over a new set of 20 subjects aged over 60. The subject's operation time (normalized) in completing designated tasks was evaluated. The ANOVA and the subsequent post-hoc Tukey tests showed the differences in interaction performance between the three interfaces were significant, with the touch-screen plus auditory message, touch-screen alone, and menu/button panel finishing the tasks in descending order. A questionnaire followed in which the subject was asked to rank their preference on the display components of the three different interfaces. Freidman rank-sum tests indicated a data pattern consistent with the performance analysis. The subject significantly favored the touch-screen plus auditory message most, with the menu/button panel receiving the least weight. The ease to learn, to remember, and to operate are the top major reasons that outstands the dual-modality, button-free interface over the other two counterparts.

It is suggested that older adults appear to have their unique needs as contrast to those of ordinary people. Mobile phone designers should seriously take older adults' uniqueness into account. Provision of multimodal interaction (visual and auditory) equipped with direct manipulation on the information items (touch screen) should receive higher weights in order for a mobile phone to be accessible and friendly for users of the older populations.

REFERENCES

Craik, F.I.M. and Byrd, M., 1982, Aging and cognitive deficits: The role of attentional resources. In *Aging and cognitive processing*, edited by Craik, F.I.M. and Trehub, S. (Plenum Press), pp. 191–211.

Hasher, L., Stoltzfus, E.R., Zacks, R.T. and Rypma, B., 1991, Age and inhibition, *Journal of Experimental Psychology: Learning, Memory and Cognition, 17*, pp. 163–169.

Lin, D-Y. M. and Hsieh, C-T. J., 2006, The role of multimedia in training the elderly to acquire operational skills of a digital camera, Gerontechnology, *5*(2), pp. 68–77.

Lin, D-Y. M., 2004, Evaluating older adults' retention in hypertext perusal: Impacts of presentation media as a function of text topology, *Computers in Human Behavior, 20*(4), pp. 491–503.

Lin, D-Y. M., 2003, Age difference in the performance of hypertext perusal as a function of text topology, *Behaviour & Information Technology, 22*(4), pp. 219–226.

Lindenberger, U. and Baltes, P.B., 1994, Sensory functioning and intelligence in old age: A strong connection, *Psychology and Aging, 9*, pp. 339–355.

Park, D.C., 2000, The basic mechanisms accounting for age-related decline in cognitive function. In *Cognitive Aging: A Primer,* edited by Park, D.C. and Schwarz, N. (Psychology Press), pp. 3–21.

Salthouse, T.A., 1996, The processing-speed theory of adult age differences in cognition, *Psychological Review, 103*, pp. 403–428.

Chapter 16   Ergonomics in manufacturing

Ergonomic Trends from the East – Kumashiro (ed)
© 2010 Taylor & Francis Group, London, ISBN 978-0-415-88178-4

Effects of bright-pixel defect on the customer perception of TFT-LCD image quality

Sheau-Farn Max Liang & Pei-Chen Lin
Department of Industrial Engineering and Management, National Taipei University of Technology, Taiwan, ROC

ABSTRACT: Current standards regarding image quality of flat panel displays are usually lack of the involvement of customers. Therefore, the perceived image quality from customers was the focus of this study. While the classification of pixel faults of ISO 13406-2 standard was reviewed, an eye-tracking experiment was conducted to understand the distribution of gaze duration on the screen. Two 17-inch LCD computer monitors were used for displaying different levels of image quality in terms of various numbers and locations of bright-pixel defects to 30 subjects. Subjects made their preference comparisons between the two screens through a questionnaire. Results showed that the perceived image quality was decreased as the number of defects was increased, and as the defects were near the central area of the screen. A partial interaction effect was found between the number and the location of defects. The findings of this study reveal that the location of defects should be taken into account along with the number of defects for the establishment of ergonomic image-quality requirements for the design and evaluation of flat panel displays.

Keywords: TFT-LCD, Customer Requirements, Image Quality, ISO 13406-2, Gaze Tracking

1 INTRODUCTION

Thin Film Transistor-Liquid Crystal Display (TFT-LCD) is the mainstream sector of the Flat Panel Display (FPD) industry. Most of TFT-LCD panels are manufactured by East Asian countries, such as Taiwan, Korea, Japan and China. To ensure the image quality of TFT-LCD panels, several standards have been proposed for establishing quality requirements and specifying measurement methods (see Downen, 2006 and Stewart, 2000 for a review). However, these standards were usually made through a committee and without the involvement of customers. As a result, these technology-oriented and manufacturing-driven standards did not address much about customer requirements. Since customers are the end users who make purchase decisions, it is necessary to listen to the voices of customers and understand their perception about the image quality of TFT-LCD panels.

Pixel faults are classified and defined with the fault types and fault classes in the ISO 13406-2 standard. The classification implies that the less the defects are, the better the image quality is. However, it is reasonable to consider that the location of defects is another factor to affect the image quality. Users may pay more attention to the central area of the screen than the peripheral area. Posent (1980) suggested that attention is associated with gaze location and duration. Most of gaze tracking studies are about reading or information search performance (e.g., Backs and Walrath, 1992; Findlay and Gilchrist, 1998; Rayner, 1998). Hence, a gaze tracking experiment especially for this study is necessary. A screen was divided into three concentric areas with the same area size and labeled as Area 1, 2, and 3 shown as in Figure 1.

A gaze tracking system was applied to measure gaze durations of 30 subjects. Results showed that the gaze durations were significantly decreased from Area 1 to Area 2, and from Area 2 to Area 3. We concluded that subjects paid different levels of attention to these three areas and these three areas of defects were used for a further experiment. With the review of ISO 13406-2 standard and the results of the gaze tracking experiment, three hypotheses were proposed as follows:

Hypothesis 1: The number of bright-pixel defects affects the perception of the image quality of TFT-LCD panels.

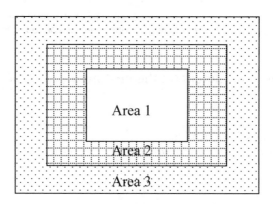

Figure 1.   Three Levels of the Location of Defects.

Table 1.   Labels of Nine Image Quality Conditions

| | | Location of Defects | | |
		Area 1	Area 2	Area 3
Number of Defects	Level X	X1	X2	X3
	Level Y	Y1	Y2	Y3
	Level Z	Z1	Z2	Z3

Hypothesis 2:   The location of bright-pixel defects affects the perception of the image quality of TFT-LCD panels.

Hypothesis 3:   The interaction effect exists between the number and location of bright-pixel defects on the perception of the image quality of TFT-LCD panels.

2   METHODS

The number of defects and the location of defects were the two independent variables. While the number of defects was designed into three levels (Level X, Y, and Z) according to the ISO 13406-2 standard, the screen was divided into three defect location areas (Area 1, 2, and 3) based on results from the gaze tracking experiment. Since a 17-inch LCD computer monitor with resolution 1280×1024 was used, the total number of pixels on the screen was about 1.3 million. We then defined that the number of bright-pixel defects is 2, 4, and 6 in Level X, Level Y, and Level Z, respectively. Based on the combination of the levels of two independent variables, the image quality of a TFT-LCD panel was categorized into 9 conditions. Each condition was labeled in Table 1.

Two 17-inch LCD computer monitors labeled as Screen A and B were used to display different conditions of the image quality in terms of different levels of the number and location of defects. Subjects made 36 (C_2^9) comparisons between these two screens and rated their preferences about the image quality of the screens through a series of six-level bipolar scale questions.

3   RESULTS

Hypothesis 1 was tested by a one-sample T-test among Level X, Y, and Z within the same areas of defect location. All the means were significantly different to zero at p-value <0.001 level. That is, the less the number of defects was, the better the image quality was. Thus, Hypothesis 1 was supported.

Hypothesis 2 was tested by a one-sample T-test among Area 1, 2, and 3 with the same number of defects. All the means were significantly different to zero at p-value <0.001 level. That is, the more central the defects located on, the worse the image quality was. Thus, Hypothesis 2 was supported.

Hypothesis 3 was tested by a one-sample T-test among the different numbers and locations of defects. All the means were significantly different to zero at p-value <0.05 level except the comparisons between X1 and Y3, and between X1 and Z3. That is, no differences in preference were found between a screen with 2 defects on the central area (Area 1) and screens with 4 and 6 defects on the peripheral area (Area 3). Thus, a weak interaction effect was found. Furthermore, the means of the comparisons between Y1 and Z3, and Y2 and Z3 were negatives. That is, subjects preferred a screen with 6 defects on the peripheral area (Area 3) to a screen with 4 defects on the central area (Area 1) and the middle area (Area 2). Thus, a strong interaction effect was found. In summary, the effects of the number and the location of defects on the perceived image quality were partially interacted.

4   CONCLUSION

Overall, results showed that the perceived image quality was decreased as the number of defects was increased, and as the defects were near the central area of the screen. A partial interaction effect was found between the number and the location of the defects. The perceived image quality of some screens with more number of defects but located on the peripheral area were better than the ones with less number of defects but located on the central area. The findings of this study reveal that the location of defects should be taken into account along with the number of defects for the establishment of ergonomic image-quality requirements for the design and evaluation of flat panel displays.

ACKNOWLEDGEMENTS

This research was funded by the National Science Council of Taiwan (NSC 95-2221-E-027-081-MY3).

REFERENCES

Backs, R. W. and Walrath, L. C. (1992). Eye movement and pupillary response indices of mental workload during visual search of symbolic displays, *Applied Ergonomics*, 23, 4, 243–54.
Downen, P. (2006). A closer look at flat-panel-display measurement standards and trends, *Information Display*, 1/06, 16–21.
Findlay, J. M. and Gilchrist, I. D. (1998). Eye Guidance and Visual Search, In G. Underwood (Ed.), *Eye Guidance in Reading and Scene Perception*, pp. 295–312.
ISO (International Organization for Standardization) (2001). ISO 13406-2: *Ergonomics Requirements for Work with Visual Displays based on Flat Panels – Part 2: Ergonomic Requirements for Flat Panel Displays*. ISO, Geneva.
Posner, M. I. (1980). Orienting of attention, *Quarterly Journal of Experimental Psychology*, 32, 1, 3–25.
Rayner, K. (1998). Eye movements in reading and information processing: 20 years of research, *Psychological Bulletin*, 124, 3, 372–422.
Stewart, T. (2000). Ergonomics user interface standards: Are they more trouble than they are worth? *Ergonomics*, 43, 7, 1030–1044.

Ergonomic Trends from the East – Kumashiro (ed)
© 2010 Taylor & Francis Group, London, ISBN 978-0-415-88178-4

Progress of finger skills on forming process for ceramic beginners

Tadao Makizuka
School of Humanity-Oriented Science and Engineering, Kinki University, Iizuka, Japan

1 INTRODUCTION

With the progress of high-technology industries, the advanced skills of craftsmen will disappear from the many factories of Japan. This craftsmanship may be impossible to revive once lost. It is an important problem to be transferred the refined, accumulation of the craftsmanship for many years experiences (Mori, et al., 2003, Kimotsuki, 1988). Accordingly, this study was planned as a fundamental research on the transference of finger skills.

Finger skills represent one of the typical abilities of a ceramic artist. Finger movements in the forming process consist of many important elements. Estimating the acquirement levels of these skills is useful in assessing the progress of the pupil in learning the forming process from the teacher (ceramic craftsman).

In this report, we sought to quantitatively analyze the development of finger skills of a ceramic arts student, during the initial stages of learning.

2 MATERIALS AND METHODS

The 2 subjects (A and B) were normal male students, aged 21 years old, who had the same desire to study ceramic art, but were inexperienced. Subject A was an ingenious and subject B was in awkward for his scrupulousness, accuracy and infinitesimal comparatively. Each subject learnt ceramic art separately from a professional ceramist (35-year-old male) who has been teaching at a ceramic art school.

Subject A had learnt first, followed by subject B. They studied the forming process for a tea mug using an electric wheel. Each subject studied the forming skills for 16 days over a period of 2 months, with random intervals. After the learning period ended, an examination was conducted to assess the mug-forming skills.

The examination and the classes conducted during the learning period were recorded in VTR. The understanding of the forming process, refined motions, and control of motion space were analyzed. The ameliorations of their works were evaluated. Their imaginative abilities were assessed through interviews. The total skills were integrated using scores for the various parameters.

The mental load of subjects and the activities of the central nervous system during the learning period were analyzed using a portable brain wave recorder. Electroencephalograms (EEGs) of the subjects were recorded, at the frontal (Fz), central (Cz), occipital (Oz), and left temporal regions (T3), which were induced by a monopole (A1 + 2) of the 10–20 system (Kadobayashi, 1983) on the 1st, 2nd, 4th, 8th, and 16th day of the exercise.

3 RESULTS AND DISCUSSION

3.1 Technical progress

The coaching method included verbal advice and practical help at any time by the teacher. Subject B made various shapes in ceramic until the 5th exercise and spent the rest of the time completely on mug production.

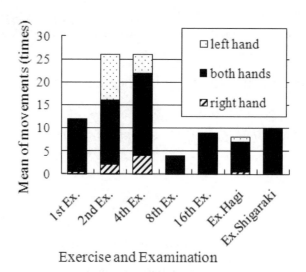

Figure 1.   Progress in movements of both hands in subject B.

Table 1.   Results of the examination after 16 exercises.

Items Conditions	Subject A		Subject B	
	Hagi	Shigaraki.	Hagi	Shigaraki
Completion rate	100 %	100 %	55.6 %	100 %
Height (mm)	84.3 ± 1.9	87.6 ± 0.9	82.9 ± 3.2	84.8 ± 1.7
Width (mm)	63.6 ± 0.5	65.3 ± 1.4	67.1 ± 1.7	67.4 ± 1.3
Thickness (mm)	3.0 ± 0.5	3.5 ± 0.5	6.0 ± 0.8	7.3 ± 0.8

In the first half, many right-handed and standing movements while making the body of the mug, and the advice of the teacher were mostly on finger motions. Later, he used a standard procedure, as did the teacher, using both hands simultaneously (Figure 1) and it was possible to control motion space.

The operating procedure for his forming process was fixed. The delay time was decreased. The verbal advices from the teacher for mental attitude were increased. The standing posture of subject B disappeared.

During the examination (Table 1) the completion rate of an unaccustomed soil (Hagi) was lower than that of the accustomed ones (Shigaraki) using in this experiments (Naitoh, 1979). It was suggested that the subject had acquired the fundamental abilities, but lacked experience (Mori, 1995).

Accordingly, the first half of the course involved acquiring the fundamental finger skills (configuring the image and making adjustment) and the process of making a mug. The latter half was about acquiring the advanced skills of revision in order to make a mug of a given form (Figure 2).

Subject A also made ceramics of various shapes, but moved on to mug production earlier than subject B. In all experiments, subject A used both hands simultaneously while making the body of the mug. His right and left hand movements and both hands operations appeared proportional. His handling space become fixed as the experiments advanced. No standing posture of subject A was apparent. The operating procedure of his forming process was fixed, from the beginning to the end. No delay time appeared.

Subject A received more verbal advices than subject B from the master.

During the examination (Table 1), subject A finished making 5 mugs of 2 kinds (Hagi, unaccustomed and Shigaraki, accustomed). It was suggested that the subject A had acquired fundamental

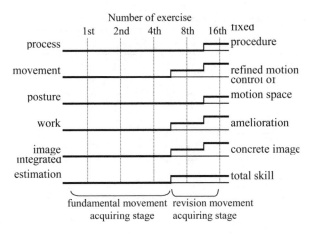

Figure 2. Integrated estimation of finger skills of subject B in the fundamental stage.

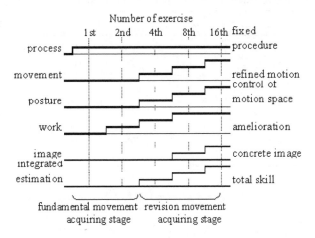

Figure 3. Integrated estimation of fundamental finger skills in Subject A.

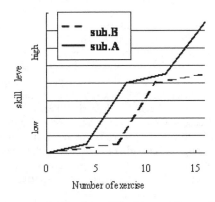

Figure 4. Comparison of the integrated estimations of fundamental finger skills of both the Subjects.

abilities better than subject B. As with subject B, 2 stages of learning in subject A's finger motions were recognized. Subject A was quick to learn as compared to subject B (Figure 3).

The integrated estimations of the development of the fundamental stages of finger skills in both the subjects are shown in Figure 4.

347

Table 2-1. Changes in EEG powers before and after exercise in subject B.

Locations	Fz				Cz				Oz				T3			
No. of exercise	1	2	4	16	1	2	4	16	1	2	4	16	1	2	4	16
Total power	+	+	+	−	+	+	+	−	+	+	+	−	+	+	+	+
Delta' power	+	−	−	+	+	−	−	+	+	+	−	−	+	+	−	−
Theta power	+	+	−	−	+	+	−	−	+	+	−	−	+	+	−	−
Alpha power	+	+	+	+	+	+	+	+	+	+	+	−	+	+	+	+
Alpha1 power	+	+	+	+	+	+	+	+	+	+	+	+	+	+	+	+
Alpha2 power	+	+	+	+	+	+	+	+	+	+	+	−	+	+	+	+
Beta' power	−	−	+	−	−	+	−	−	−	+	−	−	−	+	+	+
Beta1 power	−	−	+	−	−	+	+	−	−	−	−	−	−	−	+	+
Beta2 power	−	−	−	−	−	−	−	−	−	+	−	−	−	+	+	+

Table 2-2. Changes in EEG relative(%) powers before and after exercise of subject B.

Locations	Fz				Cz				Oz				T3			
No. of exercise	1	2	4	16	1	2	4	16	1	2	4	16	1	2	4	16
% of delta'	+	+	−	−	+	+	−	−	+	+	−	−	+	+	−	−
% of theta	+	−	−	+	+	−	+	+	+	−	+	+	+	+	−	−
% of alpha	+	−	+	+	+	−	+	+	+	+	+	+	+	−	+	−
% of alpha1	+	+	+	+	+	+	+	+	+	+	+	+	+	+	+	+
% of alpha2	+	−	+	+	+	−	+	+	+	−	+	−	+	−	+	−
% of beta'	−	−	−	−	−	−	−	−	−	−	−	+	−	−	−	+
% of beta1	−	−	+	+	−	−	−	+	−	−	−	−	−	−	−	+
% of beta2	−	−	−	−	−	−	−	−	−	−	+	−	−	−	+	+

3.2 Mental loads and images

The mental loads of subjects during an exercise were analyzed by recording the activities of the central nervous system.

EEGs of the subjects showed specific patterns according to their consciousness levels. The increase in alpha power on the EEG after exercise indicated high wakefulness of the subjects and the moderate psycho-physiological load (Ohkuma, 2006, Andreassi, 1985 and Hisahara, 1981).

A tendency for slow wave activities to increase after exercise in the early stages has been recognized. It was considered to be associated with a feeling of unpleasantness (Knott, 1976). The experimental apparatus, restrictions on the experiment schedule, and fatigue in acquiring the fundamental movement skills were affected (Andreassi, 1985).

It is considered that an increase in the fast beta power component in beginner subjects continues to influence the mental load, with higher nervous activity required for assembling the motion procedure and imagining the mug form (Tsuji, et al., 1988, Kadobayashi, et al., 1983).

The slow wave and fast wave activities in motor and somatic sensory areas of the subjects during the middle of their learning period showed the process of experience and of acquiring the basic finger movements.

The beta activation in the auditory areas at the beginning of the exercises indicated activity related to revising and adjusting the ceramic piece to the subject's satisfaction.

Two mental phases in learning the forming process were also recognized.

In the beginning, unpleasant feelings and restrictions due to memorizing the sensory and motor skills, which had not been experienced before, influenced the EEGs, with slow wave activation.

Concentrating on devising movement with central nervous system involvement influenced the latter half, with fast wave activation.

4 CONCLUSIONS

The ability to learn the forming process of ceramic mugs was assessed 16 times in highly motivated, normal male students, using moderate, psycho-physiological exercises under the supervision of a professional ceramist.

In the learning of finger motions, 2 stages were recognized. The first involved gaining experience and acquiring basic finger movements. The second involved revising and adjusting the ceramic piece to the subject's satisfaction. The second stage appeared after the image of the ceramic piece was fixed.

Obvious differences between the 2 individuals in learning and acquiring the finger skills were also recognized.

REFERENCES

Andreassi, J.L., 1985, In Psychophysiology, Nakanishiya shuppan, pp. 46–50, (Japan).
Hisahara, K. et al., 1981, In Japanese Journal of EEG and EMG, Vol.9, pp.95, (Japan).
Kadobayashi, I. et al., 1983, In Noha, KINPODO Inc., pp. 6–88, (Japan).
Kimotsuki,K., 1988, In Japanese Journal of Ergonomics, Vol. 24, pp. 343, (Japan).
Knott, J.R., 1976, In Electroencephalography and Clinical Neurophysiology, 6A, 69.
Mori, K. et al., 1995. In Japanese Journal of Ergonomics, Vol. 31, pp. 131–139, (Japan).
Mori, K. et al., 2003, In Rodo-no-Kagaku, Vol.58(5), pp. 261–276, (Japan).
Naitoh,T., 1979, In Shintei-Kotoji-no-Kagaku, Heibunsha, pp. 98–127, (Japan).
Ohkuma, T., 2006, In Noha-Handoku Nyumon , Igaku-shoin Ltd., pp. 468, (Japan).
Tsuji, Y. et al., 1988, In Electroencephalography and Clinical Neurophysiology, 70, 110–117.

Ergonomic Trends from the East – Kumashiro (ed)
© 2010 Taylor & Francis Group, London, ISBN 978-0-415-88178-4

Prapen: Sustained site of blacksmith profession in households

Ketut Tirtayasa[1]
Postgraduate Program of Ergonomics and Work Physiology,
Udayana University, Denpasar, Bali, Indonesia

1 INTRODUCTION

Bali is one of the many islands of Indonesia with a population of around four million, with a majority (92.3%) of Hindus (Wikipedia, 2007). In Bali, there exist certain segments of the community called *soroh* or clans such as *brahmana, dewa, gusti, pasek, pande* etc. A person can be identified as regard of his *soroh* from his name such as Ida Bagus Jinar (signifying a brahmana), I Nyoman Pasek Budiman (pasek clan), Pande Suteja (pande clan), etc. However, it is often difficult to know a person's *soroh* or clan because nowadays many people do not put their *soroh* in their names explicitly, as for example I Wayan Budiman, I Ketut Indra, etc.

A person's name can often be connected with a specific profession that has been descended from generation to generation. For example, the *soroh* of *pande* that has been well known in Indonesia and Bali in particular signifies the profession of blacksmith with a main task of producing household tools or weapons made of iron (Anonym, Haryono, 2006). The work or skill related to this particular profession is called *memande* in the Balinese language.

The articles made by this particular *soroh* consist of weapons or armouries like spears, kris, bullets etc. Also are made household and working tools such as pacols or hoes, sickles, and other carpentry tools. Other than that, *pande* clan often also make jewellery made of silver, gold, and brass. In other word, the *pande soroh* or clan is concerned with the work of producing goods or articles that are made of metal.

2 PRAPEN AS WORKPLACE

The workplace where a person or persons work for producing metal articles is called *prapen*; a contracted form of the word *perapian* (derived from the word *api* = *fire*). The width of *prapen* should be spacious enough to accommodate several working tools. Among the tools are *pengububan* for blowing air to keep the fire burning. With the heat of the fire the iron is made soft in order to be able to shape it according to the pande's design. The iron base used for shaping the inflamed iron by beating it repeatedly by a hammer is called *paron*. While *palungan* is the big water container made of sandstone into which the already beaten hot iron is immerged to cool down the heat. Other tools needed for the work include an iron hammer and holder.

The workload of a *pande* is generally hard. For the work, significant muscle strength is necessary for beating the inflamed iron repeatedly in great speed in order to avoid the inflamed iron to quickly become cool and hard. In addition, the hard workload is further made greater by improper body working posture (Tirtayasa, et al). The working environment is hot as it is exposed to heat of fire in the *prapen*.

In the old *prapen* sites, there exist workplaces that are designed not ergonomically, which result in poor working environment. In this situation, we can still see *pande* workers carry out their profession in a squatting position (Figure 1).

Since around 1980 when electricity was first brought into the villages, the traditional *pengububan* used to blow air onto the fire has been replaced by the modern electric blower.

---

[1] Corresponding author: tirtayasa@dps.centrin.net.id

Figure 1. Prapen with sitting or squatting working position.

Figure 2. Prapen with ergonomic working posture. A: working floor lowered. B: working pad elevated.

3 CHANGE OF PEROFESSION

Along with the increase of population and hardship of life, people who previously have *memande* as their profession may change it to another profession. Becoming a government or private corporate servicemen or running their own businesses seems to be more promising. Fewer and fewer of the young generation hold *memande* as their profession as their ancestors did. On the other hand, *prapen* or workplaces in households who maintain *memande* as their profession are now becoming more ergonomic thanks to the information or guidance given by certain group of ergonomists of Udayana University in the 1980s. In the several places and in particular in Batusanghiang village in Tabanan regency where the education was carried out, the working environment and position of the *pande* have become more ergonomic by making a special hole and lowering the working platform (Figure 2A) to make a more convenient working posture or by elevating the working pad to change the working posture to a standing position (Figure 2B).

4 RESERVATION

Whether still maintaining memande as their profession or having shifted to other professions, *pande* families always respect their ancestors' old profession as *pande*. Therefore, *memande* is now more as an attribute of profession of the *pande* clan. As this attribute is a pride to this clan, it is maintained

352

Figure 3.   Prapen site with its tools.

Figure 4.   A prapen with beautiful, high quality building.

in certain ways. The pride attached to *memande* profession is expressed as maintaining the *prapen* sites in households as workplaces with all the necessary tools for working with metal. The sizes of the tools may be the same as the original old ones or now more frequently smaller. As shown in Figure 3, a *prapen* of the size of 180 × 180 cm is equipped with four essential tools namely *pengububan, paron, palungan* and *palu* (hammer). Although now most of the *prapen* are not in use any more, their sites in the households are still preserved well.

The sites may vary as regard with their quality from very modest buildings to elaborate or beautiful ones with terracotta roof and finely carved and gilded wooden poles (Figure 4).

Moreover, people of the *pande* clan consider the *prapen* sites in the households sacred as evidenced by a specific ritual ceremony periodically and regularly carried out i.e. in every special day of *Tumpek Landep* according to the Balinese calendar. In conclusion, although now most *prapen* are not being used, they are still made or preserved well in order to keep the old profession of *memande* sustained.

5 DISCUSSION

To all Balinese Hindus, the factors of ancestries or decencies have a very significant value in their life. We have seen quite many instances in which families who have not yet found their root of decencies would make any efforts possible to find their origins or ancestors. The ultimate goal of such efforts is the creation of peace of mind and emotional feeling of oneness or completeness of the relevant families.

From the socio-cultural viewpoint, it is clear that the identity of the *pande* clan is the *memande* (blacksmith) profession of their ancestors, involving the making of articles or goods made of iron. As mentioned above, although now most *pande* families do not hold blacksmith profession any more, the *prapen* sites in the households are still maintained and preserved or even upgraded regarding the construction, all to pay homage to their ancestors. This act of preservation brings about the feeling of oneness and completeness.

The relationship of humans with their workplaces and the emotional bond resulted with their workplaces descended from their ancestors have formed a mind set with an implication of what is known now as *cultural ergonomics* (Kaplan, 2004). In this relation, since the 1980s, workers have gradually changed their workplaces and tools to be more ergonomic such as from squatting working position to standing position. In Bali, the segments of community that still hold the blacksmith profession are mostly those belonging to *pande* clan or caste. In other word, it is almost unlikely to find a person or persons having blacksmith or *memande* as their profession who are not of the *pande* ancestry. Therefore, this aspect should be considered as one important factor when recruiting new workers in Bali.

6 CONCLUSIONS

From this review, some conclusions are made, as follows.

1. The building of *prapen* or workplace in households, although not any more in use, is strongly related to paying respect to the ancestors of the Pande clan.
2. In the old *prapen* sites still exist working tools or work posture which are not ergonomic, but in the newer ones more ergonomic ones are already in existence.

REFERENCES

Anonim. *Six advices of Mpu Siwa Saguna to Pande Clan through Brahmana Dwala (Indonesian)* (Denpasar: Maha Semaya Warga Pande Propinsi Bali).

Bali Wikipedia Indonesia (cited 1 July 2007) available at: http://localhost.

Haryono, H., 2006, *Keris Jawa between mystique and rationality (Indonesian)* (Jakarta:PT Indonesia Kebanggaanku).

Kaplan, M.2004, *Cultural Ergonomics,* (Amsterdam: Elsevier)., 2004

Tirtayasa,K., Adiputra N., Djestawana, IG.A., 2003, The change of working posture in manggur decreases cardiovascular load and musculoskeletal complaints among Balinese gamelan craftsmen. *Journal of Human Ergology,* Vol.32, No. 2, 2003.

Author index